Perspectives in Animal Ecology and Reproduction

Perspectives in Animal Ecology and Reproduction

— *Volume 6* —

Editor-in Chief
Dr. V.K. Gupta

Editors
Dr. Anil K. Verma
Dr. G.D. Singh

2010
DAYA PUBLISHING HOUSE
Delhi - 110 035

ISBN 81-7035-635-0
ISBN 978-81-7035-635-6

Published by : **Daya Publishing House**
1123/74, Deva Ram Park
Tri Nagar, Delhi - 110 035
Phone: 27383999
Fax: (011) 23260116
e-mail : dayabooks@vsnl.com
website : www.dayabooks.com

Showroom : 4760-61/23, Ansari Road, Darya Ganj,
New Delhi - 110 002
Phone: 23245578, 23244987

Laser Typesetting : **Classic Computer Services**
Delhi - 110 035

Printed at : **Chawla Offset Printers**
Delhi - 110 052

PRINTED IN INDIA

Editorial Board

– Members –

Foreword

Drs. V. K. Gupta and Anil. K. Verma are to be congratulated on the publication of Volume 6 in the series on *Perspectives in Animal Ecology and Reproduction*. In this latest task, Dr. G.D. Singh has joined their editorial team to contribute to the diligent work undertaken by these editors of earlier volumes. This volume, like its predecessors, covers a wide range of topics in both the fields of animal ecology and animal reproduction. The term 'animals' is interpreted liberally and, as a result, some attention is also given to biophysical studies of some environments. Furthermore, some studies give attention to conservation management of wildlife and to pest control issues. The animal reproduction papers are, on the whole, more specialized than the animal ecology ones but no less important.

Volumes in this series mainly focus on issues and research in India but many of the findings are of considerable interest to the scientific community generally. As a result of this series, the latest results of those working in the field can be made available to a worldwide audience in a timely manner.

Many of the articles bring results that have potential applications that will enable animals and their environments to be better managed for economic gain or to ensure conservation results. The latter is very important given the rapid rate of biodiversity loss that is occurring mainly as a result of human economic activities. All branches of ecology can directly or indirectly contribute to the improved management or care for animals. In the end, it is important that a holistic approach be adopted in these tasks. The volumes in this series contribute to this approach and some even extend their reach to ecological economies.

It is to be hoped that this valuable series will continue to add new volumes and further advance the study of animal ecology and reproduction. I believe Dr. B.L. Kaul's comments on Volume 5 are also applicable to Volume 6. He said he was sure that "the current volume will be well received like the previous ones and serve as a reference book for future workers in the field of animal ecology and reproduction". May the editors continue for long time to come to make new scientific discoveries known to the whole world.

Clem Tisdell *Ph.D.; FASSA*
Professor Emeritus
The University of Queensland
Brisbane, Australia

Preface

Life having originated in water, has evolved itself into an enchanting world of rich and vastly diverse flora and fauna. The dependence of animals man included on the biological wealth of ecosystem can not be, therefore, over emphasized. Various anthropogenic activities have altered the physico-chemical and biological processes within ecosystem and failure to restore them would result in sharply increased costs later; in the extinction of the species or ecosystem types and in permanent ecological damage. As such, there is a dire need to depend and increase our understanding of the environment with regard to its physical, chemical and biological characteristics together with various interactions *viz.* animal ecology and reproductive strategies taking place in the diverse environments.

India is losing at an alarming rate its vast genetic heritage, the animal wealth of the country, even before they are properly studied and understood. The nature has gifted the biological diversity to be enjoyed and exploited gainfully for the progress of humanity and not to be destroyed and written off from the face of earth altogether. We must, therefore, understand that our own survival depends on their existence and not on their extinction. For the economic and technological development of the country, it is necessary that we should have an adequate knowledge of our natural resources and this knowledge is necessary so that our future developmental programmes should not be at the cost of our faunal wealth.

Animal's reproductive biology together with the ecology, as a discipline of study and research has, for the last quarter century or so emerged as an important discourse in many educational institutions and research is going on in many universities, colleges and laboratories. Many books on this subject are being published from various quarters which unfold different aspects of the said discipline. But in most cases, animal ecology v/s reproduction did not get due attention. It is with the purpose of focusing our attention on those areas where some deficiency or lacuna exists and there is still a need and scope for further work or expertise developed, that we have utilized our own experience in this series which is indeed a unique feature. We are confident that this volume in concept and contents, shall satisfy the needs of the students, researchers, environmentalists, planners and policy makers who are interested in this discipline.

V.K. Gupta

Anil K. Verma

G.D. Singh

Contents

Section I
Animal Ecology

Chapter 1

Impact of Supplementary Feeds on the Growth and Excretory Metabolite Levels in *Clarias batrachus*

Meenakshi Jindal*

Department of Zoology and Aquaculture,
C.C.S. Haryana Agricultural University, Hisar – 125 004, Haryana

ABSTRACT

The fingerlings of *Clarias batrachus* were fed on 6 formulated diets (protein levels ranged between 35-45 per cent) over a period of 100 days to study their protein requirements and effect of diets on the excretory levels of ammonical nitrogen (NH_4-N) and ortho-phosphate (o-PO_4). Growth and Specific growth rate remained low when fed on low protein levels (diets 1 and 2). Highest growth performance was observed in groups fed on 40.25 per cent dietary protein irrespective of the protein source (soybean; HPS and fishmeal; FM). An increase in dietary protein level (beyond 40.25 per cent) repressed growth. The values of DO and excretion of NH_4-N and o-PO_4 remained low where fish were fed on 40.25 per cent dietary protein irrespective of the protein source. But these values remained high for groups fed on FM based diets. The results clearly showed that out of two protein sources used in these

* E-mail: meenakshi@hau.ernet.in; Ph. +919416139139.

studies, soybean gives the most promising results. Therefore, the use of soybean (at 40.25 per cent protein level) is recommended as an eco-friendly and cost effective dietary protein source in the diets of *Clarias batrachus*.

Keywords: *Ammonical nitrogen, Clarias batrachus, Excretion, Fishmeal, Growth, Ortho-phosphate, Protein levels, Soybean.*

Introduction

Fish excrete phosphorus in soluble and particulate forms. The soluble fraction called ortho-phosphate (*o*-PO_4), is most available for plant growth.However the main loading of phosphorus to the environment was reported to be *via* faecal pellets (Vielman *et al.*, 2000).

Nitrogen in wastewater from aquaculture effluents is often considered a pollutant. In freshwater systems, nitrogen is sometimes a limiting nutrient, so adding it stimulates plant and algal growth. A majority of the excess nitrogen in either tank or pond culture systems originates as ammonia excreted by fish (Kibria *et al.*, 1998). The ammonia, as a waste product, is formed during the breakdown of proteins and excess amino acids not incorporated into tissues by the fish.

Therefore, there is a need to research alternate protein sources that may also reduce the input of nitrogen and phosphorus into the environment. Feeds for cultivated species contain a significant amount of fishmeal (FM). But, the increasing costs and unpredictable availability of FM necessitates the search for its replacement with cheaply and abundantly available plant protein feed stuffs. Due to world wide dominance of soybean and its appreciation as quality protein, several workers have attempted to replace FM with soybean meal (SBM) in diets formulated for several fish species (Jindal, 2001; Jindal and Garg, 2005; Jindal *et al.*, 2007a,b; Jindal,2007; Robinson and Menghe, 2007 and Jindal, 2008).

Hence, the aim of present investigation was to search an alternate protein source of plant origin (*i.e.* soybean) that will not be only cost effective, but certainly would reduce excretion of nitrogenous wastes and total organic matter, possibly also of phosphorous, and alleviate pollution problems in intensive aquacultural systems.

Material and Methods

Specimens of *Clarias batrachus* were obtained from Sultan Singh Fish Farm, Nilokheri, Haryana, India. Specimens with mean body weight (5.2 g) were used in the studies. Fish were placed in the transparent glass aquaria (60×30×30 cm) kept in the laboratory where the temperature was maintained at 25±1°C and the lighting scheduled at 12h of light alternating with 12h of darkness. The fish were acclimatized for a minimum of 15 days prior to the initiation of experimental treatments. The water was renewed daily with chlorine free water.

Fishmeal and soybean seeds were used as the protein sources. Groundnut oil cake and rice bran were used as the basic feed ingredients. Soybean seeds were cleaned, autoclaved for half an hour at 121.6°C at 15 lb pressure to remove antinutritional factors (ANFs) such as trypsin inhibitors, haemoglutinins, lectins and phytic acid(Garg *et al.*, 2002). After oven drying at 60°C, it was ground into fine powder.

Groundnut oil cake, rice bran, fishmeal and processed soybean were finely ground to pass through 0.5 mm sieve. All the ingredients were subjected to proximate analysis (AOAC, 2000) prior to the preparation of diets (Table 1.1). All the ingredients were mixed according to Table 1.2 and dough was made using distilled water. Thereafter, the dough was passed to a mechanical palletizer to obtain pellets (0.5 mm thick) which were dried in an oven and used in the studies for 100 days.

The experimental diets were fed to duplicate groups of fish to satiation twice a day for a one week acclimatization period before starting the study. After this period, the fish were individually weighed and their initial weights recorded. The fish were then offered the test diets (1-6) twice a day (9:00h and 16:00h) to satiation, for 100 days. This period was considered enough to produce the effect of feeding on daily excretory pattern in the test species. Faeces were siphoned from culture aquaria every morning before fish feeding. In addition, about 20-50 per cent of the culture water was replaced daily with new, fresh, dechlorinated water. The pooled faecal samples were dried in an oven maintained at 60°C for subsequent

Table 1.1: Proximate Analysis (Per cent Dry Weight Basis) of the Ingredients Prior to the Preparation of Diets

Ingredients	*Proximate Composition (per cent)*					
	Crude Protein	*Crude Fat*	*Crude Fiber*	*Total Ash*	*Nitrogen Free Extract*	*Gross Energy* (KJg^{-1})
Groundnut Oil Cake (GNOC)	35.266 ± 0.005	6.250 ± 0.003	6.000 ± 0.005	7.000 ± 0.000	45.496 ± 0.006	18.607 ± 0.003
Rice Bran (RB)	14.100 ± 0.005	10.066 ± 0.005	11.003 ± 0.005	20.556 ± 0.003	44.440 ± 0.005	14.906 ± 0.007
Fish Meal (FM)	42.926 ± 0.002	10.996 ± 0.003	3.493 ± 0.003	29.653 ± 0.003	12.930 ± 0.001	16.713 ± 0.000
Processed Soybean* (HPS)	43.733 ± 0.008	25.603 ± 0.000	4.496 ± 0.003	3.796 ± 0.000	22.370 ± 0.001	24.298 ± 0.001

* Raw soybeans were hydrothermically processed in an autoclave at 121°C at 15 lbs for 30 min. to remove anti-nutritional factors (ANFs) (Garg *et al.*, 2002).

All values are mean ± S.E. of means of 3 observations.

Table 1.2: Ingredient Composition and Proximate Analysis (Per cent dry weight basis) of 6 Compounded Diets (Diets 1-6) with Different Protein Levels

	Diets					
	1	*2*	*3*	*4*	*5*	*6*
Ingredients (per cent)						
Groundnut Oil Cake[a] (GNOC)	60	60	60	60	58	57.9
Rice Bran[b] (RB)	18	17.5	5	7.6	–	–
Fish Meal[c] (FM)	15	–	28	–	35	–
Processed soybean[d] (HPS)	–	15.5	–	28.4	–	35.1
Binder[e]	5	5	5	5	5	5
Chromic Oxide[f] (Cr_2O_3)	1	1	1	1	1	1
Mineral premix[g] and amino acids (MPA)	1	1	1	1	1	1
Proximate Analysis (per cent)						
Crude Protein	35.00	35.00	40.25	40.25	45.50	45.50
Crude Fat	5.50	7.00	7.00	9.50	8.50	12.00
Crude Fibre	3.50	6.50	3.50	7.25	4.50	5.00
Ash	9.15	9.05	7.30	6.50	6.65	7.15
Nitrogen Free Extract (NFE)	46.85	42.45	42.95	36.50	34.85	30.35
Gross Energy KJg^{-1}	18.48	18.32	19.65	19.53	20.09	20.71

a, b Used as basic feed ingredients; c: Used as reference protein source of animal origin; d: Used as main protein source of plant origin; e: Used is Carboxyl methyl cellulose to make the diets water stable; f: Used as an external indigestible marker for estimating apparent digestibility; g: Added to supplement the diets with minerals and amino acids.

Each Kg contains Copper: 312mg; Cobalt: 45mg; Magnesium: 2.114g; Iron: 979mg; Zinc: 2.13g; Iodine: 156mg; DL-Methionine: 1.92g; L-lysine mono hydrochloride: 4.4g; Calcium: 30 per cent and Phosphorous: 8.25 per cent.

analysis. Individual weight of the fish was recorded at the end of the experiment.

Analytical Techniques

The feed ingredients, experimental diets, faecal samples and fish carcass were analysed (AOAC, 2000). Chromide oxide levels in the diets as well as in the faecal samples were estimated spectrophotometrically (Spyridakis *et al.*, 1989). Live weight gain (g), growth percent gain, specific growth rate (SGR, per cent d^{-1}), protein efficiency ratio (PER) and gross conversion efficiency (GCE) were calculated using standard methods (Steffens, 1989). Apparent nutrient digestibility (APD) of the diets were calculated (Cho *et al.*, 1982) as follows:

$$APD = 100 - \frac{100 \times \%\ Cr_2O_3 \text{ in diet} \times \%\ \text{nutrient in faeces}}{\%\ Cr_2O_3 \text{ in faeces} \times \%\ \text{nutrient in diet}}$$

Water Quality Parameters

On the last day of experiment offer the same feed to the fish in sufficient quantity so that the same is consumed, wait for 2 hours. Maintain a fixed level of water in each aquarium (say 30-40 L). Remove the excess of feed. The various water quality parameters like dissolved oxygen, pH, temperature, conductivity, free carbon dioxide, total alkalinity and total hardness of aquaria water were analyzed (APHA, 1998). Start collecting water samples from each aquarium in replicate of 2 for the determination of ammonical nitrogen (NH_4-N) and ortho-phosphate (*o*-PO_4) (APHA, 1998) to see the influence of compounded feeds on pollution status of receiving water in the aquaria.

Calculate the excretory levels of NH_4-N and *o*-PO_4 in treated water as follows:

$$\begin{array}{c}NH_4\text{-N excretion}\\ \text{(mg/100g BW of fish)}\end{array} = \frac{NH_4\text{-N (mg l}^{-1}\text{) in aquarium water}}{\text{Fish weight (mg) per L of water}}$$

$$\begin{array}{c}o\text{-}PO_4\text{ excretion}\\ \text{(mg/100g BW of fish)}\end{array} = \frac{o\text{-}PO_4\text{ (mg l}^{-1}\text{) in aquarium water}}{\text{Fish weight (mg) per L of water}}$$

Statistical Analysis

Data was analysed following ANOVA at 5 per cent probability level. Group means were compared by student 't' test.

Results and Discussions

Water Quality Parameters

The effects of 6 different diets are very clearly reflected on the physico-chemical characteristics of aquaria water (Table 1.3).

The table showed that DO fluctuated between 4.01 to 5.00 mg/l, conductivity between 0.48 to 0.55 micro mhos/cm, Free CO_2 between 16.40 to 17.60 mg/l, total alkalinity between 207.33 to 252.66 mg/l and total hardness between 210.00 to 254.66 mg/l. The pH remained alkaline (7.4 to 7.8). The water temperature fluctuated between 24.5 to 25.5°C.

Although DO levels, remained at optimum levels, but low DO values in aquaria where the fish were fed on 40.25 per cent dietary protein (HPS based diets) clearly indicated its utilization by the growing fish (Jindal, 2007; Jindal, 2008 and Jindal *et al.*, 2008).

Post-prandial Excretory Levels of Ammonical Nitrogen (NH_4-N) and ortho-phosphate (O-PO_4)

A significant (P<0.05) decrease in NH_4-N excretion and *o*-PO_4 production was observed in the aquaria waters up to the dietary protein levels 40.25 per cent, irrespective of the protein source (Table 1.3). But as the protein level increased above 40.25 per cent, increase in NH_4-N excretion and *o*-PO_4 production was observed. These studies showed that digestibility of protein in the test species is only up to 40.25 per cent protein level and above this level, they will starts polluting the aquaria water resulting in poor growth of fish (Jindal, 2001; Jindal, 2007; Jindal, 2008 and Kalla *et al.*, 2003).

The results further indicated that the level of NH_4-N excretion and *o*-PO_4 production were found significantly (P<0.05) low in fish fed on HPS based diets (plant proteins) as compared to FM based diets (animal protein). Reduction in NH_4-N excretion and *o*-PO_4 production with the use of HPS in the diets have been reported by Jindal, 2001; Jindal and Garg, 2005; Singh *et al.*, 2003 and Kalla and Garg, 2004. These results further showed that plant protein based

Table 1.3: Water Quality Parameters of Different Aquaria Stocked with *C. batrachus* Fingerlings Fed on Experimental Diets (Diets 1–6) at Different Dietary Protein Levels

Parameters	*Diets*					
	1	*2*	*3*	*4*	*5*	*6*
Dissolved Oxygen (DO) mg/l	4.400± 0.115	4.363±0.008	4.166±0.006	4.011±0.115	5.000±0.115	4.900±0.120
pH	7.8	7.4	7.5	7.7	7.4	7.6
Water temperature (°C)	25.0	25.3	24.5	25.5	25.0	25.5
Conductivity micro (μ) mhos cm^{-1}	0.500± 0.001	0.493±0.001	0.480±0.001	0.550±0.002	0.513±0.001	0.540±0.002
Free Carbon dioxide (Free CO_2) mg/l	16.40±0.115	17.40±0.115	16.86±0.120	16.60±0.120	17.03±0.003	17.60±0.119
Total alkalinity (mg/l)	220.000±5.773	252.66±1.764	207.333±1.764	229.33±2.906	246.666±1.764	225.33±2.906
Total hardness (mg/l)	219.333±1.763	226.66±2.404	226.666±1.764	254.66±2.404	249.333±4.807	210.00±2.309
Ammonical nitrogen (NH_4-N) excretion (mg/100g BW of fish)	0.724±0.000	0.708±0.000	0.316±0.000	0.245±0.002	0.406±0.002	0.392±0.000
Ortho-phosphate (*o*-PO_4) excretion (mg/100g BW of fish)	0.573±0.002	0.522±0.000	0.406±0.002	0.383±0.003	0.474±0.001	0.434±0.002

*: All values are mean ± S.E. of mean of 3 observations.

diets excrete less ammonia and phosphorous and thus less pollute the water.

In the present studies, Maximum NH_4-N excretion occurred after 8h post feeding (Figure 1.1) and maximum *o*-PO_4 production after 6h post feeding (Figure 1.2) irrespective of the protein source (Jindal, 2008; Kalla and Garg, 2004 and Jindal *et al.*, 2009).

Growth Parameters

Survival in different treatments were high and varied between 89 to 95 per cent (Table 1.4). Fingerlings in different treatments were fed on one of the 6 formulated diets containing protein levels ranging from 35.0 to 45.5 per cent. ANOVA revealed that a significant ($P<0.05$) high live weight gain, per cent weight gain and SGR were observed in the fingerlings fed on diets containing 40.25 per cent dietary protein, irrespective of the protein source. On the other hand, FCR values at 40.25 per cent dietary protein level remained significantly ($P<0.05$) low in comparison to other dietary treatments. The data further indicates that an increase in dietary protein contents (beyond 40.25 per cent) not only repressed growth performance but also

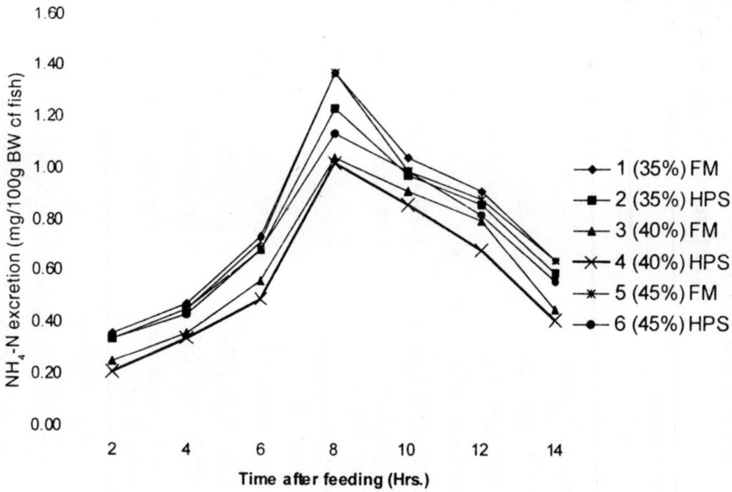

Figure 1.1: Diurnal Excretory Pattern of Ammonical Nitrogen (NH_4-N) in Treated Waters in Fish *C. batrachus* Fed on Experimental Diets (1–6)

Table 28.4: Effect of Six Dietary Protein Levels on Growth Performance, Digestibility and Nutrient Retention in *C. batrachus* Fingerlings Fed on Experimental Diets (1–6)

Parameters	*Diets*					
	1	*2*	*3*	*4*	*5*	*6*
Survival (per cent)	89	90	90	95	92	93
Live Weight gain (g)	21.675±0.005	22.980±0.002	23.030±0.210	24.985±0.505	21.945±0.002	23.625±0.135
Fish length gain (cm)	7.350±0.005	7.550±0.005	7.900±0.100	8.200±0.100	7.700±0.100	8.000±0.000
Growth per cent gain in body wt	421.121±4.394	413.899±9.679	440.845±8.930	437.536±5.294	411.331±4.900	459.243±6.000
Growth/day per cent body wt	1.356±0.004	1.347±0.001	1.375±0.012	1.372±0.000	1.345±0.001	1.393±0.002
Specific growth rate SGR (per cent d^{-1})	0.717±0.004	0.711±0.001	0.733±0.001	0.731±0.001	0.708±0.008	0.747±0.000
Food Consumption/day per cent Body weight	3.969±0.004	3.982±0.001	4.003±0.016	3.964±0.001	3.999±0.001	3.975±0.003
Feed Conversion ratio (FCR)	2.927±0.001	2.656±0.000	2.911±0.003	2.888±0.001	2.973±0.003	2.853±0.000
Protein efficiency ratio (PER)	0.647±0.001	0.338±0.003	0.611±0.003	0.637±0.000	0.514±0.002	0.549±0.002
Gross conversion ratio (GCE)	0.342±0.001	0.342±0.005	0.343±0.004	0.346±0.001	0.336±0.000	0.350±0.000
Apparent protein digestibility (APD) per cent	81.680±0.250	82.910±0.005	84.545±0.185	87.580±0.008	80.045±0.004	85.755±0.003

Duration of experiment = 100 days.

Mean with the same letter in the same column are not significantly ($P>0.05$) different.

All values are Mean ± S.E. of mean of three observations.

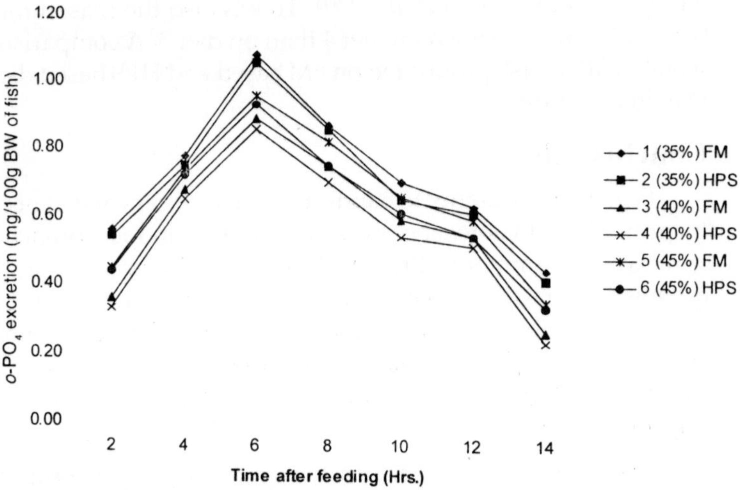

Figure 1.2: Diurnal Excretory Pattern of Ortho-phosphate (*o*-PO_4) in Treated Waters in Fish *C. batrachus* Fed on Experimental Diets (1–6)

significantly (P<0.05) increased FCR values. The data further showed a slight high growth performance at 40.25 per cent dietary protein from HPS based diets may be attributed to the high digestibility of plant protein.

Low growth at other dietary treatments (treatments 1, 2, 5 and 6) may indicated that optimum protein requirements of *C. batrachus* fingerlings is about 40.25 per cent. Feeding the fingerlings on high dietary protein level (45 per cent) not only repressed growth performance, but also deteriorated water quality as evident from high NH_4-N and o-PO_4 excretion. Protein levels above optimum requirements may results in decreased growth rates because of a reduction in dietary energy available for growth due to energy required to deaminate and excrete excess absorbed amino acids. These resultrs are in agreement with those of Jindal (2001); Kalla and Garg (2004), Jindal (2007); Robinson and Menghe (2007); Jindal (2008) and Jindal *et al.* (2009).

The studies further showed that growth and digestibility parameters were found to be negatively correlated with NH_4- N and

o-PO_4 excretion (Jindal *et al.*,2009). This is also the reason for the better growth of fish fed on diet 4 than on diet 3. A comparison of weight gain in fish groups fed on FM based and HPS based diets is shown in Figure 1.3.

Conclusions

Fish have the ability to handle protein in excess of that needed for growth and maintenance by deaminizing amino acid bronchially and excreting ammonia. Present studies on *C. batrachus,* thus, have established that when the protein levels in the diets exceeds the limits of digestibility (above 40.25 per cent) the excess are deaminized and are excreted as ammonia in the aquaria water, which sometimes may be stressful for the fish especially in some impoundments and impedes growth.

Further, the use of plant proteins make the aquaculture feed more environmental friendly as the metabolism of such feeds results in the lower excretion of nitrogen, phosphorous and other organic wastes in the environment. Higher the ingestion of proteins by the fish, more ammonia, urea and phosphorous are excreted in the

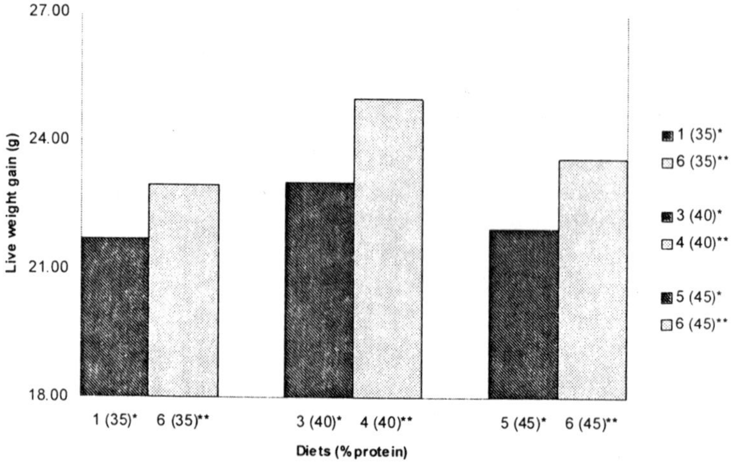

Figure 1.3: Comparison of Feeding 6 Diets (1–5 FM* and 6–10 HPS) Containing 3 Different Dietary Protein Levels on Live Weight Gain in *C. batrachus* Fingerlings**

environment, therefore, determination of nutrient budgets and daily pattern of excretion of metabolites by the fish is important for evaluating the potential waste load of the fish farm effluents. Studies have revealed that the use of plant proteins in fish feeds not only reduces the cost of feed formulation but also reduces the post prandial excretion of ammonia and phosphorous in the treated waters.

The results of this study clearly demonstrated that HPS supplemented with MPA can be recommended as a replacement of FM based diet for *C. batrachus*, up to the protein level of 40.25 per cent. This not only saves total feed cost, but certainly would also reduce excretion of nitrogenous and total organic matter, possibly also of phosphorous and alleviate pollution problems in the intensive aquacultural systems.

Acknowledgements

The author acknowledges funding received under the scheme "Women Scientist Scholarship Scheme for Societal Programmes (WOS-B), Department of Science and Technology, Government of India" for carrying out this research.

Mr. Sultan Singh, owner Sultan Fish Seed Farm, Nilokheri, Haryana, India is fully acknowledged for providing *C. batrachus* fingerlings for carrying out the research work.

References

AOAC (Association of Official Analytical Chemists), 2000. *Official Methods of Analysis*. Assoc. Off. Anal. Chem. Washington, Sc, USA.

APHA (American Public Health Association), 1998. *Standard Methods for the Examination of Water and Wastewater*, 20th Edn., APHA, AWWA, EPFC, New York.

Cho, C.Y., Slonger, S.J. and Bayley, H.S., 1982. Bioenergetics of salmonid fishes. Energy intake, expenditure and productivity. *Comp. Biochem. Physiol.*, 73B: 25–41.

Garg, S.K., Kalla, Alok and Bhatnagar, Anita, 2002. Evaluation of raw and hydrothermically processed leguminous seeds as supplementary feed for the growth of two Indian Major carp species. *Aquaculture Res.*, 33: 151–163.

Jindal, Meenakshi and Garg, S.K., 2005. Effect of replacement of fishmeal with defatted canola on growth performance and nutrient retention in the fingerlings of *Channa punctatus* (Bloch.). *Pb. Univ. Res. J. (Sci.)*, 55: 183–189.

Jindal, Meenakshi, 2001. Effect of feeding of plant origin proteins on nutrient utilization and growth of *Channa punctatus* (Bloch.). *Ph.D. Thesis*, CCS HAU, Hisar, p. 141.

Jindal, Meenakshi, 2007. Use of supplementary feeds for the development of sustainable aquaculture technology. A report submitted to Science and Society Division, Department of Science and Technology, New Delhi, p. 47.

Jindal, Meenakshi, 2008. Studies on protein requirements of catfish *Clarias batrachus* fingerlings for sustainable aquaculture. In: *National Seminar on Technical Advances in Environment Management and Applied Zoology*. Department of Zoology, Kurukshetra University, Kurukshetra. Sponsored by Council of Scientific and Industrial Research, New Delhi, India, pp. 59–62

Jindal, Meenakshi, Garg, S.K. and Yadava, N.K., 2007a. Effect of replacement of fishmeal with dietary protein sources of plant origin on the growth performance and nutrient retention in the fingerlings of *Channa punctatus* (Bloch.) for sustainable aquaculture. *Pb. Univ. Res. J. (Sci.)*, 57: 133–138.

Jindal, Meenakshi, Garg, S.K. and Yadava, N.K., 2009. Effect of feeding defatted canola on daily excretion of ammonical nitrogen (NH_4-N) and ortho-phosphate (o-PO_4) in *Channa punctatus* (Bloch.). (Accepted for publication in *Livestock Research for Rural Development*, USA).

Jindal, Meenakshi, Garg, S.K., Yadava, N.K. and Gupta, R.K., 2007b. Effect of replacement of fishmeal with processed soybean on growth performance and nutrient retention in *Channa punctatus* (Bloch.) fingerlings. Livestock Research for Rural Development, USA, Volume 19, Article #165. Retrieved from http://www.cipav.org.co/lrrd/lrrd19/11/jind19165.htm.

Kalla, Alok and Garg, S.K., 2004. Use of plant proteins in supplementary diets for sustainable aquaculture. In: *National Workshop on Rational Use of Water Resources for Aquaculture*, (Eds.)

S.K. Garg and K.L. Jain. CCS HAU Press, Hisar, India, pp. 31–47.

Kalla, Alok, Garg, S.K. and Kaushik, C.P., 2003. Effect of dietary protein source on growth, digestibility and body composition in the fingerlings of *Cirrhinus mrigala* (Ham.). In: Proceedings of 3rd Interaction Workshop on Fish Production Using Brackish Water in Arid Ecosystem, (Eds.) S.K. Garg and A.R.T. Arasu. S.S. Tonk Press Manager at CCS HAU Press, Hisar, India, pp. 139–145.

Kibria, G., Nugegoda, D., Fairclough, R. and Lam, P., 1998. Can nitrogen pollution from aquaculture be reduced? *NAGA, ICLARM,* 21: 17–25.

Robinson, Edwin H. and Menghe, H. Li., 2007. Catfish Protein Nutrition (Revised). Bulletin 1153. Office of Agricultural Communications, Mississippi State University, USA.

Singh, K., 2001. Effect of supplementary feeds on survival and growth performance of Indian major carp fry and fingerlings. *M.Sc. Thesis,* CCS HAU, Hisar, India, p. 124.

Singh, K., Garg, S.K., Kalla, Alok and Bhatnagar, Anita, 2003. Oil cakes as protein source in supplementary diets for the growth of *Cirrhinus mrigala* (Ham.) fingerlings: Laboratory and field studies. *Bioresource Technology,* 86: 283–291.

Spyridakes, P., Metailler, R., Gabandan, J. and Riaza, A., 1989. Studies on nutrient digestibility in European sea bass *Dicentrarchus labrax*. I. Methodical aspects concerning faeces collection. *Aquaculture,* 77: 61–70.

Steffens, W., 1989. *Principles of Fish Nutrition*. Ellis Horwood, Chichester.

Vielman, J., Makinen, T., Ekholm, P. and Koskela, J., 2000. Influence of dietary soybean and phytase levels on performance and body composition of large rainbow trout (*Oncorhynchus mykiss*) and algal availability of phosphorus load. *Aquaculture,* 183: 349–362.

Chapter 2

Physico-Chemical and Microbial Studies of Lake Mansar in Jammu Shiwaliks

J.P. Sharma and V.K. Gupta*

Indian Institute of Integrative Medicine (IIIM), CSIR, Canal Road, Jammu – 180 001

ABSTRACT

During the present investigations the seasonal fluctuation in the physico-chemical factors and the bacterial populations were recorded from lake Mansar in order to evaluate their intrinsic relationship and suitability for pisciculture and for human consumption. The physico-chemical factors *viz.* temperature ranges from 14–32°C, pH 6.9–8.8, dissolved oxygen 3.4–9.8 ppm, free CO_2 0–3 ppm, total alkalinity 56–132 ppm, DOM 2–19 mg/l and chloride ranges from 24–1.8 ppm. The bacterial count *viz.* TBC ranges from 3×10^2–2.16×10^{11}, SPC 1.0×10^2–8.0×10^5 cfu/ml and coliform count range from 17–1100/100 ml as recorded from four spots of lake Mansar. The lake water was found to be congenial for the growth of plants and animals but for human consumption water can be used only after proper treatment and filtration.

Keywords: Physico-chemical, Microbiological, Lake, Water.

* Corresponding author: E-mail: drsharma77@yahoo.com.

Introduction

Lake Mansar is one of the oldest lake in Jammu (J&K State) which is perennial water body having an area of 3.29 Sq kms, situated 56 km in the east of Jammu city at an altitude of 710 m above sea level located between latitude 32°, 45′.5′ north to 33°42′.36′ and longitude 75°8′,32′ and 73°9′.8′ east. The lake is a big basin of Shiwalik formation and the origin of the lake is lost in antiquity, there being several myths about its origin.

The common geological belief is that the lake owes its origin due to the damming of the river, which was flowing along the lower Shiwalik ranges in the main Shiwalik belt. The lake has a circumference of 4.3 km and is primary fed by surface run off and also has some submerged spring resources. There is large thermal gradient, trophogenic and tropholytic zones and single winter overturn. The lake is subtropical monomictic having high population of plants and animals.

Although, Mansar lake is one of the most important lake of this region yet it has been least subjected to detailed microbiological investigations. Although limnological studies on Kashmir lakes are numerous (Hutchinson, 1937; Kant and Kachro, 1971; Zutshi and Khan, 1971, 1973; Zutshi *et al.*, 1972; Zutshi, 1975), the studies on the lakes of Jammu are scanty except for Zutshi and Khan (1977), Malhotra *et al.* (1979) and Jyoti and Sehgal (1980). The present study is aimed to fulfill the existing lacuna in this regard. The seasonal study of physico-chemical parameters like temperature, pH, dissolved oxygen, free CO_2, alkalinity of carbonate and bi-carbonate, chloride, organic matter and the bacterial population *i.e.* total bacterial count, standard plate count and coliform count of lake water has been done in order to determine the ecological condition of the lake water and its suitability for pisciculture and also for human consumption.

Materials and Methods

For physico-chemical and microbiological parameters the water samples were collected during the period 2002–2003 from four fixed spots 25 feet inside the lake. The analysis of various physico-chemical parameters, like pH, temperature, dissolved oxygen, alkalinity, chloride, dissolved organic matter was done on the spot and also in the laboratory as per manual of Jhingran *et al.* (1988) and APHA

(1998). Fish samples were collected by trapping the fish in a net or trap. Bacterial samples from fish were taken from the skin, gills and intestine of the fish in sterilized condition. Bacterial species from lake water, soil and fishes were isolated by streaking loop full of sample on Nutrient agar plates and the incubating at 37°C for 24 hours. The isolated colonies were further sub-cultured on Nutrient agar slants and their characterization was recorded. The bacterial load *i.e.* total bacterial count (TBC/ml), standard plate count (SPC cfu/ml) and MPN of coliform (MPN /100 ml) were also recorded. For the identification of bacterial isolates biochemical tests *viz.* carbohydrates utilization, H_2S production, No_3 reduction, indole production, MR-VP test, citrate utilization, urease activity, gelatin liquefaction, starch hydrolysis, lipid hydrolysis and arginine hydrolysis were performed. The identification of the culture was confirmed by PCR techniques.

Results and Discussion

The Mansar lake being rich in nutrient has high population of plants and animals. The common plants are *Nitella, Typha, Potamogeton, Myriophyllum, Hydrilla, Chara, Ipomea* and *Nymphea* sps.; animal include fish like *Channa puntatus, Rasbora rasbora, Danio rerio, Puntius sophore, P. ticto, P. conconius* and *Cyprinus carpio,* amphibian like *Rana tigrana, R. cyanocephlyctis* and reptiles like *Natrix piscator, Trionyx gangeticus* and *Lissemys punctata punctata.*

The seasonal fluctuations of the physico-chemical factors reveals that the temperature ranges from 14–32°C, pH 6.9–8.8, dissolved oxygen 3.4–9.8 ppm, free CO_2 0–3 ppm, total alkalinity 56–132 ppm, dissolved organic matter 2.0–19 mg/L and chloride 0.24–1.8 ppm. There were not many variations of physico-chemical factors as recorded from four different spots. Dissolved organic matter however, was found to be maximum *i.e.* 19 mg/L from spot A may be due to the influx of more organic load because of more exploitation by the people at this spot as compared to other spots (Tables 2.1–2.4). Sharma and Gupta (2005) recorded the significant relationship of organic matter and the aerobic and facultative bacteria from a fresh water pond but no such observation was recorded from the present study.

Table 2.1(a): Physico-chemical Studies of Lake Mansar (SPOT-A) Year–2002

Month	Temp°C		pH	Dissolved Oxygen (ppm)	Free CO_2 (ppm)	Alkalinity (ppm)		DOM mg/l	Chloride (ppm)
	Air	Water				CO_2	HCO_3		
Jan	21.2	15.4	8.48	6.4	2.0	Nil	100	7.8	0.6
Feb	230	14.0	7.80	8.0	1.0	Nil	104	9.0	1.1
Mar	27.0	23.0	7.20	8.0	2.0	Nil	88	11.0	1.2
Apr	28.2	24.0	7.33	7.2	2.0	Nil	62	19.0	1.9
May	32.5	30.6	7.45	8.4	Nil	16	60	16.0	1.6
Jun	33.1	31.6	8.37	90	Nil	6.0	60	18.0	0.9
Jul	28.7	30.5	8.20	8.8	Nil	6.0	50	8.0	1.0
Aug	31.0	30.0	8.00	7.4	Nil	12	52	4.6	1.0
Sept	28.2	28.0	8.20	9.6	Nil	6.0	50	3.8	0.7
Oct	26.0	25.0	8.00	9.4	Nil	52	62	6.0	1.1
Nov	22.4	18.0	7.80	7.6	0.8	10	76	7.6	0.4
Dec	17.1	16.1	7.74	6.6	1.1	Nil	66	3.2	0.8

Table 2.1(b): Physico-chemical Studies of Lake Mansar (SPOT-A) Tear–2003

Month	Temp°C		pH	Dissolved Oxygen (ppm)	Free CO_2 (ppm)	Alkalinity (ppm)		DOM mg/l	Chloride (ppm)
	Air	Water				CO_2	HCO_3		
Jan	16.4	15.4	7.70	4.6	10	14	70	4.8	0.7
Feb	22.7	18.1	7.94	5.0	2.0	Nil	60	2.6	0.8
Mar	24.7	21.6	8.00	9.0	1.0	Nil	100	4.0	0.7
April	31.0	28.0	7.80	9.4	Nil	8.0	86	7.0	0.7
May	36.0	30.0	7.40	7.0	Nil	12.8	88	5.6	0.7
Jun	35.0	32.0	7.20	6.8	10	18	90	4.8	0.6
Jul	315	31.0	8.04	8.0	0.6	22	68	84	0.7
Aug	31.3	30.9	8.001	8.0	2.0	26	52	4.2	1.1
Sept	30.0	29.0	8.00	6.8	Nil	16	56	3.2	0.6
Oct	27.0	26.0	8.04	8.4	Nil	8.0	58	7.0	1.4
Nov	28.0	18.5	7.60	7.8	1.0	8	48	8.4	0.6
Dec	17.0	16.0	7.80	6.4	1.0	Nil	66	3.8	0.8

Table 2.2(a): Physico-chemical Studies of Lake Mansar (SPOT-B) Year-2002

Month	Temp°C		pH	Dissolved Oxygen (ppm)	Free CO_2 (ppm)	Alkalinity (ppm)		DOM mg/l	Chloride (ppm)
	Air	Water				CO_2	HCO_3		
Jan	20.5	153	8.18	4.0	2.0	Nil	90	8.0	0.9
Feb	22.8	14.5	7.80	6.4	0.8	Nil	110	8.2	1.3
Mar	27.0	23.0	6.90	8.8	Nil	6.0	98	10.4	1.3
Apr	28.5	23.5	7.60	8.0	2.0	Nil	86	15.0	1.8
May	32.5	29.5	7.20	8.4	Nil	10	90	17.0	1.8
Jun	32.9	31.0	7.10	9.6	Nil	8.0	66	8.6	0.8
Jul	28.7	30.5	8.20	8.8	Nil	6.0	50	8.8	1.0
Aug	31.0	30.0	7.80	7.8	Nil	18	62	5.8	12
Sept	28.0	28.1	8.80	9.2	Nil	5.0	52	3.9	0.7
Oct	26.0	25.0	8.10	9.2	Nil	5.0	56	6.2	12
Nov	22.0	18.0	8.80	7.0	0.8	8.0	70	7.0	0.5
Dec	17.2	16.2	7.88	6.2	0.6	50	70	3.2	0.7

Table 2.2(b): Physico-chemical Studies of Lake Mansar (SPOT-B) Year–2003

Month	Temp°C		pH	Dissolved Oxygen (ppm)	Free CO_2 (ppm)	Alkalinity (ppm)		DOM mg/l	Chloride (ppm)
	Air	Water				CO_2	HCO_3		
Jan	16.4	14.9	7.56	3.4	3.0	11	72	3.9	0.24
Feb	22.8	17.6	7.92	5.2	1.0	Nil	104	2.2	0.90
Mar	24.7	21.3	8.1	9.0	1.0	Nil	100	4.0	0.65
Apr	31.2	29.2	7.7	9.0	Nil	10	94	6.9	0.75
May	37.0	30.5	7.0	6.8	1.0	10	60	6.4	0.68
Jun	36	31.5	7.8	5.6	Nil	10	72	6.3	0.70
Jul	32	31	8.1	8.2	1.0	14	60	9.2	0.70
Aug	30.8	30.8	7.95	8.0	2.2	20	42	5.4	1.00
Sept	29	28	8.12	6.8	Nil	5.0	56	7.0	0.70
Oct	26	26	7.8	8.0	1.0	2.0	56	6.8	1.10
Nov	22	18	7.8	6.8	0.8	10	56	7.8	0.40
Dec	17.2	16	7.8	6.2	0.8	4.0	68	3.6	0.70

Table 2.3(a): Physico-chemical Studies of Lake Mansar (SPOT-C) Year–2002

Month	Temp°C		pH	Dissolved Oxygen (ppm)	Free CO_2 (ppm)	Alkalinity (ppm)		DOM mg/l	Chloride (ppm)
	Air	Water				CO_2	HCO_3		
Jan	20.5	17.0	8.57	9.6	Nil	12	120	6.8	1.0
Feb	23.0	16.7	7.18	8.2	2.0	Nil	108	7.4	1.0
Mar	27.0	24.0	7.2	8.0	Nil	16	90	8.0	1.3
Apr	28.0	25.0	7.3	7.0	2.0	Nil	62	16.0	1.8
May	32.0	30.0	7.2	8.8	Nil	12	86	14.0	1.4
June	33.0	32.0	8.3	8.6	Nil	8.0	80	8.8	0.9
Jul	28.0	30.0	8.5	8.8	Nil	6.0	54	8.0	1.0
Aug	31.2	30.7	8.1	8.8	Nil	7.0	46	4.0	0.7
Sept	29.0	28.5	8.2	9.0	Nil	16	68	6.8	0.9
Oct	27.2	26.0	8.4	9.2	Nil	8.0	60	8.2	1.8
Nov	23.8	18.0	8.3	8.2	1.2	10	70	8.4	0.5
Dec	22.5	18.3	8.2	6.6	1.9	Nil	64	3.4	0.7

Table 2.3(b): Physico-chemical Studies of Lake Mansar (SPOT-C) Year–2003

Month	Temp°C		pH	Dissolved Oxygen (ppm)	Free CO_2 (ppm)	Alkalinity (ppm)		DOM mg/l	Chloride (ppm)
	Air	Water				CO_2	HCO_3		
Jan	19.5	17.6	7.82	7.2	2.0	23	70	6.7	0.28
Feb	22.2	18.0	7.9	6.0	1.5	Nil	98	2.2	0.9
Mar	26.0	24.0	7.8	7.6	1.0	Nil	96	4.6	0.7
Apr	32.0	28.6	7.8	7.8	2.0	8	74	6.9	0.7
May	34.6	30.2	8.2	7.8	1.0	8	60	4.0	0.7
Jun	35.0	32.0	7.6	5.8	1.0	Nil	80	5.2	0.68
Jul	31.0	31.4	8.11	7.4	2.0	44	78	10.2	0.75
Aug	31.0	30.0	7.8	7.8	2.4	42	58	4.0	1.2
Sept	29.0	28.0	8.2	7.0	Nil	12	62	5.6	0.7
Oct	26.0	25.0	8.2	6.8	1.0	Nil	60	8.0	0.8
Nov	23.0	19.0	8.2	6.8	Nil	12	78	6.9	0.8
Dec	21.0	18.0	8.4	6.4	1.8	Nil	56	3.8	1.0

Table 2.4(a): Physico-chemical Studies of Lake Mansar (SPOT-D) Year–2002

Month	*Temp°C*		*pH*	*Dissolved Oxygen (ppm)*	*Free CO_2 (ppm)*	*Alkalinity (ppm)*		*DOM mg/l*	*Chloride (ppm)*
	Air	*Water*				*CO_2*	*HCO_3*		
Jan	20.0	17.5	8.0	9.0	Nil	10	80	6.0	0.8
Feb	22.0	15.5	7.6	6.8	1.0	Nil	110	8.0	1.3
Mar	27.0	24.0	6.9	8.8	Nil	6	90	10.2	1.2
Apr	28.0	23.0	7.0	8.6	2.0	Nil	56	11.4	1.7
May	32.0	31.0	8.2	10.0	Nil	12	58	8.2	0.8
Jun	33.0	30.0	7.8	9.0	Nil	12	66	8.6	0.8
Jul	31.0	30.0	8.3	8.8	Nil	15	65	7.8	1.0
Aug	31.0	30.6	8.14	8.6	Nil	7	38	3.6	0.7
Sept	29.5	28.5	8.0	8.6	Nil	10	76	7.2	1.0
Oct	26.0	25.0	8.0	8.6	1.0	Nil	86	8.6	1.6
Nov	25.5	19.0	8.1	7.8	0.8	10	80	7.8	0.5
Dec	22.0	18	7.8	5.2	1.1	3	46	3.4	0.8

Table 2.4(b): Physico-chemical Studies of Lake Mansar (SPOT-D) Year–2003

Month	*Temp°C*		*pH*	*Dissolved Oxygen (ppm)*	*Free CO_2 (ppm)*	*Alkalinity (ppm)*		*DOM mg/l*	*Chloride (ppm)*
	Air	*Water*				*CO_2*	*HCO_3*		
Jan	19.5	18.9	7.9	6.4	2.0	8.0	55	7.8	1.24
Feb	24.0	19.4	7.72	3.6	0.6	10	108	2.0	0.8
Mar	23.9	21.5	7.9	7.6	1.0	Nil	96	4.4	0.7
Apr	32.4	21.3	7.75	7.6	2.0	4.0	100	6.8	0.75
May	33.8	29.6	8.0	7.6	Nil	8.0	96	4.6	0.8
Jun	34.8	32.0	7.8	5.6	Nil	12	84	5.8	0.8
Jul	29.6	31.4	8.24	7.4	1.0	36	62	8.6	0.65
Aug	31.6	30.6	7.8	8.0	2.0	40	62	5.6	0.1
Sept	29.0	28.5	8.0	7.6	Nil	10	76	6.8	2.0
Oct	26.5	29.0	8.0	6.2	Nil	8.0	68	7.6	1.2
Nov	25.0	16.0	8.1	7.2	Nil	10.0	62	5.8	1.1
Dec	21.0	19.0	8.4	5.0	1.0	4.0	58	3.8	0.8

pH of the surface water was slightly alkaline with maximum of 8.8 recorded in January. Zutshi and Khan (1977) also recorded the pH as slightly alkaline with maximum of 8.8 in the month of November from this lake. pH of the water effect the growth of bacteria. Thimann (1964) suggested that most of the bacteria could grow only in pH range of 4 to 9. The pH during the present study ranged from 6.9–8.8. No marked seasonal fluctuation in pH was recorded so it did not have much bearing upon the seasonal counts of bacteria. Dissolved oxygen was found to be maximum *i.e.* 9.8 mg/l during the month of July and August where as minimum *i.e.* 3.4 mg/l was recorded in the month of January from surface layers at spot-B. Zutshi and Khan (1977) recorded 8.4 mg/l of DO from Mansar lake and 10.4 mg/l from surface water of Surinsar lake, however deeper water of both the lakes were having low oxygen concentration with minimum DO of 0.48 mg/l. Malhotra *et al.* (1979) also recorded DO concentration from surface layer as 9.2 mg/l which confirmed our findings that surface layer has maximum dissolved oxygen as compared to deeper layers.

Free carbon dioxide (F.CO_2) ranges from 0–3 ppm recorded during the present investigation from surface water of the lake. Free CO_2 concentration was recorded to be maximum 3 ppm from spot-B in the month of January, whereas DO concentration was recorded to be less (3.4 ppm) during this month. During this period, mortality in some fish species like *Puntius sophore* and *Rasbora* sps. was recorded. On comparing the values of DO and free CO_2 on this occasion, rise in free CO_2 would appear to be more related to the mortality of fishes in lake Mansar. Malhotra *et al.* (1977) recorded varied F.CO_2 concentration at different depth in lake surinsar and Mansar. The fish kill was recorded when F.CO_2 concentration was 7 ppm in January 1977 and 8.5 ppm in February 1977 by Malhotra *et al.* (1977).

The bacterial load is total bacterial count (TBC/ml), standard plate count (SPC cfu/ml) and the coliform count (MPN/100 ml) recorded from four different spots of the lake. During the present study TBC ranges from 3×10^2–2.16×10^{11}/ml, standard plate count SPC ranges from 1×10^2–8×10^5 cfu/ml where as coliform count MPN/100 ml was recorded to be 17–1100$^+$/100 ml. The coliform count was found to be maximum at spot D during the period of study, which may be because of greater human interference at this spot as compared to other spots (Tables 2.5–2.8). During the period of investigations four number of isolates belonging to family Enterobactericeae were identified as *Echerichia coli, Klebesiella, Salmonella* and *Shigella* which indicated that the lake water contains the coliform organisms which may act as fish pathogen and can also be as intestinal parasite of human and other animals.

Table 2.5(a): Monthly Microbial Load of Lake Mansar (SPOT-A) Year–2002

Month	*TBC/ml*	*SPC Count (cfu/ml)*	*MPN/100 ml*
Jan	7×10^5	2×10^3	25
Feb	2×10^5	3×10^2	13
Mar	7×10^2	4×10^3	25
Apr	5×10^2	1×10^2	93
May	3×10^2	2×10^2	15
Jun	2×10^4	2×10^2	20
Jul	3×10^2	2×10^2	20
Aug	2.16×10^{11}	5×10^3	21
Sept	8×10^8	6×10^3	9
Oct	7×10^8	4×10^3	93
Nov	5.2×10^7	6×10^2	75
Dec	3.9×10^7	7×10^2	23

From the published literature it is apparent that bacterial counts will be maximum during summer in eutrophic water body (Sharma, 1993). Carney *et al.* (1975) recorded 10^3 to 10^5 cfu/ml of aerobic heterotrophic bacteria and also reported higher counts from January to March. However, Geesey and Costerton (1979) reported the higher concentration in spring and autumn and lowest in August. During the present study period total bacterial count range from 3×10^2–2.16×10^{11}/ml and the standard plate count of aerobic facultative bacteria ranges from 1×10^2–8×10^5 cfu/ml was found to be maximum during August and September from all the four spots of lake Mansar. These finding are in confirmation with Austin and Allan-Austin (1985) who emphasized that in reservoir fishery there was no apparent seasonality of bacterial count in water although lower number is found in winter months and higher in summer months.

Table 2.5(b): Monthly Microbial Load of Lake Mansar (SPOT-A) Year–2003

Month	*TBC/ml*	*SPC Count (cfu/ml)*	*MPN/100 ml*
Jan	4.3×10^7	142×10^2	14
Feb	3.5×10^6	3.5×10^3	15
Mar	2.4×10^7	5.3×10^3	29
Apr	7.2×10^6	1.4×10^3	23
May	3.8×10^6	1.9×10^3	43
Jun	3.2×10^6	12×10^3	23
Jul	1.0×10^9	1.27×10^3	240
Aug	2.8×10^7	2.7×10^3	21
Sept	7×10^8	5.6×10^3	21
Oct	2.8×10^8	3.0×10^3	93
Nov	4.2×10^7	4.8×10^3	75
Dec	2.8×10^8	12×10^3	93

Table 2.6(a): Monthly Microbial Load of Lake Mansar (SPOT-B) Year–2002

Month	*TBC/ml*	*SPC Count (cfu/ml)*	*MPN/100 ml*
Jan	2×10^4	2×10^4	17
Feb	2×10^5	3×10^2	15
Mar	2×10^3	4×10^3	15
Apr	5×10^3	1×10^2	1100
May	3×10^3	2×10^2	210
Jun	4×10^4	2×10^3	290
Jul	4×10^2	2×10^2	43
Aug	2.4×10^{11}	4×10^3	24
Sept	1.6×10^9	4.3×10^3	21
Oct	2×10^9	9.6×10^3	43
Nov	2.5×10^8	1.1×10^3	9
Dec	2.0×10^7	7.3×10^2	15

Table 2.6(b): Monthly Microbial Load of Lake Mansar (SPOT-B) Year–2003

Month	*TBC/ml*	*SPC Count (cfu/ml)*	*MPN/100 ml*
Jan	8×10^7	15.4×10^2	75
Feb	2.4×10^6	1.8×10^3	20
Mar	8×10^6	3×10^3	21
Apr	2×10^6	2×10^3	7
May	4.2×10^5	4×10^2	93
Jun	2.8×10^6	3.2×10^3	75
Jul	1.2×10^8	1.4×10^3	44
Aug	3.4×10^7	1.2×10^4	21
Sept	1.5×10^8	2.4×10^3	21
Oct	2.0×10^8	8.6×10^3	43
Nov	2.5×10^7	1.8×10^2	21
Dec	2.4×10^8	8.2×10^2	75

Table 2.7(a): Monthly Microbial Load of Lake Mansar (SPOT-C) Year–2002

Month	*TBC/ml*	*SPC Count (cfu/ml)*	*MPN/100 ml*
Jan	7×10^4	6×10^4	17
Feb	3×10^4	2×10^4	210
Mar	2×10^4	3×10^3	1100
Apr	1×10^3	4×10^3	1100+
May	7×10^3	2×10^3	93
Jun	4×10^3	4×10^3	210
Jul	9×10^2	3×10^2	36
Aug	2.4×10^9	2×10^5	1100
Sept	2.08×10^9	7×10^3	150
Oct	1.7×10^9	5.5×10^3	440
Nov	1.8×10^8	7.5×10^3	460
Dec	5.6×10^8	7.2×10^3	1100

Table 2.7(b): Monthly Microbial Load of Lake Mansar (SPOT-C) Year–2003

Month	*TBC/ml*	*SPC Count (cfu/ml)*	*MPN/100 ml*
Jan	2.4×10^6	3.7×10^2	15
Feb	8.8×10^6	19.8×10^3	210
Mar	1.7×10^7	1.2×10^3	240
Apr	3.4×10^6	1.2×10^3	460
May	2.3×10^7	1.1×10^2	460
Jun	3.7×10^6	2×10^3	240
Jul	1.1×10^9	4×10^3	1100
Aug	3.6×10^7	3.5×10^5	460
Sept	2.3×10^9	6×10^3	150
Oct	1.7×10^9	5.3×10^3	460
Nov	5.8×10^8	2.8×10^3	240
Dec	2.5×10^9	2.5×10^3	240

Table 2.8(a): Monthly Microbial Load of Lake Mansar (SPOT-D) Year–2002

Month	*TBC/ml*	*SPC Count (cfu/ml)*	*MPN/100 ml*
Jan	3×10^4	8×103^9	1100+
Feb	4×10^4	8×10^5	1100
Mar	5×10^2	2×10^2	1100
Apr	2×10^4	3×10^2	1100+
May	5×10^3	6×10^2	240
Jun	1×10^5	3×10^2	240
Jul	4×10^2	3×10^2	1100+
Aug	1.64×10^9	4.2×10^5	240
Sept	2.4×10^9	1.7×10^4	460
Oct	1.85×10^9	3.1×10^3	240
Nov	3.5×10^8	3.5×10^3	240
Dec	1.6×10^8	4×10^3	93

Table 2.8(b): Monthly Microbial Load of Lake Mansar (SPOT-D) Year–2003

Month	*TBC/ml*	*SPC Count (cfu/ml)*	*MPN/100 ml*
Jan	2×10^6	1.8×10^2	37
Feb	7.4×10^7	11.6×10^4	460
Mar	8.6×10^7	2.3×10^4	53
Apr	1.2×10^7	4.2×10^3	1100
May	2.3×10^7	1.2×10^3	460
Jun	2.8×10^6	8×10^3	240
Jul	1.4×10^9	8.5×10^3	1100
Aug	3.8×10^7	4.9×10^5	75
Sept	2.5×10^9	3.8×10^7	460
Oct	2.8×10^8	2.7×10^3	240
Nov	3.2×10^6	3.5×10^4	75
Dec	3.0×10^6	2.8×10^3	240

From the above observations it has been concluded that the lake is eutrophic with high trophic level and the ecological factors are conducive for the growth of aquatic flora and fauna. The inflow of sewage and organic load should be restricted to save the lake from eutrophication and water quality degradation. The lake water was not found to be fit for drinking and hence suggested that it can be used only after proper treatment and filtration.

Acknowledgements

The authors are thankful to the Director, Indian Institute of Integrative Medicine, Jammu for his keen interest and encouragement. Our thanks to Ministry of Environment and Forests, New Delhi for financing the project.

References

APHA, 1989. *Standard Method of Examination of Water and Wastewater*. American Public Health Association, Washington, D.C., 1134 pp.

Austin, B. and Allan-Austin, D., 1985. Microbial quality of water in intensive fish rearing. *Journal of Applied Bacteriology Symposium Supplement*, p. 2075–2265.

Carney, J.F., Carty, C.E. and Colwell, R.R., 1975. Seasonal occurrences and distribution of microbial indicators and pathogens in Rhode river of Chesapeake Bay. *Applied and Environmental Microbiology*, 30: 771–780.

Geesey, G.G. and Costertion, J.W., 1979. Microbiology of a Northern river bacterial distribution and relationship to suspended sediment and organic carbon. *Canadian Journal of Microbiology*, 25: 1058–1102.

Hutchinson, G.E., 1937. Limnological Studies in Indian Tibet. *Int. Rev. Hydrobiol.*, 35: 137–177.

Jhingran, V.G., Natrajan, A.V., Banargee, S.M. and David, A., 1988. Methodology on reservoir fisheries investigation. *India Bull.*, 12: 8–26.

Jyoti, M.K. and Sehgal, Harjeet, 1980. Rotifer fauna of Jammu (J&K) India Part–1: Loricates. *Limnologia (Berlin)*, 12(1): 121–126.

Malhotra, Y.R., Jyoti, M.K. and Sehgal, Harjeet, 1977. Causes of size and sex restricted kill of *Puntius conchnius* in subtropical lake in Jammu. *Indian Journal of Experimental Biology*, 17(8): 836–837.

Sharma, J.P., 1993. Some microbial relations in fishes of Jammu. *Ph.D. Thesis*, 282 pp.

Sharma, J.P. and Gupta, V.K., 2005. Seasonal population dynamics of bacteria and their relationship with abiotic and biotic factors in a subtropical fish pond. In: *Proceeding National Seminar on Aquatic Resource Management in Hills*, p. 47–54.

Thimann, K.V., 1964. *Daas Lebebder Bakterien Jana, Fisher*, 875 pp..

Zutshi, D.P. and Khan, M.A., 1977. Limnological investigations of two subtropical lakes. *Geobios*. 4(2): 45–48.

Zutshi, D.P and Vass, K.K., 1971. Ecology and production of *Salvinia natans* Hoffin in Kashmir. *Hydrobiol.*, 38: 303–320.

Zutshi, D.P. and Vass, K.K., 1973. Variation in the water quality of some Kashmir lakes. *Trop Ecol.*, 14: 182–186.

Zutshi, D.P., Kaul, V. and Vass, K.K., 1972. Limnology of high altitude Kashmir lakes. *Verh. Enforna. t verein Limnol.*, 18: 599–804.

Chapter 3

Entomopathogenic Nematode–A Natural Enemy of Insect Pests: Its Ecology, Mass Production, Storage and Applications

A.K. Chaubey[1], Vidhi Tyagi[1], Zakwan Ahmed[2] and Mohommad Arif[2]

[1]*Nematology Laboratory, Department of Zoology, C.C.S. University, Meerut – 250 004*

[2]*Defence Agriculture Research Laboratory, (DRDO), Haldwani*

ABSTRACT

Entomopathogenic nematodes have been found in all inhabited continents and a range of ecologically diverse habitats ranging from cultivated fields to deserts. *Steinernema* and *Heterorhabditis* are the most commonly studied genera that are in use to control the insect pests. Although phylogenetically these are not closely related but share similar life histories. The cycle begins with an infective juvenile, whose only function is to infect new hosts. After entering an insect, infective juveniles release an associated mutualistic bacterium. These bacteria are of the genus *Xenorhabdus* and *Photorhabdus* for *Steinernema* and

* E-mail: kumar_ac2001@rediffmail.com; akc.nema@gmail.com.

Heterorhabditis, respectively and cause host mortality with in 48 hours.

Steinemematids can be stored at 4–10°C for 6–12 months without much loss of activity while heterorhabditids can be stored for 2–4 months at the same temperature. Steinemematids and heterorhabditids can also be cultured on a variety of substrates *viz.*, potato ash, ground veal pulp and dog food. Currently, a medium based on chicken offal is commonly used.

Steinernema sp. and *Heterorhabditis sp.* provide the best control when environmental conditions keep them healthy under optimal environmental conditions. Pest mortality is temperature dependent. Cooler soil temperature may slow nematode activity and delay pest mortality whereas higher optimal soil temperatures may hasten mortality.

Application of EPNs is usually easy. In most cases, there is no need for special application equipment. Most nematode species are compatible with pressurized, mist, electrostatic, fan and aerial sprayers. Hose-end sprayers, pump sprayers, and watering cans are effective applicators as well. Nematodes can even be applied through irrigation systems on some crops.

Keywords: *EPN, Ecology, Mass Production, Applications.*

Introduction

Nematodes are usually considered harmful because of the diseases they cause in humans and animals and the economic impact they have on many agricultural products. There are, however, a small but significant number of beneficial nematodes, *i.e.* nematodes associated, often parasitically, with insects. Some of these nematodes are of considerable interest because of their potential as biological control agents of insects pest (Entomopathogenic Nematodes).

Entomopathogenic nematodes (EPNs) are soil-inhabiting, lethal insect parasites that belong to the phylum Nematoda, commonly called roundworms. The term Entomopathogenic comes from the Greek word *entomon*, meaning insect, and *pathogenic*, which means causing disease. EPN s live inside the body of insect and so they are designated endoparasitic. They infect many different types of soil insects, including the larval forms of butterflies, moths, beetles and flies, as well as adult crickets and grasshoppers. "EPNs are everywhere; they have been found in all inhabited continents and a range of ecologiclly diverse habitats, from cultivated fields to deserts. The most commonly studied genera (*Steinernema* and *Heterorhabditis*) are those that are useful in the biological control of insect pests (Gaugler, 2006).

Poinar (1979) listed 9 families of nematodes containing members that are facultative or obligate parasites of insects. Nine families (Allantonematidae, Diplogasteridae, Heterorhabditidae, Mermithidae, Neotylenchidae, Rhabditidae, Sphaerularidae, Steinemematidae, and Tetradonematidae) include species that attack insects and kill, sterilize, or alter host development. Both the genera *viz. Steinernema* and *Heterorhabditis* belong to family Steinemematidae and Heterorhabditidae respectively. Bacterial symbionts of EPN in the families Steinemematidae, Heterorhabditidae are members of the family Enterobacteriaceae and belong to the genus *Xenorhabdus* and *Photorhabdus* respectively. These bacteria are carried in the intestinal vesicle of the nonfeeding infective stage of members of the Steinemematidae, and through out the whole intestine of infective juveniles of members of the family Heterorhabditidae. The nematodes release their bacterial symbionts into the hemocoel of the insects, where growth induces a lethal septicemia and contributes to the symbiotic relationship by providing nutrients required by nematode partners during reproduction in insect cadavers.

Most of our knowledge of this group has been developed in the last 20 years and has been stimulated by increased interest in biological control. EPNs are particularly attractive because:

1. They offer little or no environmental hazard;
2. They offer no threat from competitive displacement of other desirable organisms because of their life cycle; and
3. The potential exists for inundative release to give high initial host reduction, or inoculative release to establish the nematode and give partial control for an indefinite period.

General Morphology: Diagnostic Characters

Females

1st Generation

Cuticle smooth, head rounded, not offset from rest of body; six lips, each with a papilla. Fm cephalic papillae occur further back on the head, located in sub-medial positions. Amphids distinc located behind lateral labial papillae. Stoma partially collapsed. Esophagus extending nearly to mouth opening. Cheilorhabdions well sclerotized. Below this is another sclerotized ring that represents the prorhabdions. Other parts indistinct. Esophagus muscular with a cylindrical procorpus, nonvalvated metacorpus, isthmus, and basal bulb with a valve. The nerve ring often surrounded the anterior portion of the basal bulb. The excretory pore opening variable, mostly posterior to metacarpus. Excretory duct unusually prominent forming an elliptically-shaped structure seemingly with a hole at the center. Lateral fields and phasmids not observed. Gonads didelphic, reflexed. Vulva a transverse slit, usuall slightly protruding with a prominent double-flapped epiptygma. Vagina short. Eggs deposited initially but they later hatch inside the females. First generation adults larger than those of the second generation and with a wide tail with a rounded wedge-like projection on the tip. Anteriorly, female similar to male but much larger.

2nd Generation

Second generation female similar morphologically to that of the first generation with the following exceptions: about one-half as long and two-third as wide, valve in basal bulb of esophagus more prominent, elliptically-shaped structure less prominent. Tail of second generation female straight and pointed, sometimes with a slight postanal ventral swelling. Pigmy forms occurred in some instances (Figures 3.1A–C).

Males

1st Generation

First generation male much smaller than first generation female, but anatomically the two are similar anteriorly. Body usually plump. Head rounded continuous with body. Lips not seen but 6 labial papillae prominent, and 4 cephalic papillae well developed, much larger than labial papillae. Procorpus cylindrical, metacorpus slightly swollen, isthmus distinct, basal bulb swollen with small valve. Nerve ring located in isthmus region of esophagus but position variable. Excretory pore located anterior to mid-esophagus. Excretory duct not forming elliptically-shaped structure present as in females. Posterior part of body curved ventrally. Testis .single, reflexed. Spicules paired and strongly curved with blunt tips. Spicule head short, truncate anteriorly. Shaft present but short. Blade with anterior part well curved, posterior part slightly curved or almost straight. Velum large not covering spicule tip. Gubernaculum large. Posterior region with twenty three genital papillae, variable in position. The ventral portion of the tail is usually concave. This character seems to be fairly consistent although the degree of concavity varies somewhat. A thin flap of sclerotized cuticle occurs over the cloaca and serves to close the cloacal opening when the spicules are in a resting position. The cloaca opening is slit-like and on both sides is a pair of small lateral papillae-like structures (Figures 3.1, D–G). Whether these are innervated could not be determined but if so they could function as sensory organs to locate the vulva.

2nd Generation

Second generation males are similar morphologically to that of the first generation except that it is about two-third as long and one-half as wide and the spicules have an elongate head.

Infective Juveniles

These are narrower than the corresponding parasitic juveniles. Mouth and anus closed, esophagus and intestine collapsed usually constricted on the dorsal side. Lateral field with six to eight incisures depending on region of body observed. A bacterial pouch present in anterior portion in intestine containing symbiotic bacteria (Figure 3.2). There was no retractile spine in the tail of the infective (Figure 3.1, H–J).

Biology

Life Cycle

Since EPNs are economically important, the life cycles of the genera heterorhabditidae and steinernematidae are particularly well known. Although not closely related, phylogenetically, both share similar life histories (Poinar, 1993). The cycle begins with an infective juvenile, whose only function is to seek out and infect new hosts.

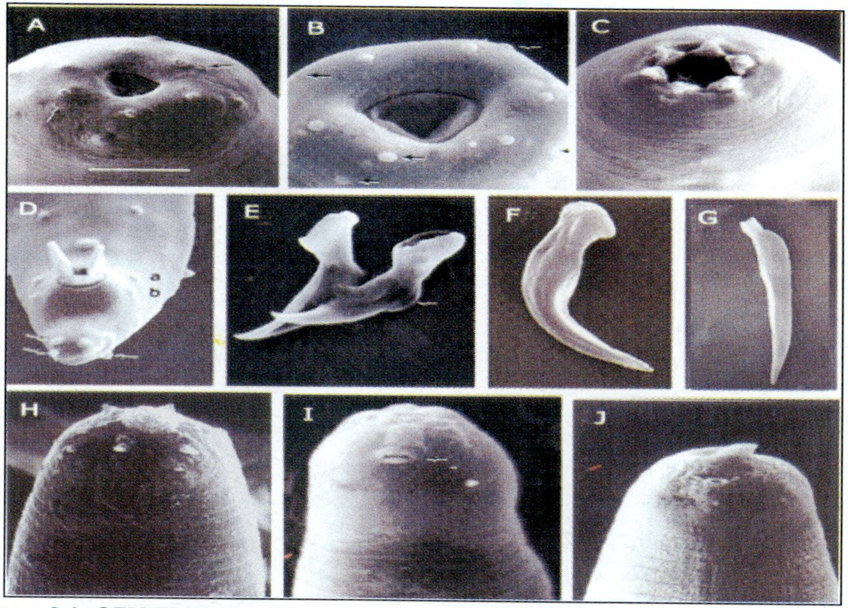

Figure 3.1: SEM Photographs of *Steinernema, Neosteinernema* and *Heterorhabditis*

A–C: Female heads: A. *Steinernema glaseri*; B: *Neosteinernema*; C: *Heterorhabditis*; D–G: Male structures; D: Posterior region of *Steinernema* with papillae; E: Spicules and gubernaculum of *Neosteinernema*; F: Spicule of *Steinernema*; G: Spicule of *Heterorhabditis;* H–J: Heads of infective juvenile; H: *Steinernema scapterisci*; I: *Neosteinernema longicurvicauda*; J: *Heterorhabditis bacteriophora.*

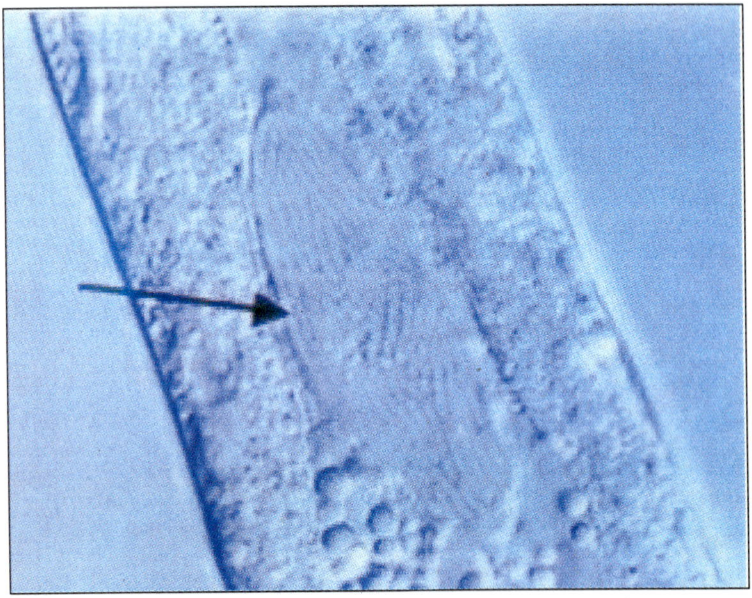

Figure 3.2: Position of Symbiotic Bacteria in *Steinernema* spp.

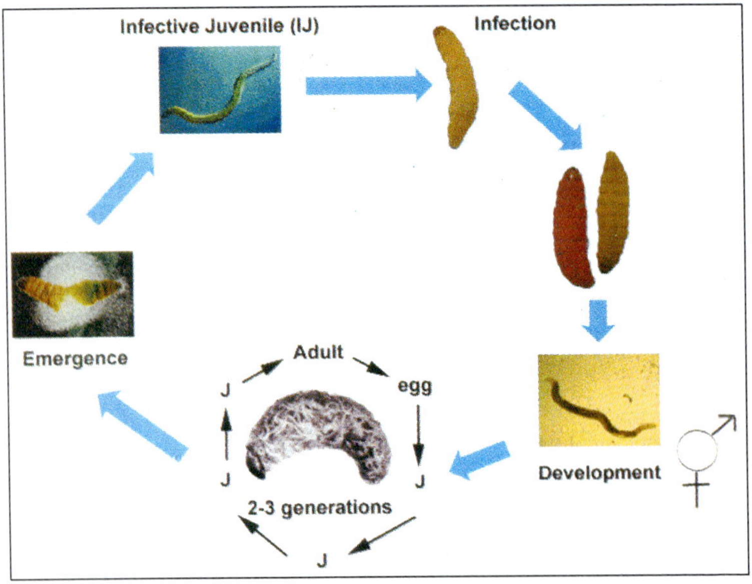

Figure 3.3: Life Cycle Pattern of Entomopathogenic Nematodes

After entering an insect, infective juveniles release an associated mutualistic bacterium. These bacteria are of the genus *Xenorhabdus* and *Photorhabdus* for *Steinernema* and *Heterorhabditis* respectively, cause host mortality within 48 hours. The nematodes provide shelter to the bacteria, which, in return, kill the insect host and provide nutrients to the nematode. Together, the nematode and bacteria feed on the liquefying host, and reproduce for several generations inside the cadaver (Figure 3.3). Steinernematid infective juveniles may become males or females, whereas heterorhabditids develop in to self fertilizing hermaphrodites with later generations producing two sexes. When food resources in the host become scarce, the adults produce new infective juveniles adapted to withstand the outside environment. After about a week, hundreds of thousands of infective juveniles emerge and leave in search of new hosts, carrying with them an inoculation of mutualistic bacteria, received from the internal host environment (Gaugler, 2006).

Foraging Strategies

The foraging strategies of EPNs vary between species, influencing their soil depth distributions and host preferences. Infective juveniles use strategies to find hosts that vary from ambush and cruise foraging (Campbell and Gaugler, 1997). In order to ambush prey, some *Steinernema* species nictate, or raise their bodies off the soil surface so they are better poised to attach to passing insects, which are much larger in size (Campbell and Gaugler, 1993) whereas many *Steinernema* are able to jump by forming a loop with their bodies that creates stored energy which, when released, propels them through the air (Campbell and Kaya, 2000). Other species adopt a cruising strategy and rarely nictate. Instead, they roam through the soil searching for potential hosts. These foraging strategies influence which hosts the nematodes infect. For example, ambush predators such as *Steinernema carpcapsae* infect more insects on the surface, while cruising predators like *Heterorhabditis bacteriophora* infect insects that live deep in the soil (Campbell and Gaugler, 1993).

Population Ecology

Competition and Coexistence

Inside their insect hosts, EPNs experience both intra- and interspecific competition. Intraspecific competition takes place among nematodes of the same species when the number of infective juveniles penetrating a host exceeds the amount of resources available. Interspecific competition occurs when different species compete for resources. In both cases, the individual nematodes compete with each other indirectly by consuming the same resource, which reduces their fitness and may result in the local extinction of one species compete directly, can also occur. For example, a Steinemematid species that infects a host first usually excludes a *Heterorhabditis* species. The mechanism for this superiority may be antibiotics prevent the symbiotic bacterium of the heterorhabditid from multiplying (Kaya and Koppenhofer, 1996). In order to avoid competition, some species of infective juveniles of *S. carpocapsae* are repelled by 24-hr-old infections, likely by the smell of their own species' mutualistic bacteria (Grewal *et al.*, 1997).

Interspecific competition between nematode species can also occur in the soil environment outside of hosts. Millar and Barbercheck (2002) showed that the introduced nematode *Steinernema riobrave* survived and persisted in the environment for up to a year after its release. *S. riobrave* significantly depressed detection of the endemic nematodes *H. bacteriophora*, but never completely displaced it, even after 2 years of continued introductions. *S. riobrave* had no effect on populations of *S. carpocaosae*, which suggests that coexistence is possible. Niche differentiation appears to limit competition between nematodes. Different foraging strategies separate the nematodes in space and enable them to infect different hosts. EPNs also occur in patchy distributions, which may limit their interactions and further support coexistence (Koppenhofer and Kaya, 1996.).

Population Distribution

Entomopathogenic nematodes are typically found in patchy distributions, which vary in space and time, although the degree of patchiness varies between species (Lewis, 2002). Factors responsible for this aggregated distribution may include behavior, as well as the spatial and temporal variability of the nematode's natural enemies like nematode trapping fungus. Nematodes also have limited dispersal ability. Many infective juveniles are produced from a single host could also produce aggregates. Patchy EPN distributions may also reflect the uneven distribution of host and nutrients in the soil (Lewis *et al.*, 1998; Stuart and Gaugler, 1994; Campbell *et al.*, 1998). EPNs may persist as meta-populations, in which local population fragments are highly vulnerable to extinction and fluctuate asynchronously. The meta-population as a whole can persist as long as the rate of colonization is greater or equal to the rate of population extinction (Lewis *et al.*, 1998). Recent studies suggest that EPNs may also use non-host animals, such as isopods and earthworms for transport (Eng *et al.*, 2005).

Community Ecology

Parasites can significantly affect their hosts, as well as the structure of the communities to which they and their hosts belong (Minchella and Scott, 1991). Entomopathogenic nematodes have the potential to shape the populations of plants and host insects, as well as the species composition of the surrounding animal soil community. EPNs affect population of their insect hosts by killing and consuming individuals. If EPNs suppress the population of insect root herbivores, they indirectly benefit plants by freeing them from grazing pressure. This is a trophic cascade in which consumers at the top of the food web (nematodes) exert an influence on the abundance of resources (plants) at the bottom. The idea that plants can benefit from the application of their herbivore's enemies is the principle behind biological control.

Not only entomopathogenic nematodes affect their host insects, they can also change the species composition of the soil community. Many familiar animals like earthworms and insect grubs live in the soil, but smaller invertebrates such as mites, collembolans and nematodes are also common. Aside from EPNs, the soil ecosystem includes predatory, bacteriovores, fungivores and plant parasitic nematode species. EPNs have not adverse effect on mite and collembolan populations (Georgis *et al.*, 1991), yet there is strong evidence that they affect the species diversity of other nematodes (Somaseker *et al.*, 2002). EPNs have no effect on free-living nematodes. However, there may be a reduction in the number of genera and abundance of plant parasitic nematodes, which often remain enclosed within growths on plant root. The mechanism by which insect parasitic nematodes have an effect on plant parasitic nematodes remains unknown. Although this effect is considered beneficial for agricultural systems where plant parasitic nematodes cause crop damage, it raises the question of what other effects are possible. Future research on the impacts EPNs have on soil communities will lead to greater understanding of these interactions.

In above ground communities, EPNs have few side effects on other animals. Study conducted by Bathon (1996) indicated that *Steinernema feltiae* and *Heterorhabditis megidis,* when applied in a range of agricultural and natural habitats, had little impact on non-pest arthropods. Some minimal impacts did occur, however, on non-pest species of beetles and flies. Unlike chemical pesticides, EPNs are considered safe for humans and other vertebrates.

Disturbance

Frequent disturbance often perturbs agricultural habitats and response to disturbance varies among EPN species. In traditional agricultural systems, tilling disturbs the soil ecosystem, affecting biotic and abiotic factors *e.g.* tilled soils have lower microbial, arthropods, and nematode species diversity (Lupwayi *et al.*, 1998). Tilled soil also has less moisture and higher temperatures. In a study, conducted by Millar and Barbercheck (2002), examining the tolerances of different EPN species to tillage indicated that the density of a native nematode, *H bacteriophora,* was unaffected by tillage, while the density of an introduced nematode, S. *carpocapsae,* decreased. The density of a third nematode introduced to the system, S. *riobrave,* increased with tillage The response of EPNs to other forms of disturbance is less well defined. Nematodes were not affected by certain pesticides and are able to survive flooding. The effects of natural disturbances such as fire have not been examined.

Ecological Barriers

Although the biological control industry has acknowledged entomopathogenic nematodes since 1980s, relatively little is understood about their biology in natural and managed ecosystems (Georgis, 2002). Nematode-host interactions are poorly understood, and more than half of the natural hosts for recognized *Steinernema* and *Heterorhabditis* species remain unknown (Akhurst and Smith, 2002). Information is lacking because isolates of naturally infected hosts are rare, so native nematode are often baited using *Galleria mellonella,* a lepidopteran that is highly susceptible to parasitic infection. Laboratory studies showing wide host ranges for EPNs were often overestimates, because in a laboratory, contact with a host is assured and environmental conditions are ideal; there are no "ecological barriers" to infection (Kaya and Gaugler, 1993; Gaugler *et al.*, 1997). Therefore, the broad host range initially predicted by assay results has not always translated into insecticidal success.

The lack of knowledge about nematode ecology has resulted in unanticipated failures to control pests in the field. For example, parasitic nematodes were found to be completely ineffective against blackflies and mosquitoes due to their inability to swim (Lewis *et al.* 1998). Efforts to control foliage-feeding pests with EPNs were equally unsuccessful, because nematodes are highly sensitive to UV light and desiccation. Comparing the life histories of nematode species has a unique array of characteristics, including different environmental tolerances, dispersal tendencies, and foraging behaviors (Lewis *et al.*, 1998). Increased knowledge about the factors that influence EPN populations and the impacts they have in their communities will likely increase their efficacy as biological control agents.

Mass Propagation

Relatively efficient and economical methods for large-scale culturing of steinernematids and heterorhabditids are available. Workable procedure for storage and transport of the nematodes has been developed. Moreover, recent evidences indicate that these nematodes can be stored anhydrobiotically, which will increase the feasibility of using the nematodes in integrated pest management systems.

In vivo Propagation

Since *Steinernematids* and *Heterorhabditids* infect and reproduce in a broad spectrum of insects, they may be readily reared *in vivo* in the laboratory. *Galleria mellonella* (4th–5th instar larva) is often used as a host because it is widely available, easily reared, very susceptible and an excellent host for nematode reproduction. Up to 2,00,000–3,50,000 EPNs can be harvested from one *Galleria* (Dutky *et al.*, 1964, Milstead and Poinar, 1978).

Baiting

Galleria larvae (5th instar) are generally used for baiting. Place 5–7 larvae in beaker containing soil sample and incubated at 28°C. Regular screening is required for dead larvae. The infective juveniles (IJ_2) of the EPN can be collected from the dead larvae. Keep the IJ suspension to room temperature (20–24°C) and examine the nematodes briefly under dissecting microscope. Dead dauers will generally be straight. While the live dauers will actively move about.

Evenly distribute 1 ml of the homogeneous nematode suspension on a 9.0 cm Whatman No. 1 filter paper in the lid of a 100 × 14 mm plastic petri dish. Add 10 conditioned *Galleria* larvae. The goal is to have about 20 nematodes per larva. Too many nematodes per larva produce few progeny due to competition and foreign bacteria. Cover the lid (containing nematodes and *Galleria*) with the inverted petri dish bottom and store them at room temperature. Place infected larvae in to White traps after 5–7 days of infection (White, 1927). *Steinernema* infected larvae turned yellowish brown whereas *Heterorhabditis* infected larvae turn brick-red (Figure 3.4). Heavily contaminated insects or insects dying of other causes are usually blackish and smell putrid. It is best not to harvest from them as their production is poor and they contaminate the whole batch.

Harvesting

To make White traps (White, 1927) place a 9.0 cm Whatman No.1 filter paper in a concave-side-up watch glass in a large glass petri dish (150 × 20 mm) and autoclave for 20 minutes at 121°C. Pour about 70 ml sterilize distilled water (sdw) or 0.1 per cent formalin into the petridish. Do not put any water in to the watch glass. Drape the filter paper over the watch glass so that it comes in to contact with the liquid surface. Place typically infected larvae (10–30 as they fit) on the filter paper over the edge of the watch glass. IJs will start to exit 10–12 days after infection (Figures 3.5, 3.6). The goal is for active juveniles to migrate in to the water or formalin while most of the host tissues slough in to the concavity of the watch glass. Once nematodes begin to appear, they should be harvested daily until production drops (3-4 days). To harvest, remove the watch glass with dead larvae, pour the IJs in to a beaker (rinse the petri dish to collect all the IJs), add 70 ml of sdw or 0.1 per cent formalin and replace watch glass.

Preparation for Storage

Examine the Infective juveniles (IJs) for activity. Host tissues and noninfective stages should not be present. If extraneous matter or large numbers of inactive IJs are collected, a separator step should be performed before the final rinsing procedures.

Separation of active IJs from other material is performed by allowing the IJs to migrate through a filter chosen to retain the host

Figure 3.4: Arrangement of Dead Wax Worms on White Traps to Collect EPN

A: *Steinernema*; B–D: *Heterorhabditis*

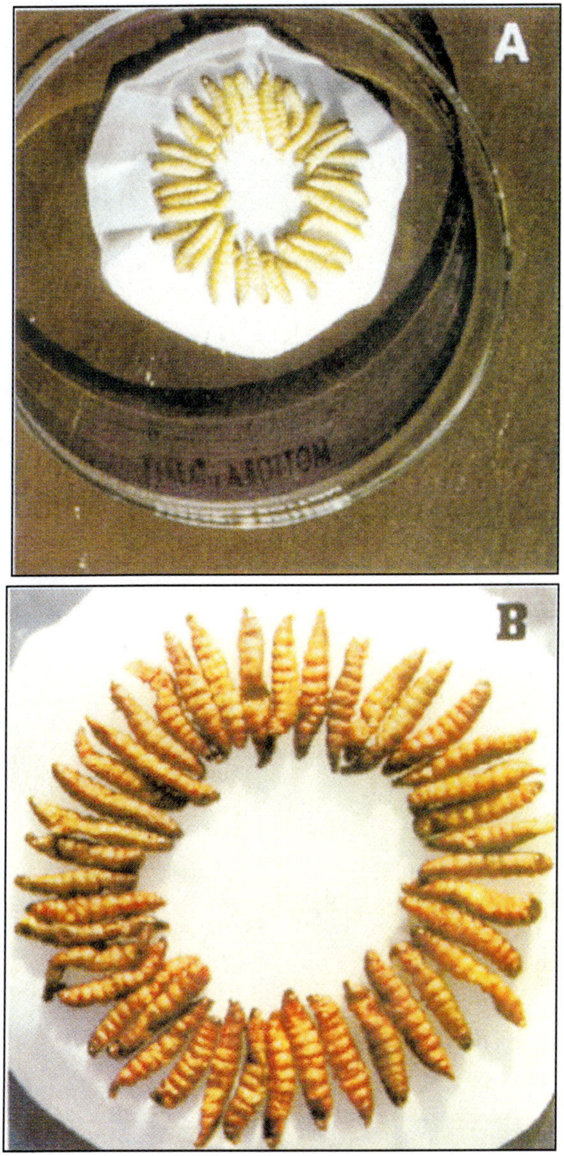

Contd...

Figure 3.4–Contd...

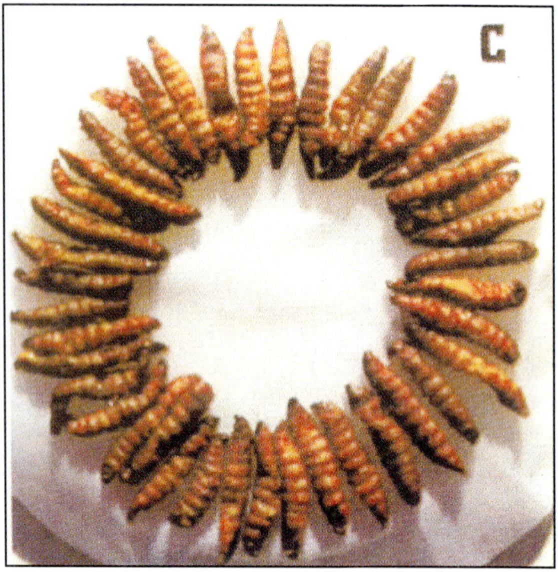

Figure 3.5: Emergence of Entomopathogenic Nematodes from the Dead Larvae

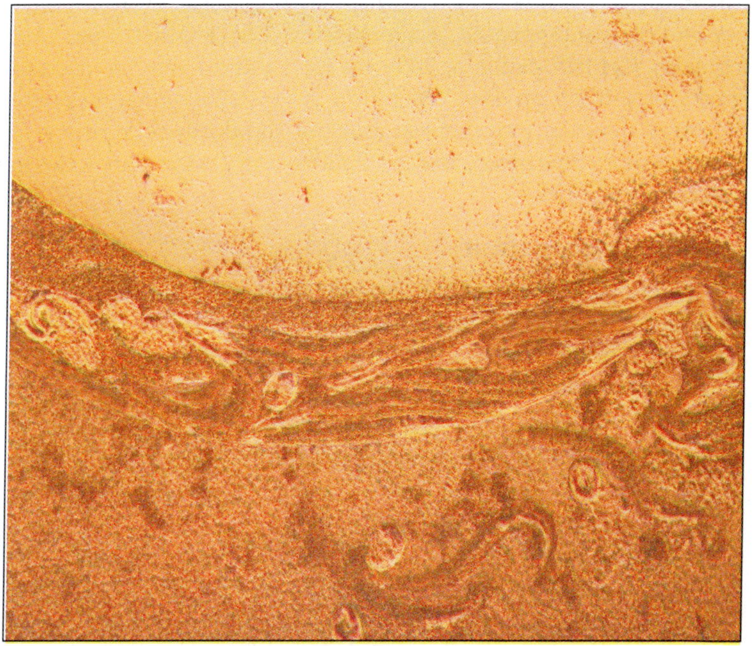

Figure 3.6: Infective Juveniles are coming out from the Anterior Region of Female EPN

tissues and noninfective stages. Noninfective stages of *Steinernematids* may first be killed by rinsing the nematode in 0.4 per cent "Hyamine-I" solution (Methylbenzethonium chloride) for 15 minutes. If desired, noninfective stages of both *Steinernematids* and *Heterorhabditids* may be killed by allowing the nematodes to sit at room temperature for a few days. In some cases, however it will be simpler to use a fine mesh filter to retain the larger stages. Filters in use include:

1. 3–4 stacked "Kimwipesl" or a milk filter supported on a wire mesh in a Baermann funnel.
2. A 500-mesh screen in a separator funnel.
3. A 500-mesh sieve.

In each case the filter is placed in contact with the collection container. Depending upon the numbers, most IJs should have migrated within 8 hours. To rinse IJs, allow them to settle in a beaker. Then aspirate or decant the supernatant, and add more sdw until the suspension is clean (2-4 times). If the suspension appears particularly contaminated, it may be rinsed once with 0.1 per cent formalin. Centrifugation at 300 rpm for 1 minute may be used to speed the settling process. When time is short, the nematodes may be stored overnight in a refrigerator but should then be rinsed at least one more time before storage. Finally, transfer the nematodes to a storage container.

In vitro Propagation

In the past, *Steinernematids* and *Heterorhabditids* have been cultured on a variety of substrates: Potato mash (McCoy and Glaser, 1936), ground veal pulp (McCoy and Girth, 1938) and dog food (House *et al.*, 1965, Hara *et al.*, 1981). Currently, a medium based on chicken offal (Bedding, 1981) is common.

Bedding (1981) has developed a technique whereby huge numbers of nematodes may be economically produced using a chicken offal medium on a porous foam substrate. Polyether polyurethane provides the largest surface-to-volume ratio while providing adequate interstitial space (Bedding, 1986). Glass flasks or large autoclavable bags serve as rearing containers. Proficiency with the flask system is considered a prerequisite to successful culturing in the bags, with average production of over 1 billion *S. feltiae* IJs/bag (Bedding, 1986).

Preparation of Rearing Flasks

Wearing rubber gloves during this procedure is highly desirable. Impregnate small foam pieces (1 cm diameter) with chicken, duck or turkey offal homogenate is required. Bedding (1981) recommends 12.5 parts medium to 1 part foam, by weight. The pores of the foam should still be clearly visible. Fill the flask with foam homogenate mixture to the cotton wrapped in cheese cloth and autoclave for 20 minutes at 121° C. When the offal homogenate has been frozen prior to use, it can develop an odor, so an autoclave deodorizer should be applied.

Inoculation with Bacteria

The day before the flasks will be prepared, liquid cultures of the primary form of the appropriate *Xenorhabdus or Photorhabdus bacteria* should-be-incubated. Cells of the bacterium should be aseptically transferred to 5 ml of nutrient broth in a test tube. If possible, leave the tubes overnight on a shaker or at least vortex the tubes before incubating.

Allow the autoclaved flasks to cool to room temperature. Inoculate by pouring the contents of one tube of bacteria in to each flask. Shake to mix the broth and bacteria throughout the foam substrate. Store for 2-3 days at 25°C to allow the respective bacterial population build up.

Inoculation with Nematodes

When monoxenic cultures are already available, foam from them is used to septically inoculate the new flasks. One flask can be divided in to about seven new ones. Care should be taken to maintain monoxenicity during the transfers. Best results are obtained if the flask is not shaken after introduction of the nematodes. The flask will be ready to harvest in about 2 weeks. It is possible to employ a much smaller inoculum such as one or two pieces of foam. However, harvest will then be delayed to 4 weeks, and the final yield will be reduced.

When monogenic cultures are not available, or when their purity is suspected, surface-sterilized infective may serve as the inoculum. Do not add too much liquid with the nematodes (< 5 ml). When beginning with newly sterilized IJs, the purity of the new cultures should be verified with in 2–3 days. Aseptically streaking from the flask onto MacConkey agar or NBTA (Nutrient agar, bromothymol blue and tetrazolium chloride) and then checking for appropriate colony colour and morphology should generally be sufficient to verify purity. Cell morphology (rod-shaped) and motility may also be examined, if desired.

Harvesting

The foam may be piled 5 cm deep on a 20-mesh sieve (20 meshes/inch). Place sieve in a pan of tap water level adjusted so that the foam is just submerged. If a mist chamber is available, the pan and sieve may be placed in it for 2–24 hours. Longer periods tend to be homogenate in to the water. With in 2 hours 9.5 per cent of the IJs will migrate in to the water (Bedding, 1981).

The nematodes may be sedimented and rinsed if necessary to remove particulate matter and inactive IJs. The IJs may then be permitted to migrate through a 500-mesh sieve. Nematodes rinsed from the inside of the flask should also be allowed to migrate through the 500-mesh sieve to remove particulate matter. (For steinemematids, the non-IJs may be killed in a "Hyaminel" solution before migration through the sieve.) the nematodes should be rinsed several times until the water appears clear. An antibiotic rinse may be seemed appropriate. Rather than rinsing several times right after harvesting, Bedding (1986) first aerated steinemematids for 1 week to break down small particles of medium then rinsed the IJs.

Troubleshooting

A drop in production could indicate contamination, a reversion to the secondary phase of the bacteria (*Xenorhabdus* and *Photorhabdus*), unsuitable incubation temperatures, improper moisture content or many other problems. Contamination will sometimes be visually evident in the form of fungal or bacterial colonies or 'unusual' colouration or exudates. At other times an 'unusual' odor may indicate that contamination is a problem. In any case, purity should be routinely verified by streaking on to NBTA or MacConkey agar. Using media with specific dyes also permits determination of the proportion of primary and secondary form bacteria. By the end of two weeks, it is normal for a flask to contain a high proportion of secondary bacteria; however, it is important then the original bacterial inoculation for each flask be done with primary-phase bacteria.

As to incubation temperatures, 25°C seems optimal for *S. feltiae*, but the optimum may be different for different species. Room temperatures of 20–25°C should generally be adequate for good production of nematodes.

Moisture content can sometimes be difficult to adjust. It may be necessary to try adding more or less liquid at various steps in the procedure. If the medium is too liquid, the nematodes will be trapped in standing water and dye. If the medium is too dry, the cultures will dry out before large number of nematodes is produced. Maintaining high humidity, where flasks are incubated, may be helpful.

Production levels can remain adequate while quality, measured in terms of infectivity or activity, drops. *In vivo* passage is recommended every four to six months to avoid reduced virulence and pathogenicity of the nematodes to their insect hosts. Otherwise, strict attention to maintaining optimum conditions for growth, development and storage will reduced the potential for problems with nematode quality.

It is impossible to foresee every potential problem. However, each technician will become familiar with his/her own system and its responses to procedural changes. With some knowledge of nematode biology, logical causes and solutions for most troubles should not are too difficult to find. Even so, *in vitro* production is not easy, and trial and error will be needed for the novice to perfect the system.

Primary-phase Bacterial Cultures

Primary-phase bacteria may be isolated and recognized by methods adopted after Poinar (1996). Primary colonies may be suspended in 17per cent glycerol-nutrient broth in Martney bottles and stored at 18°C (Akhurst, 1980). In this case the suspension is rapidly thawed in a water bath at 60°C prior to use (Bedding, 1986). Alternately, the colonies can be stored at 12–25°C and routinely sub cultured (monthly at 12°C or weekly at 25°C). It is best to use NBTA or, MacConkey agar so that colonies reverting to the secondary form will be easily recognized and will not be used.

Storage

The nematodes may be stored in distilled water with a drop if Triton X-100 (a wetting agent that prevents nematodes from sticking to the side of container) or 0.1 per cent formalin (use this only if continuing or recurring contamination seems to be a problem). If stored without aeration, the nematodes should be concentrated to no more than 10,000–20,000 nematodes/ml, and the water depth should be 1 cm or less. Tissue culture flasks are ideal for this type of storage since there is a large surface to volume ratio.

Higher nematode concentrations (1,00,000/ml) will not be detrimental in aerated suspension. An aquarium pump or any forced air supply can be attached to an aquarium stone (a porous structure that creates many fine bubbles as the air is force through) in the nematodes suspension. *Steinernematids* can be stored at 4–10° C for 6–12 months without much loss of activity. The *Heterorhabditids* do not store as well, but may be stored for 2–4 months at 4–10°C is considered good.

Bedding (1986) also proposed the following storage procedure: For up to 1 year, a well-rinsed semisolid paste of steinemematid IJs may be stored on moist autoclaved polyether polyurethan foam. Coat the foam with 10 times its weight IJs. Place the foam in a sterilized container and force air through bacterial filters (0.45 microns). The antiseptic precautions are important to minimize fungal and bacterial contamination. The IJs can be extracted later by migration through a sieve as during the harvesting process. Or, foam may be squeezed in water until all the tan 'nematode' colouration is gone. Extraction is time consuming for large scale commercial uses but pose few problems for research purposes.

Nematode Species and Target Insect Pests

Steinemema and Heterorhabditis nematodes provide the best control when environmental conditions keep them healthy under optimal environmental conditions, nematode infected target pests. Pest mortality is temperature dependent. Cooler soil temperature may slow nematode activity and delay pest mortality whereas higher optimal soil temperatures may hasten mortality. Certain species (*S. carpocapsae)* and relatively inactive, remain near the soil surface and use an 'ambusher' strategy in which they stand on their tails and await passing insects. Others (*H. bacteriophora*) penetrate more deeply in to the soil matrix and use an active 'cruiser' strategy to locate and infect sedentary insects. Still others (*S. feltiae*) use an intermediate or mixed strategy. Species such as *H. marelatus* and *H. megidis* are cold adapted whereas *S. riobrave* is warm temperature adapted. Therefore, the selection of appropriate nematode species to watch the biology and environment of the target pest species is very important to achieve effective control. Although they have received limited research attention for over a century, nematode known as Entomopathogenic nematodes that infect and kill insects have been developed seriously as biological control agent only since the early 1980s. The nematodes consider at present to be environmentally safe and acceptable, and they actively seek out additional host insects.

The entomopathogenic nematodes are lethal to an extraordinarily broad range of insect pests in the laboratory. Field host range is considerably more restricted, with some species being quite narrow in host specificity. When considered as a group of nearly 30 species, however, EPNs are useful against a large number of insect pests, many of which are listed in the Table 3.1.

Applications

Application is usually easy and in most cases, there is no need for special equipment. Most nematode species are compatible with pressurized, mist, electrostatic, fan and aerial sprayers. Hose-end sprayers pump sprayers and watering cans are effective applicators as well. Nematodes can even be applied through irrigation systems on some crops. Check the label of the nematode species to use the best application method. Repeat applications if the insect is in the soil for a longer period of time.

There is no need for masks or specialized safety equipments. Insect parasitic nematodes are safe for plants, animals and even the human beings (worms, other non-targeted organisms, birds, pets, children). Dosage could vary according to the pest or product, but normally about 2.5 billion/hectare showing the satisfactory results. Following important points should follow while applying nematodes:

1. Nematodes must be applied during early morning or late evening when intensity of lethal sunlight is minimal.

Table 3.1: Entomopathogenic Nematodes and their Respective Target Pests

Sl.No.	EPN Species	Target Pests
1.	*Steinernema carpocapsae*	Artichoke plume moth, Root weevils, Cranberry girdler, Wood borers, Billbugs, armyworm, Cutworm, webworm, Annual bluegrass weevil, Bluegrass billbug, Hunting billbug, Black cutworm, Dog/cat flea larvae, Europearn crane fly, Armyworms, Sod webworms
2.	*Steinernema riobravae*	Root weevils, Mole crickets, Tawny mole cricket, Southern Mole-cricket
3.	*Steinernema feltiae*	Sciarids, Fungus gnats
4.	*Steinernema scapterisci*	Mole-crickets, Tawny Mole-cricket, southern Mole- cricket
5.	*Heterorhabditis bacteriophora*	Root weevils, Wood borers, Scarabs, Billbugs, Cranberry girdler, Blackvine weevil, Strawberry root weevil, Black turfgrass ataenius, European chafer, Green June beetle, Japanese beetle, Beetles appearing in May/June, NO masked chafer, SO Masked chafer, SW masked chafer, West Masked chafer
6.	*Heterorhabditis megidis*	Root weevils
7.	*Heterorhabditis marelatus*	Cranberry girdler, Blackvine weevil, Strawberry root weevil

2. Nematodes should not be applied to hot and dry soil. If the soil is dry and hot, apply irrigation of at least 0.1 inch of water before applying the nematodes.
3. Agitation must be provided in the spray tank to ensure proper mixing and dispersion of the product during spraying.
4. Sprayer screen sizing must be 50 mesh or coarser or screens can be removed to prevent clogging.

If nematodes are to be applied by fustigation equipment following additional precautions should be followed:

1. Pre-mix the required amount of product in sufficient water to uniformly inject the entire irrigation system.
2. Inject nematode during the second watering. Add nematode to a clean feeder tank. Suspension must be continuously agitated during injection to prevent nematodes from settling. After injection rinse feeder tank with clean water and inject into system to purge nematodes from fustigation system.
3. Nematode should not be applied if soil temperature is below 65°F or above 85° F at 1–2 inches below the soil surface.
4. Apply at least ¼ inch of irrigation immediately following application to wash nematodes off the foliage and facilitate penetration through the thatch and into the soil.

Acknowledgements

Authors are thankful to the Head, Department of Zoology, C.C.S. University, Meerut. Life Science Research Board, Defence Research and Development Organisation, New Delhi is also thankfully acknowledged for the financial assistance by providing research project and fellowship (LSRB–12I/FS&B/2007) to AKC and VT respectively.

References

Akhurst, R.J., 1980. Morphology and functional dimorphism in *Xenorhabdus* spp. bacteria symbiotically associated with insect pathogenic nematodes *Neoaplectana and Heterorhabditis. J. Gen. Microbiol.,* 121: 303–309.

Akhurst, R. and Smith, K., 2002. Regulation and safety. In: *Entomopathogenic Nematology*, (Ed.) I. Gaugler. CABI Publishing, New Jersey, p. 311–332.

Bathon, H., 1996. Impact of entomopathogenic nematodes on non-target hosts. *Biocont. Sci. Technol.*, 6: 421–434.

Bedding, R.A., 1981. Low cast *in vitro* mass production of *Neoaplectanata* and *Heterhabditis* species (Nematoda) for control of insect pests. *Nematologica*, 27: 109–114.

Bedding, R.A., 1986. Mass rearing, storage and transport of entomopathogenic nematodes. *Proc. 4th Intl. Colloq. Invert. Pathol.*, Veldhoven, The Netherlands, p. 308–311.

Dutky, S.R., Thompson, J.V. and Cantwell, G.E., 1964. A technique for the mass propagation of the DD-136 nematode. *J. Insect Pathol.*, 6: 417–422.

Campbell, J.F. and Gaugler, R., 1993. Nictation behavior and its ecological implications in the host search strategies of Entomopathogenic nematodes. *Behavior*, 126(3–4): 155–169.

Campbell, J.F. and Gaugler, R.R., 1997. Inter-specific variation in Entomopathogenic nematode foraging strategy: Dichotomy or variation along a continuum? *Fund. Appl. Nematol.*, 20(4): 393–398.

Campbell, J.F., Orza, G., Yoder, F., Lewis, E. and Gaugler, R., 1998. Spatial and temporal distribution of endemic and released entomopathogenic nematodes populations in turfgrass. *Entomologia Experimentalis et Applicata*, 86: 1–11.

Campbell, J.F. and Kaya, H.K., 2000. Influence of insect associated cues on the jumping behavior of Entomopathogenic nematodes (*Steinernema* spp.). *Behavior*, 137(5): 591–609.

Eng, M.S., Preisser, E.L. and Strong, D.R., 2005. Phoresy of the entomopathogenic nematode *Heterorhabditis marelatus* by a non-host organism, the isopod *Porcellio scaber*. *J. Inverte. Pathol.*, 88(2): 173–176.

Gaugler, R., 2006. *Nematodes Biological Control.* Cornell University Press, Cornell.

Gaugler, R., Lewis, E. and Stuart, R.J., 1997. Ecology in the service of biological control: The case of entomopathogenic nematodes. *Oecologia*, 109: 438–489.

Georgis, R., Kaya, H.K. and Gaugler, R., 1991. Effects of *Steinemematid* and *Heterorhabditid* nematodes (Rhabditida : Steinemematidae and Heterorhabditidae) on non-target Arthropods. *Environ. Entomol.,* 20(3): 815–822.

Georgis, R., 2002. The biosys experiment: An insider's perspective. In: *Entomopathogenic Nematology,* (Ed.) I. Gaugler. CABI Publishing, New Jersey, p. 357–371.

Grewal, P.S., Lewis, E.E. and Gaugler, R., 1997. Response of infective stage parasites (Nematoda : Steinemematidae) to volatile cues from infected hosts. *J. Chem. Ecol.,* 23(2): 503–515.

Hara, A.H., Lindegren, J.E. and Kaya, H.K., 1981. Monoxenic mass production of the entomopathogenic nematode, *Neoaplectana carpocapsae* Weiser on dog food/agar medium. *USDA/SEA, AAT–16, Oakland, CA.*

House, H.L., Welch, H.E. and Clough, T.R., 1965. Food medium of prepared dog biscuit for the mass production of the nematode DDI 36 (Nematoda : Steinemematidae). *Nature,* 296: 847.

Koppenhofer, A.M. and Kaya, H.K., 1996. Coexistence of two *Steinemematid* nematode species (Rhabditida : Steinemematidae) in the presence of two host species. *Appl. Soil Ecol.,* 4(3): 221–230.

Kaya, H.K. and Gaugler, R., 1993. Entomopathogenic Nematodes. *Ann. Rev. Entomol.,* 38: 181–206.

Kaya, H.K. and Koppenhofer, A.M., 1996. Effects of microbial and other antagonistic organism and competition on entomopathogenic nematodes. *Biocont. Sci. Technol.,* 6(3): 357–371.

Lewis, E.E., Campbell, J.F. and Gaugler, R., 1998. A conservation approach to using entomopathogenic nematodes in turf and landscapes. In: *Conservation Biological Control,* (Ed.) P. Barbosa. Academic Press, San Deigo, p. 235–254.

Lewis, E.E., 2002. Behavioral Ecology. In: *Entomopathogenic Nematology,* (Ed.) I. Gaugler. CABI Publishing, New Jersey, p. 205–224.

Lupwayi, N.Z., Rice, W.A. and Clayton, G.W., 1998. Soil microbial diversity and community structure under wheat as influenced by tillage and crop rotation. *Soil Biol. Biochem.,* 30: 1733–1741.

McCoy, E.E. and Glaser, R.W., 1936. Nematode culture for Japanese beetle control. *New Jersey Department of Agriculture Circular,* No. 265.

McCoy, E.E. and Girth, H.B., 1938. The culture of Neoplectana glaseri on veal pulp. *New Jersey Department Bur. Plant Indus. Circ.,* 285: 12

Millar, L.C. and Barbercheck, M.E., 2001. Interaction between endemic and introduced entomopathogenic nematodes in conventional till and no till corn. *Biol. Cont.,* 22: 235–245.

Millar, L.C. and Barbercheck, M.E., 2002. Effects of tillage practices on entomopathogenic nematode in a corn agroecosystem. *Biol. Cont.,* 25: 1–11.

Milstead, J.E. and Poinar, G.O. Jr., 1978. A new entomopgagus nematode for pest management system. *Calif Agric.,* 32: 12.

Minchella, D.J. and Scott, M.E., 1991. Parasitism: A cryptic determinant of animal community structure. *Trends in Ecology and Evolution,* 6(8): 250–254.

Poinar, G., 1996. Fossil velvet worm in Baltic and Domoinican amber: Onychophoran evolution and biogeography. *Science,* 273: 1370–1371.

Poinar, G.O., 1993. Origins and phylogenetic relationships of the entomophilic rhabditis, *Heterorhabditis* and *Steinernema. Fund. Appl. Nematol.,* 16(4): 333–338.

Poinar, G.O. Jr., 1979. *Nematode for Biological Control of Insects.* Boca Raton, CRC Press, Florida.

Somasekar, N., Grewal, P.S., De Nardo, E.A.B. and Stinner, B.R., 2002. Non-target effects of entomopathogenic nematodes on the soil community. *J. Appl. Ecol.,* 39: 735–744.

Stuart, R.J. and Gaugler, R., 1994. Patchiness in populations of entomopathogenic nematodes. *J. Inverte. Pathol.,* 64: 39–45.

White, G., 1927. A method for obtaining infective nematode larvae from culture. *Science,* 66: 302–303.

Chapter 4

Taxonomic Description and Ecological Notes on Some Prosobranchiate Molluscs from Jammu Region, J&K State

Fareed Ahmad[1], *Shakti Prashar*[2] *and Anil K. Verma*[3]

[1]*P.G. Department of Zoology, University of Jammu, Jammu – 180 001*

[2]*Village Dharana, Tehsil Mendhar, Distt. Poonch – 185 211, J&K*

[3]*Department of Zoology, Govt. College for Women, Gandhi Nagar, Jammu – 180 001*

ABSTRACT

Freshwater molluscs are known to play a significant role in the public and veterinary health and thus needs to be explored more extensively and scientifically. Never the less very little information is available on their biodiversity in Jammu region and present study documents the biodiversity of the molluscs inhabiting lotic freshwater of Jammu and adjoining areas.

Keywords: Gastropod molluscs, Viviparidae, Bithyniidae, Thiaridae.

Introduction

The phylum mollusca is a large assemblage of animals having diverse shapes, sizes, habits and occupy different habitats (Subba

Rao, 1993), and the molluscs are second only to Arthropoda in numerical abundance (Abbott, 1989; Bouchet, 1991). The number of species identified under phylum mollusca vary between 80,000 (Boss, 1973) to 2,00,000 (Ponder and Lindberg, 2008). There are around 100,000 extant species within the phylum (Barnes *et al.*, 2001) with an estimated 70,000 extinct species (Brusca and Brusca, 2003). Of these 31,000–100,000 are marine, 14,000–35,000 terrestrial and about 5,000 freshwater species (Abbott, 1989; Seddon, 2000).

Jammu region of the J&K state has hitherto remained poorly surveyed regarding molluscan faunal elements. Some exploratory work has been done by Duda and Verma (1992) and hence this investigation has been undertaken to consolidate further data in this regard.

The main objective of this study has been based on the periodic surveys conducted for the year 2007–2008 to search out various freshwater environments for the diversified malacofaunal elements and to collect the diversified specimens from their natural habitats by methods adopted after Parker and Haswell (1967), Tonapi (1980), Subba-Rao (1989) and Duda and Verma (1993).

Material and Methods

The flowing water bodies including rivers, stream, nallahs and irrigation canals were the main collection sites of these molluscs from the aquatic. Molluscs were mainly collected by hand picking method and by using nets: The hand digging is traditional, hard and man oriented method and is more preferable technique without damaging the habitat and its nearer area (Duda and Verma, 1995) (Plate 4.1, Figures a,b,c,d,e,f,g).

Nets used were of two types:

Hand Net

A hand net made up of a fine mosquito net fixed to round iron-ring and fitted to a wooden handle was used, and

Scoop-Net

A scoop-net usually had a metallic rod with 30 cm × 30 cm frame of steel bars and wire netting and the scoops 10 cm, deep with a 8 cm wide blade soldered to the frame was also closed. A wooden handle was attached at the other end.

The hand net or scoop net was dragged over the aquatic weeds, the contents was poured out on a spread-out cloth piece or enamel tray. The molluscs were picked up with a pair of forceps from the weeds and some time picked by hand from underneath of the stones, attached to rocks, wooden logs, vegetables matter etc.

Identification of Gastropods

The shell characters such as shape, spire length and shape, mouth opening, opercular shape, umbilicus shape and size, colour and ornamentation of the shell were used mainly for the identification of gastropods.

When the specimens were allocated to particular family, identification of the genus was made with aid of key given by Malek and Cheng (1974), Duda and Verma (1995), Subba Rao (1989), Tonapi (1980), and their status was ascertained by analyzing collected samples.

Description of the Taxa and Ecological Notes

Gastropods includes the animal with a univalve shell; derived from Greek words 'Gastros' means 'stomach or gut' and 'podos' means 'Foot'; Gastropoda is the largest and most varied molluscan class and its members occupy a wide variety of marine, fresh water and terrestrial habitats.

The prosobranchiates belonging to order: Mesogastropoda, collected during this work were identified and were representing three families *viz.* Viviparidae, Bithyniidae and Thiaridae, of the total five recognized families including Pilidae and Hydrobiidae which remain unrepresented in present collection. The taxonomic description of the collected molluscan elements has been described as under:

Prosobranchia (or Stereptoneura or Ctenobranchiata or Pectinobranchiata) is a subclass of the class Gastropod in which specimens possess following characters:

The Prosobranchiates are dioecious Gastropods in which the visceral loop is twisted in the form of figure of '8' due to torsion, and they are almost always with the shell and an operculum, shell aperture is usually with a calcareous or horny operculum, the ctenidia lie in front of the heart and mantle cavity opens anteriorly. This group is distinguished by a small radula with few teeth in a transverse series. Respiration occurs through gills (except in few terrestrial species) and presence of single internal gill with one row of leaflets and sex organs are separate.

Order: Mesogastropoda (or Pectinobranchiata or Monocardia): Single monopectinate ctenidium, one auricle and one Nephridium; Nervous system more concentrated; sexes are separate. This order included five families *viz.*, Family: (Viviparidae; Pilidae (Ampullariidae); Hydrobiidae; Bithyniidae; and Thiaridae)

Key to Family Viviparidae

Operculum horny; shell pyramidal; whorls regularly increasing in size; animal with gill only; no labial palps .... Viviparidae.

Family–Viviparidae

Characters

Shell is generally olive brown or green, moderately large, turbin form, perforate or imperforate; whorls inflated; aperture ovate and slightly retracted below; shell of females larger than that of males; operculum horny, concentric with a sub central nucleus.

Respiration entirely aquatic, monopectinate ctenidium or gill; viviparous and sexually dimorphic. In male the right tentacle stout, truncated and modified as a copulatory organ containing the penis; the left tentacle in male and both the tentacle in female long and slender; after copulation fertilization take place internally; Pallial-oviduct enlarged and acts as a uterus. It contains embryos in various stages of development; young ones are shed into water.

Subfamily–Bellamyinae

Characters

In males a big clean shaped complex of testes on the right side of the roof of mantle cavity; females with complex seminal receptacle; shells unbanded in the embryonic stage.

Key to Genus–*Bellamya*

Shell smooth, with out traces of distinct ridges or sculpture, adult shell medium sized, with or without coloured bands and when no bands the shell feebly ridged or keeled, edge of mantle not so thickened and without a conspicuous sphincter, superior margin of gill lamella never thrown into fold, embryonic shell with three

Plate 4.1: Showing the Different Lotic Water Bodies of Jammu Region (a) River Tawi (Jammu); (b) River Chenab (Akhnoor); (c) Jhajjar Kotli Stream (Udhampur); (d) Devika Stream (Udhampur); (e) Basanter River (Samba); (f) Nallah (Kathua); (g) Aik Nallah (Arnia)

Plate 4.2: Showing Morphology and Site of Collection of *Bellamya bengalensis*

(a) Dorsal View; (b) Ventral View; (c) and (d) Irrigation Canal (Arnia)

primary ridges bearing chaetae and a number of secondary ridges .... *Bellamya.*

Genus–*Bellamya* (= *Viviparus*) Jousseaume, 1886.
Jousseaume, 1886, *Bull. Soc. Zool. France,* II: 478.
Type species: *Paludina bellamya* Jousseaume.

Characters

Shells are well developed and of large size. Shell more or less oblong in outline, obtusely or strongly keeled at the periphery, scarcely complex or rather flattened whorls, thinner columellar margin. The spire and body whorl of about equal height, whorls tumid, body whorl evenly convex in profile, umbilicus narrow; dark spiral bands variable and irregular, though the coloration and the bands on the surface in some of the cases are much faded, other characters of the shells agree well with those of form typica. In adults, shell usually without ridges or spines. Most of the shells are of an olive green or golden yellow colour, with the rim of aperture being typically black.

Head small in relation to the size of the animal. Margin of the mantle moderately thick with three short processes corresponding in position to the three rows of chaetae on the embryonic shell. Uterus usually with a large number of embryos in various stages of development.

Bellamya bengalensis (Lamarck, 1882) (Plate 4.2, Figures a, b, c and d)

Key to Species

Shell with dark spiral bands; suture not greatly impressed .... *Bellamya bengalensis.*

Habitat

Specimens were mostly collected from irrigation canals of the paddy fields having depth 4–5 feet and along the bank of the streams.

Key to Family–Bithyniidae

Operculum calcareous and concentric, as large as aperture and cannot be withdrawn inside, verge bifurcate .... Bithyniidae.

Family–Bithyniidae

Characters

Shell ovate to elongate, turrated, narrowly umbilicate to imperforate, moderately convex whorls. Aperture ovate to round, some times expended basally, peristome continues and thickened. columellar fold ridge like. Operculum as large as aperture, thick calcareous and concentric with the small subcentral, spiral nucleus. Tentacles long with thickened bases bearing eyes on lateral side.

Radula 2 : 1, central tooth of radula with or without a series of later basal denticulation on either side. Sexes separate. Verge located on the dorsal side of the neck, curved fleshy and bifid. A long finger like appendage on the inner concave part. A small, cup shaped skin flap attached to the right side of the head just behind the right tentacle.

Subfamily–Bithyniinae

Characters

Shell not very thick, smooth to the naked eye or with fine spiral ridges; ovate or globose, collumellar fold ridge-like, central tooth with a series of latero-basal denticulation on both sides.

Key to Genus–*Bithynia*

Operculum concentric, thick with central or subcentral nucleus, internally smooth, columellar fold ridge-like and prominent, a well defined groove running downwards from the umbilicus; shell generally more elongate, smooth or with microscopic spiral striae; Outer lip thin, non-protruded or angulate at its inner extremity *Bithynia*.

Genus–*Bithynia* Leach, 1818

Bithynia Leach, 1818 in Abel's _Narrative of journey into interior of China_, p. 362.

Type species: _Helix tentaculata_ Linnaeus, 1758.

Characters

Shell elongate, conical and acuminate. Whorls convex, usually perforate but constricted, a well defined groove extending obliquely downwards from the umbilicus to columellar fold, a sharp and prominent ridges forming a well along the inner margin of the groove.

Operculum not very thick, externally with coarse concentric ridges, nucleus central or sub central, internally, almost smooth. The denticulation of the lateral teeth greatly enlarged, denticulation of marginal minute and sharp.

Bithynia tentaculata kashmirensis (Nevill, 1884) (Plate 4.3 Figures a, b, c)

Diagnostic Characters

Shell conically ovate greenish brown, whorls five, smooth, convex, apex rather sharp, umbilicus minute or almost closed, aperture pyriformly ovate, lip generally dark edged, scarcely reflected; operculum with the central nucleus.

Habitat

Specimens were collected along the bank of a stream near Arnia with slow running water and having moderate depth. The bottom of stream was muddy with little sand.

Key to Family–Thiaridae

Shell mostly turrated or when globose operculum completely closing the aperture, mantle border fringed with a brood pouch .... Thiaridae.

Family–Thiaridae

Characters

Shell high, elongate turreted to ovate conical, whorls rounded with moderate or impressed sutures, sculpture, smooth to spiral or longitudinal ridges, ribs, knobs or tubercles imperforate, aperture generally narrowly ovate and some what canaliculated. Operculum smaller than aperture, dark brown, corneous and multi spiral with the central or subcentral nucleus of paucispiral with an excentric nucleus.

Animal with a short, broad snout slightly bifurcate and laterally "keeled". Eyes on short stalks near the external base of tentacles. Mantle margin smooth or papillate. Foot small squarish, radula with the formula 2 : 1 : 1 : 1 : 2. Cutting edges of teeth with a few numerous cusps.

Either oviparous or ovoviviparous with the brood pouch for the eggs to hatch, birth pore situated on the right side of the neck, males rare in many species.

Key to Subfamily–Thiarinae

Shell moderately elongate, usually not more than 12 whorls; mantle margin with about 14 or more papillae; aperture more or less vertical and proportionately smaller .... Thiarinae.

Subfamily–Thiarinae

Characters

Shell imperforate within thick periostracum. Ovately conical to an elongately turrated shape, smooth or spirally sculptured, tuberculated or spiny. Aperture more or less elongately ovate, acutely angulated above, usually indented below, operculum oblong with the terminal nucleus. Mantle margin with about 14 papillae.

Genus–*Thiara*

Characters

Shell generally measures 3 to 5 cm in length; whorls varying in number from 8 to 15; sutures deep to moderately deep or shallow; sculpture also variable, spiral striae to beaded and a few species with spines; aperture more or less vertical, without any sinus or siphonal canal; operculum pear shaped; with an excentric nucleus at the narrow basal end. Mantle edge bears 10 to 14 papillae; ovoviviparous with neck brood pouch.

Key to Subgenus–*Melanoides*

Spire proportionately larger than the body whorl, sculptured with spiral striae crossed by axial striae giving a tubercled appearance; shell neither very slender nor elongated, generally not more than 12 whorls, without hairs .... *Melanoides.*

Subgenus–*Melanoides* Olivier, 1807.

***Melanoides* Olivier, 1807, *Voyage a 1 'Emp. Ottoman,* 2: 40.**

Type species: *Melanoides fasciolata* Olivier= *Nerita tuberculata* Muelier, 1776.

Characters

Shell small or of moderate size with a number of regularly increasing whorls, spire usually twice the length of the aperture or more, sculpture of axial and spiral striae or thin ribs often producing a more or less granular appearance. Aperture small, ovoid and not

Plate 4.3: Showing Morphology and Site of Collection of *Bithynia tentaculata*

(a) Dorsal View; (b) Ventral View; (c) Stream near Arnia

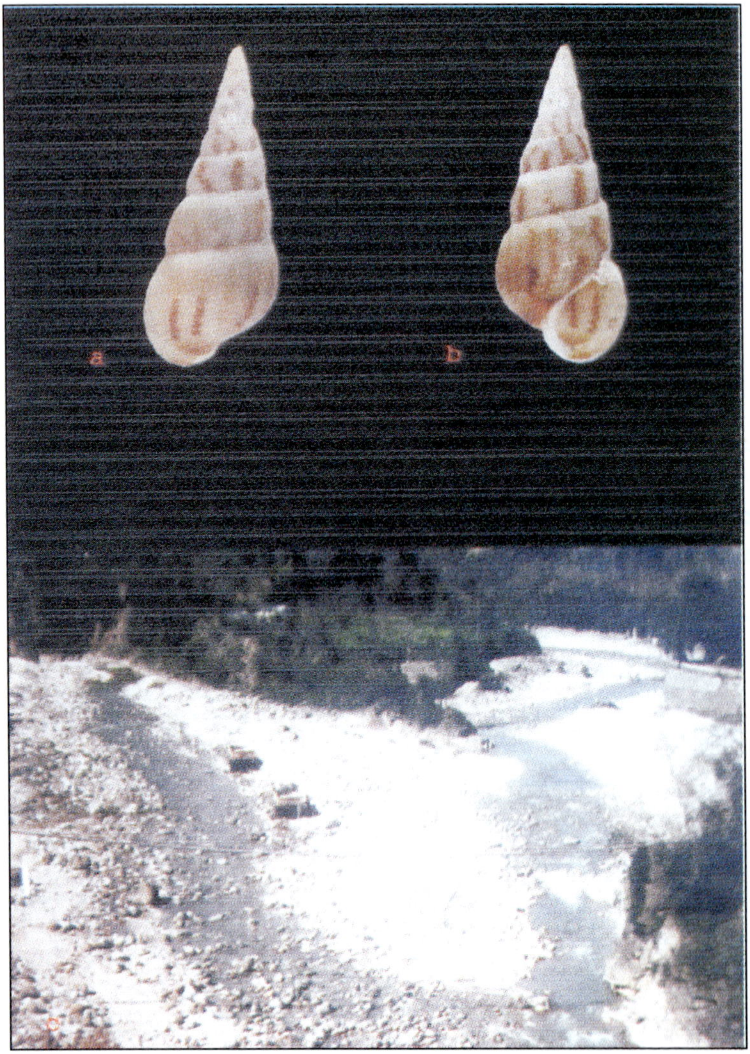

Plate 4.4: Showing Morphology and Site of Collection of *Thiara tuberculata*

(a) Dorsal View; (b) Ventral View; (c) Jhajjar Kotli Stream

greatly produced in front; collumella slightly bent and not produced anteriorly, the lip slightly or not at all thickened. Operculum relatively large with posterior extremity pointed, paucispiral, with a basal nucleus sometimes almost approaching the subspiral condition. Ovoviviparous and parthenogenesis.

Thiara (*Melanoides*) *tuberculata* (Mueller, 1774) (Plate 4.4, Figures a, b, c)

Diagnostic Characters

Shell with a high spire and moderately large body whorl, whorls evenly rounded dark red-brown dots and flames, either irregularly distributed or longitudinally arranged on the shell surface, sculptured with vertical ribs and spiral striae, distinct and raised on the upper whorls, but flatter on the lower ones.

Mantle edge bears 10–12 appendages, exterior surface of mental not pigmented. Only females known. Specimens are greatly variable in shape and size of their shells.

It was known by several names. Starmueler (1976) gave a complete synonymy. In India it was described under the names like *M. pyramis* (Hutton), *M. tigrina* (Hutton) and their varieties. Many of the names given by Preston may turn out to be synonyms of this species.

Distribution

District Udhampur (Jhajjar Kotli) and District Jammu (Jewel). It has a selective geographical distribution in the Jammu province in being restricted mostly to the plains of the districts Jammu and Udhampur.

Habitat

It is common melanid occurring in streams, rivers, irrigation canals and stagnant water ponds, often extending in to brackish waters. It is found exclusively in clear/transparent waters with low decaying organic matter and thus an indicator of pollution of water body.

References

Abbott, R.T., 1989. *Compendium of Landshells*. American Malacologists, Burlington, MA.

Barnes, R.S.K., Calow, P., Olive, P.J.W., Golding, D.W. and Spicer, J.I., 2001. *The Invertebrates: A Synthesis*, 3rd edn. Blackwell Science, UK.

Boss, K.J., 1973. Critical estimate of the number of Recent Mollusca. *Dccas. Pap. Mollusks*, 3: 81–135.

Bouchet, P., 1991. Extinction and preservation of species in the tropical world: What future for Mollusks? *American Malacologists*, 20: 20–24.

Brusca, R.C. and Brusca, G.J., 2003. *Invertebrates*, 2nd edn. Sinauer Associates Inc., MA, USA.

Duda, P.L. and Verma, A.K., 1992. Role of gastropods in the transmission of diseases among Herpetiles Part I: Amphibia–A numerical analysis. *Presented at the First International Congress of IUCN/ISRAG–SSG*, at Bhubaneswar.

Duda, P.L. and Verma, A.K., 1995. Ecology and Distribution of Gastropods Mollusca of Jammu and Kashmir State together with the Cercariae parasite. Final Tech. Report, Submitted to DO En, Govt. of India, New Delhi, p. 277.

Malek, A.E. and Cheng, C.T., 1974. *Medical and Economic Malacology*. Academic Press, New York and London, xi+398 pp.

Parker, T.A. and Haswell, 1967. *Invertebrate Zoology*. McMillan and Co. Ltd., New York, London.

Ponder, W.F. and Lindberg, D.R., 2008. *Phylogeny and Evolution of the Mollusca*. University of California Press, Berkeley, pp. 481.

Seddon, M.B., 2000. Molluscan diversity and impact of large dams. Prepared for thematic review II. 1: Dams, ecosystem functions and environmental restoration. IUCN Report.

Subba Rao, N.V., 1989. *Handbook: Freshwater Molluscs of India*, p. 290.

Subba Rao, N.V., 1993. Freshwater molluscs of India. In: *Recent Advances in Freshwater Biology*, (Ed.) K.S. Rao. Anmol Publication, New Delhi, 2: 187–202.

Tonapi, G.T., 1980. *Freshwater Animals of India: An Ecological Approach*. Oxford and IBH Publishing Co., New Delhi.

Chapter 5

A Synoptic List of Gerromorpha (Hemiptera: Insecta) from Western Ghats of Sindhudurg District, Maharashtra

T.V. Sathe and T.M. Chougale*

Biosystematics Division of Entomology, Department of Zoology, Shivaji University, Kolhapur

ABSTRACT

The suborder Gerromorpha contains 8 families namely, Gerridae, Veliidae, Hydrometridae, Mesoveliidae, Hebridae, Macroveliidae, Paraphrynoveliidae and Hermatobatidae etc. and provides aquatic form of bugs which are being recognized as important zoogeographical indicators. The very special feature of this group is their poor dispersal capacity. Western Ghat is among the 18 hot spots of the world, however little attention has been paid on this group of insects. Hence biodiversity of Gerromorpha have been studied and described with a synoptic list.

Keywords: *Gerromorpha, Aquatic bugs, Biodiversity, Synoptic list, Western Ghats.*

* Corresponding author: E-mail: drtvsathe@rediffmail.com.

Introduction

The suborder Gerromorpha contains 8 families namely, Gerridae, Veliidae, Hydrometridae, Mesovliidae, Herbidae, Macroveliidae, Paraphrynoveliidae and Harmatobatidae etc. which provides aquatic form of bugs and recognized as important zoological indicators. The very special feature of this group is their poor dispersal capacity. However, they are found in all kinds of climatic zones except coldest and driest parts. They function in its natural habitat as intermediate links specially at the base level, in the tropic structure within a fresh water habitat system and also act as ecological indicators. Western Ghats is among the 18 hot spots of the world, however, little attention is paid on the Western Ghats of Maharashtra region. Keeping in view all above facts, the present work was undertaken. In past, Ananthakrishnan (1999), Anderson (1981a, 1982, 1983, 1989), Annandale (1919), Bal and Basu (1997), Distant (1903, 1910a, 1910b), Dover (1928), Ghosh *et al.* (1989), Paira (1919a, 1919b), Pradhan (1950), Tirumalai (1986, 1989, 1994a, 1999a, 1999b, 2001, 2002), Tirumalai and Radhakrishnan (1999), Tirumali and Krishnan (2000), Tirumalai and Sharma (2002b), etc. have worked on aquatic and semi-aquatic Gerromorpha bugs.

Materials and Methods

Gerromorpha bugs have been studied from different habitats of Western Ghats, district Sindhudurg specially Sawantwadi, Amboli, Kudal, Kankavali, Vaibhavwadi and Fonda by spot observations and temporary collection of the bugs. The bugs were anesthetized for their morphological and taxonomical studies and then released in the environment from which they were collected. The collected species were identified by consulting appropriate literature (Distant, 1903, 1910a, b; Thirumalai, 1986, 1989, 1994 a, b, 1999a, b, 2001, 2002; Thirumalai and Sharma, 2002a, b, etc.) Thirumalai, 1986, 1989, 1994a, b, 1999a, b, 2001, 2002; Thirumalai and Sharma, 2002a, b, etc.) and synoptic list have been prepared.

Results and Discussion

The results recorded indicates that 36 species have been reported from Western Ghats of Sindhudurg. Out of 6 study spots, Sawantwadi and Vaibhavwadi showed rich diversity of Gerromorpha bugs. Common and rare species found in the region are tabulated in Table 5.1.

Table 5.1: Biodiversity and Abundance of Gerromorpha Bugs in Western Ghats of Sindhudurg District, Maharashtra

Sl.No.	Species	Status
	Family: Gerridae	
	Subfamily: Cylindrostethinae	
1.	*Cylindrostethus producuts* (Spinola)	Common
	Subfamily: Eotrechinae	
2.	*Amemboa* (*Amemboa*) *kumari* (Distant)	Rare
3.	*Onychotrechus major* Anderson	Common
4.	*Onychotrechus rhexenor* Kirkaldy	Common
5.	*Onychotrechus spinifer* Anderson	Common
	Subfamily: Gerrinae	
6.	*Aquarius adelaidis* (Dohrn)	Common
7.	*Neogerris parvula* (Stal)	Rare
8.	*Limnogonus* (*Limnogonus*) *fossarum fossarum* (Fabricius)	Common
9.	*Limnogonus* (*Limnogonus*) *nitidus* (Mayr)	Common
10.	*Limnometra anadyomene* (Kirkaldy)	Common
11.	*Limnometra fluviorum* (Fabricius)	Common
	Subfamily: Halobatinae	
12.	*Halobates galatea* Herring	Rare
13.	*Metrocoris dembickyi* Chen and Zettel	Rare
14.	*Metrocoris indicus* Chen and Nieser	Common
15.	*Metrocoris malabaricus* Thirumalai	Rare
16.	*Metrocoris variegans* Thirumalai	Common
17.	*Metrocoris velmentus* Chen and Nieser	Rare
18.	*Metrocoris communoides* Chen and Nieser	Common
19.	*Ventidius* (*Ventidius*) *aquarius* Distant	Common
	Subfamily: Ptilomerinae	
20.	*Pleciobates indicus* Thirumalai	Common
21.	*Stridulobates nostras* (Thirumalai)	Rare
22.	*Striduobates anderseni* Zettel and Thirumalai	Common
23.	*Ptilomera* (*Ptilomera*) *agroides* Schmidt	Common
24.	*Jucundus custodiendus* Distant	Common

Contd...

Table 5.1–Contd...

Sl.No.	Species	Status
	Subfamily: Rhagadotarsinae	
25.	*Rhagadotarsus* (*Rhagadotarsus*) *kraepelini* Breddin	Common
26.	*Cryptobatus raja* (Distant)	Rare
27.	*Naboandelus signatus* Distant	Rare
	Family: Veliidae	
	Subfamily: Microveliinae	
28.	*Microvelia* (*Microvelia*) *douglasi* Scott	Common
29.	*Microvelia* (*Microvelia*) *santala* Hatiz and Ribeiro	Rare
30.	*Pseudovelia* (*Pseudovelia*) *lingula* Gupta and Khandelwal	Common
	Subfamily: Rhagoveliinae	
31.	*Rhagovelina* (*Rhagovelina*) *ceylanica* Lundbald	Common
32.	*Rhagovelina* (*Rhagovelina*) *hibialis* Lundbald	Common
	Family: Herbidae	
	Subfamily: Hyrcaninae	
33.	*Hyrcanus reichli* Zettel	Rare
	Family: Mesoveliidae	
	Subfamily: Mesoveliinae	
34.	*Mesovelia vittigera* Horrath	Common
	Family: Hydrometridae	
	Subfamily: Hydrometrinae	
35.	*Hydrometra bulteri* Hungerford and Evans	Rare
36.	*Hydrometra greeni Kirkaldy*	Common

Systematic List of Gerromorpha from Western Ghats of Sindhudurg District, Maharashtra

Family: *Gerridae*

Subfamily: *Cylindrostethinae*

1. *Cylindrostethus productus* (Spinola, 1840)

Gerris productus Spinola, Essai sur less insects hemipteres rhynchotes on heteropteresl, 64, 1840.

Cylindrostethus productus (Spinola): Distant, Founa British India 2, 184.

Cylindrostethus productus (Spinola): Polhemus, Bishop Mus. Occ, 1905. Pap, 38, 10, 1994.

Janias elegantulus Distant, *Ann. Mag, Nat–Hist.* 5(8), 145, 1910.

Distribution

Karnataka, Kerala, Madhya Pradesh, Tamil Nadu, Punjab, Orissa, Uttar Pradesh, West Bengal, Maharashtra, Western Ghats: Sindhudurg: Amboli, Sawantwadi, Kudal, Kankavali, Vaibhavwadi.

Subfamily: *Eotrechinae*

2. *Amemboa (Amemboa) kumari* (Distant, 1910)

Onychotrechus kumari Distant, Ann. Mag. Nat. Hist, 5(8) 145, 1910.

Onychotrechus kumari (Distant): Esaki, 1928. *Ann. Mag. Nat Hist,* 2 (10), 509.

Amemboa (*Amemboa*) *kumari* (Distant): Polhemus and Anderson, *Steenstrupia,* 10(3) 85, 1984.

Amemboa parvati Pradhan, 1950b., *Rec. Indian. Musy* 48 (3 and 4) 12.

Distribution

Karnataka, Kerala, Tamil Nadu, Orissa, Maharashtra, Western Ghats: Sindhudurg: Vaibhavwadi, Kudal.

3. *Onychotrechus major* Anderson, 1980

Onychotrechus major Anderson, *Steenstrupia,* 6(10), 133, 1980.

Distribution

Kerala, Maharashtra, Western Ghats: Sindhudurg, Sawantwadi, Amboli.

4. *Onychotrechus rhexenor* Kirkaldy, 1903

Onychotrechus rhexenor Kirkaldy, *Entomologist,* 36, 44, 1903.

Onychotrechus rhexenor Kirkaldy, Anderson, *Steenstrupia,* 6 (10), 128, 1980.

Distribution

Karnataka, Kerala, Rajasthan, Tamil Nadu, Maharashtra. Western Ghats: Sindhudurg: Sawantwadi, Amboli, Kudal, Kankavali, Vaibhavwadi, Fonda.

5. *Onychotrechus spinifer* Anderson, 1980

Onychotrechus spinifer Anderson, *Steenstrupia*, 6(10), 138, 1980.

Distribution

Karnataka, Kerala, Maharasthra, Western Ghats, Sindhudurg: Sawantwadi, Amboli, Kudal, Kankawali, Vaibhavwadi.

6. *Aquarius adelaidis* (Dohm, 1860)

Aquarius adelaidis Dohm, *Stettin, ent. Ztg*. 21, 408, 1860.

Aquarius adelaidis (Dohm): Andersen, *Steenstrupia*, 16(4), 61, 1990.

Gerris Spinolae Leth and Ser, *Lat, gen Hemiptera*, 3, 63, 1896.

Gerris Spinolae (Leth and Ser.,): Distant, *Fauna British India*, 2, 180, 1903.

Distribution

Andhra Pradesh, Uttar Pradesh, Karnataka, Kerala, Rajasthan, Tamil Nadu, West Bengal, Bihar, Orissa, Maharashtra, Western Ghats: Sindhudurg: Sawantwadi, Amboli, Kudal, Vaibhavwadi, Kankavali, Fonda.

7. *Neogerris parvula* (Stal, 1859)

Gerris parvula Stal, Zoology, 4, 265, 1859.

Limnogonus parvulus (Stal): Lundbald, Arch. *Hydrobiol, Suppl.*, 12, 384, 1934.

L. (*Limnogonus*) *parvulus* (Stal): Hungerford and Matsuda. *J. Kans, Ent. Soc.*, 32(1): 41, 1959.

L.(*Neogerris*) *parvulus* (Stal): Hungertord and Matsuda. *Insecta matsum?* 24, 114, 1959.

Neogerris parvula (Stal): Andersen, *Ent. Scand. Suppl.* 7, 86, 1975.

Gerris tristan Kirkaldy, *Revue, Ent.*, 18, 88, 1899.

Distribution

Arunachal Pradesh, Assam, Kerala, Orissa, Tamil Nadu, U.P.,

West Bengal, Pondicherry, Maharashtra, Western Ghats: Sindhudurg: Amboli, Kudal.

8. *Limnagonus* (*Limnogonus*) *fossarum fossarum* (Fabricius, 1775)

Cimex fossarum Fabricius, *Syst. Ent.*, 727, 1775.

Gerris fossarum Fabricius, *Ent. syst. emem. Aucta*, IV, 188, 1794.

Limnogonus fossarum Stal, K. Svenska Vetensk, Akad, 7, 133, 1868.

L. (*Limnogonus*) *fossarum* (Fab.,): Hungertord and Matsuda. *J. Kans, Ent. Soc.*, 32(1): 40, 1959.

L. (*L.*) *fossarum* (Fab.,): Andersen, *Ent. Scand. Suppl.*, 7, 30, 1975.

Distribution

Andaman and Nicobar Islands, Arunachal Pradesh, Assam, Bihar, Delhi, Goa, Harayana, Himachal Pradesh, J&K, Karnataka, Kerala, M.P., Orissa, Rajasthan, Tamil Nadu, West Bengal, Pondicherry, Maharashtra, Western Ghats: Sindhudurg: Amboli, Sawantwadi, Kudal, Kankavali, Vaibhavwadi, Fonda.

9. *Limnogonus* (*Limnogonus*) *nitidus* (Mayr, 1865).

Hydrometra nitida Mayr, *verh, zool. bot. Ges. Wien*, 15, 443, 1865.

Gerris nitida (Mayr): Distant, *Fauna, British India*, 2, 178, 1903.

Limnogonus nitidus (Mayr): Kirkaldy, Wissenschafl, Ergebn. der. sehwed. Zool. Exped. nach. dem. Kilimandjaro, 12, 21, 1908.

L. (*Limnogonus*) *nitidus* (Mayr): Matsuda. *Kans, Univ. Sci. Bull.*, 41, 198, 1960.

L. (*Limnogonus*) *nitidus* (mayr): Andersen, *Ent. Scand. Suppl.*, 7, 62, 1975.

Distribution

Andaman and Nicobar Islands, A.P., Assam, Bihar, Delhi, Karnataka, Kerala, U.P., West Bengal, Tamil Nadu, Rajasthan, Orissa, Tripura, Maharashra, Western Ghats: Sindhudurg: Amboli, Fonda, Kudal, Vaibhavwadi, Sawantwadi, Vaibhavwadi.

10. *Limnometra anadyomene* (Kirkaldy, 1901)

Gerris, anadyomenc Kirkaldy, *Entomologist*, 34, 117, 1910.

Tenagogonus andyomene (Kirkaldy): *Bergroth, Zool, meded, Leiden,* 1(2): 122, 1915.

Unnometra anadomene (Kirkaldy): Lundbald, *Arch. Hydrobiol. Suppl.* 12, 371, 1934.

Tenagogonus (Limnometra) anadyomene (Kirkaldy): Mutsuda, *Kans. Univ. Sci. Bull.*, 41, 206, 1960.

Limnometra anadyomene (Kirkaldy): Andersen, *Steenstrupia,* 21, 117, 1955.

Tenagogonus (Limnometra) longispinalus Thirumalai, *Rec. Zool. Surv. India,* 84(1-4): 11, 1986.

Distribution

Karnataka, Kerala, Tamil Nadu, Maharashtra, Western Ghats: Sindhudurg: Amboli, Sawantwadi, Fonda, Kankavali, Kudal.

11. *Limnometra fulviorum* (Fabricius, 1798)

Cimexfulviorum Fabricius, *Ent. Syst. Suppl.*, 543, 1798.

Gerrisfulviorum (Fab.): Distant, *Fauna British, India,* 2, 177, 1903.

Limnometra fulviorum (Fab.) Lundbald, *Arch Hydrobiol. Suppl.,* 12, 371, 1934.

Tenagogonus limnometra fulviorum (Feb.) Matsuda. *Kani Univ. Sci. Bull.*, 41, 206, 1960.

Limnometra fulviorum (Fab.) Andersen, *Steenstrupia,* 21, 118, 1995.

Gerris armata Spinola: Distant, *Fauna British India,* 2, 180, 1903.

Distribution

Karnataka, Kerala, Pondichery, Tamil Nadu, West Bengal, Maharasthra, Western Ghats: Sindhudurg, Amboli, Sawantwadi, Kudal, Kankavali, Vaibhavwadi.

Subfamily: Halobatinae

12. *Halobates galatea* Herring, 1961

Halobates galatea Herring, *Pacif Insects,* 3(2-3), 294, 1961.

Distribution

Goa, Kerala, Maharashtra, Western Ghats: Sindhudurg: Kudal, Vaibhavwadi.

13. ***Metrocoris dembickyi*** **Chen and Zettel, 1999**

Metrocoris dembickyi Chen and Zettel, 1999. *Ann. Naturhist, Mus. wien,* 101 B, 2, 1999.

Distribution

Kerala, Maharashtra, Western Ghats: Sindhudurg, Sawantwadi, Kankavali.

14. ***Metrocosis indicus*** **Chen and Nieser, 1993**

Metrocosis indicus Chen and Nieser, 1993, *Steenstrupia,* 19(2), 48, 1993.

Metrocoris stali (*Dohrn*): Distant, Fauna British India, 2, 190, 1903.

Metrocoris stali: Den Boer, 2001. verh. Leideu, 74, 8, 1965.

Metrocoris stali: Thirumalai, Misc. Dee. Pap. *Rec. Zool. Surv. India,* 165, 37, 1994.

Distribution

Karnataka, Kerala, Tamil Nadu, Maharashtra, Western Ghats: Sindhudurg, Amboli, Sawantwadi, Kudal, Fonda, Vaibhavwadi.

15. ***Metrocoris malabaricus*** **Thirumalai, 1986**

Metrocoris malabaricus Thirumalai, 1986, *Rec. Zool. Surv. India,* 84 (1-4), 22, 1986.

Distribution

Karnataka, Kerala, Maharashtra, Western Ghats: Sindhudurg: Sawantwadi, Kudal.

16. ***Metrocoris variegans*** **Thirumalai, 1986**

Metrocoris variegans Thirumalai, 1986, *Rec. Zool. Surv. India,* 84, (1-4), 25, 1986.

Distribution

Karnataka, Kerala, Maharashtra, Western Ghats: Sindhudurg: Sawantwadi, Amboli, Kudal, Kankavali, Vaibhavwadi, Fonda.

17. ***Metrocoris velamentus*** **Chen and Nieser, 1993**

Metrocoris velamentus Chen and Nieser, 1993, *Steenstrupia,* 19(2) 61, 1993.

Distribution

Kerala, Tamil Nadu, Maharashtra, Western Ghats: Sindhudurg: Amboli, Sawantwadi, Kudal, Kankavali.

18. *Metrocoris cummumoides* Chen and Nieser, 1993

Metrocoris cummumoides Chen and Nieser, 1993, Steenstrupia, 19(2), 51, 1993.

Genus: Ventidius, Distant, 1910.

Distribution

Kerala, Tamil Nadu, Maharashtra, Western Ghats: Sindhudurg: Amboli, Sawantwadi, Kankavali.

Subgenus: Ventidius Distant, 1910

19. *Ventidius (Ventidius) aquarius* Distant, 1910

Ventidius aquarius, Distant, *Ann. Mag. nat. Hist.* 5(8): 150, 1910.

Ventidius (Ventidius) aquarius Distant: Hungerford and Matsuda, *Kans., Univ., Sci. Bull.*, 40(7): 324, 1910.

Distribution

Karnataka, Kerala, Tamil Nadu, Maharashtra, Western Ghats: Sindhudurg: Amboli, Kankavali, Kudal.

Subfamily: Ptilomerinae

Genus: Pleciobates Esaki, 1930.

20. *Pleciobates indicus* Thirumalai, 1986

Pleciobates indicus Thirumalai, *Rec. Zool. Surv. India*, 84(1-4): 16, 1986.

Distribution

Kerala, Maharashtra, Western Ghats: Sindhudurg: Sawantwadi, Amboli, Kudal, Vaibhavwadi.

21. *Stridulobates nostras* (Thirumalai) 1986

Plecobates nostras Thirumalai, *Rec. Zool. Surv. India*, 84(1-4): 19, 1986.

Distribution

Karnataka, Kerala, Maharashtra, Western Ghats: Sindhudurg: Amboli, Sawantwadi, Fonda, Kudal.

22. *Stridulobates anderseni* Zettel and Thirumalai, 2001

Pleciobates tuberculatus: Esaki, 1930: Thirumalai, 1992, *Hexapoda*, 4(2), 173, 2001.

Distribution

Karnataka, Kerala, Maharashtra, Western Ghats: Sindhudurg: Kudal, Amboli, Kankavali, Vaibhavwadi.

Genus: Ptilomera Amyot and Serville, 1843

Subgenus: Ptilomera Amyot and Serville, 1843.

23. *Ptilomera (Ptilomera) agroides* Schmidt, 1926

Ptilomera agroides Schmidt, *Ent. Mitt.*, 15(1): 63, 1926

Ptilomera (Ptilomera) agroides Schmidt: Matsuda, *Kant., Univ. Sci. Bull.*, 41 (2); 269, 1926.

Ptilomera (*Ptilomera*) *agroides* Schmidt: Thirumelai, *Rec. Zool. Surv. India*, 84(1-4), 15, 1986

Ptilomera laticaudata (Hardwicke): Distant, *Fauna British, India*, 2, 185 (Figure 133), 1903.

Ptilomera lachne Schmidt, *Ent.Mitt.* 15(1), 64, 1926.

Distribution

Karnataka, Kerala, Tamil Nadu, Maharashtra, Western Ghats: Sindhudurg: Sawantwadi, Amboli, Kankawali, Vaibhavwadi.

Genus: Jucundus Distant, 1910

24. *Jucundus custodiendus* Distant, 1910

Jucundus custodiendus Distant, 1910, *Ann. Mag. Nat. History* (VIII series), 5, 145, 1910; Zettel and *Thirumalai, Ann. Naturhist. Mus, wien.*, 103 B, 273, 2001.

Rheumatogonus Custodiendus (Distant, 1910): Esaki, EOS, Rev. Espan. Entomol. 3, 267, 1910; Lundblad, *Arch. Hydrabiol, Suppl.*, 12, 372, 1933.

Pleciobates tuberculatus Esaki, 1930: Thirumalai, *Hexapoda*, 4(2), 173, 1930.

Distribution

Karnataka, Kerala, Maharashtra, Western Ghats: Sindhudurg: Amboli, Sawantwadi, Vaibhavwadi, Kudal.

Subfamily: Rhagadotarsinae

Genus: Rhagadotarsus Breddin, 1905

Subgenus: Rhagadotarsus Breddin, 1905

25. *Rhagadotarsus (Rhagadotarsus) kraepelini* Breddin, 1905

Rhagadotarsus kraepelini Breddin, *Mitt. Naturhist, Mus. Hamb,* 22, 137, 1905.

R. (Rhagadotarsus) kraepelini Breddin: Matsuda, *Kans., Uni. Sci. Bull.*, 41(2), 322, 1960.

R. (Rhagadotarsus) kraepelini Breddin: Polhemus and Karunaratne, *Bull. Raffles Mus. (Zoology)*, 41(1), 100, 1993.

Nacebus dux Distant, *Ann. Mag Nat. Hist.*, 5(8), 152, 1910.

Distribution

Karnataka, Kerala, Pondicherry, Tamil Nadu, West Bengal, Maharashtra, Western Ghats: Sindhudurg: Fonda, Amboli, Sawantwadi, Kudal.

Subfamily: Trepobatinae

Genus: Cryptobates Esaki, 1929

26. *Cryptobatus raja* (Distant, 1910)

Gerris raja Distant, *Ann. Mag. Nat. Hist.*, 5(8), 142, 1910.

Cryptobatus raja (Distant): Esaki, *Ann. Mag Nat. Hist.*, 4(10), 412, 1929.

Cryptobatus raja (Distant): Polhemus and Polhemus, *Ent. Scand,* 26(1), 104, 1995.

Distribution

Kerala, Maharashtra, Western Ghats: Sindhudurg: Sawantwadi, Kudal, Amboli, Vaibhavwadi.

Genus: Naboandelus Distant, 1910

27. *Naboandelus signatus* Distant, 1910.

Naboandelus signatus Distant, 1910, *Ann. Mag. Nat. Hist.*, 5(8), 151, 1910.

Distribution

Karnataka, Tamil Nadu, Pondicherry, U.P., West Bengal, Kerala, Maharashtra, Western Ghats: Sindhudurg: Sawantwadi, Fonda, Kankavali, Amboli.

Family*: *Veliidae

Subfamily*: *Microveliidae

***Genus*: *Microvelia* Westwood, 1834**

***Subgenus: Microvelia* Westwood, 1834.**

28. *Microvelia* (*Microvelia*) *douglasi* Scott., 1874.

Microvelia douglasi Scott, *Ann. Mag. Nat. Hist.*, 44, 448, 1874.

Microvelia (*Microvelia*) *druglasi* Distant: Andersen, *Cat., Het, Palaerechi. Region;* 1, 87, 1995.

Microvelia douglasi Scott: Thirurnalai, Misc., Gee. Pap. *Rec. Zool. Surv. India,* 165, 40, 1994.

Microvelia repentiana Distant, *Fauna British India,* 3, 174, 1903.

Microvelia kumaonensis Distant, *Ann. Mag. Nat. Hist.*, 3(8), 500, 1909.

Distribution

Andaman Nicobar Islands, Arunachal Pradesh, Kerala, Karnataka, Orissa, Tamil Nadu, U.P., West Bengal, Maharashtra, Western Ghats: Sindhudurg: Amboli, Kankavali, Sawantwadi.

29. *Microvelia* (*Microvelia*) *santala* Hatiz and Ribeiro, 1939

Microvelia santala Hatiz and Ribeiro, *Rec. Indian. Mus.*, 41(4), 424, 1939.

Distribution

Bihar, Tamil Nadu, Kerala, Maharashtra, Western Ghats: Sindhudurg: Fonda, Amboli, Kankavali, Vaibhavwadi.

***Genus*: *Pseudovelia* Hoberlandt, 1950**

***Subgenus*: *Pseudovelia* Hoberlandt, 1950**

30. *Pseudovelia* (*Pseudovelia*) *lingual* Gupta and Khandelwal, 2003

Pseudovelia (*Pseudovelia*) *lingula* Gupta and Khandelwal, *Bionotes,* 5(1), 8, 2003.

Distribution

Kerala, Maharashtra, Western Ghats: Sindhudurg: Sawantwadi, Kankavali, Kudal, Vaibhavwadi.

Subfamily: Rhagovelinae

***Genus*: *Rhagovelia* Mayr, 1865**

***Subgenus: Rhagovelia* Mayr, 1865**

31. *Rhagovelia (Rhagovelia) ceylanica* Lundblad, 1936.

Rhagovelia ceylanica Lundblad, *Ark. Zool*, 28(21), 32, 1936.

Rhagovelia nigricans (Burmeister): Distant, *Fauna. British India*, 5, 132, 1980.

R. (Rhagovelia) ceylanica Lundblad; Thirumalai, Misc, Occ. Pap. *Rec. Zool. Surv. India*, 118, 51, 1989.

R. (Rhagovelia) Ceylanica Lundblad: Thirumalai, *Rec. Zool. Surv. India*, 94(24), 386, 1994.

Distribution

Assam, Bihar, Himachal Pradesh, Kerala, Meghalaya, Orissa, Punjab, Tamil Nadu, Uttar Pradesh, Maharashtra, Western Ghats: Sindhudurg: Amboli, Sawantwadi, Kudal.

32. *Rhagovelia (Rhagovelia) tibialis* Lundblad, 1936.

Rhagovelia tibialis Lundblad, *Ark. Zool.*, 2001, 28(21), 31, 1936.

R. (Rhagovelia) tibialis Lundblad; Thirumalai, *Rec. Zool. Surv. India*, 9 and (2-4), 382, 1994.

Distribution

Karnataka, Kerala, Tamil Nadu, Maharashtra, Western Ghats: Sindhudurg: Sawantwadi, Kudal, Kankavali, Vaibhavwadi.

Family*: *Herbidae

Subfamily: Hyrcaninae

***Genus: Hyrcanus* Distant, 1910.**

33. *Hyrcanus reichli* Zettel, 1998

Hyrcanus reichli Zettel, *Staptia*, 55, 597, 1998.

Distribution

Kerala, Maharashtra, Western Ghats: Sindhudurg, Sawantwadi, Amboli, Vaibhavwadi, Kudal.

Family: Mesoveliinae

Subfamily: Mesoveliinae

Genus: Mesovelia Mulsant and Ray, 1852

34. *Mesovelia vittigera* Horvath, 1895

Mesovelia-vlltigera Horvath, 1895, *Revue, Ent.*, 14, 160, 1895.

Mesovelia vittigera Horvath: Thirumalai, Misc. Occ., Pap. *Rec. Zool. Surv. India*, 165, 28, 1994.

Mesovelia mulsanti white: Distant, *Fauna British India*, 2, 169, 1903.

Mesovelia orientalis Kirkaldy, *Annali, mus, cu. star, Nat hiacom Doria*, 20, 808, 1901.

Distribution

Andaman and Nikobar Islands, Arunachal Pradesh, Bihar, Karnataka, Kerala, Tamil Nadu, Orissa, U.P., West Bengal, Maharashtra, Western Ghats: Shindhudurg: Fonda, Vaibhavwadi, Sawantiwadi, Kudal.

Family: Hydrometridae

Subfamily: Hydrometrinae

Genus: Hydrometra Latreille, 1796

35. *Hydrometra butteri* Hungerford and Evans, 1974

Hydrometra butteri Hungerford and Evans, 1974, *Ann. Mag, Nat Hist.*, 28, 71, 1974.

Hydrometra butteri Hungerford and Evans; Thirumalai, Misc. Occ., Pap. *Rec. Zool. Surv. India*, 165, 29, 1994.

Distribution

Karnataka, Kerala, Orissa, Tamil Nadu, Maharashtra, Western Ghats: Sindhudurg: Kudal, Vaibhavwadi, Kankavali, Sawantwadi.

36. *Hydrometra greeni* Kirkaldy, 1898

Hydrometra greeni Kirkaldy, 1898, *Entomologist*, 31, 2, 1898.

Hydrometra greeni Kirkaldy, 1898: Thirumalai, Misc. Occ. Pap. *Rec. Zool. Surv. India*, 165, 29, 1994.

Hydrometra vittata (Stal): Distant, *Fauna British India*, 2, 170, 1903.

Distribution

Andaman and Nicobar Islands, Arunachal Pradesh, Assam, Bihar, Gujarat, Karnataka, Kerala, Orissa, Rajasthan; Tamil Nadu, U.P., Pondicherry, West Bengal, Maharashtra, Western Ghats: Sindhudurg: Amboli, Sawantwadi, Vaibhavwadi, Kudal.

References

Ananthakrishanan, T.N., 1999. Multidimensional links in biodiversity research: An integrated exercise. *Current Science*, 77: 356–358.

Andersen, N.M., 1981. Semiaquatic bugs: Phylogeny and classification of Hebridae (Heteroptera: Gerromorpha) with a revision of *Timasius, Neotimasius* and *Hyrcanus. Systematic Entomology*, 6: 377–412.

Andersen, N.M., 1982. The semi-aquatic bugs (Hemiptera : Gerromorpha), phylogeny, adaptation, biogeography, and classification. *Entomonograph*, 3: 1–455.

Andersen, N.M., 1983. The Old World Microveliinae (Hemiptera: Veliidae). I. The status of *Pseudovelia* Hoberlandt and Perivelia Poisson, with a review of Oriental species. *Ent. Scand.*, 14: 253–268.

Anderson, N.M., 1989. The Old World Microveliinae (Hemiptera: Veliidae) II. Three new species of *Baptista* Distant and a new genus from the Oriental region. *Ent. Scand.*, 19: 363–380.

Annandale, N., 1919. The fauna of certain small streams in the Bombay Presidency. *Rec. Indian Mus.*, 16(1): 109–161.

Bal, A. and Basu, R.C., 1997. Hemiptera–Water Bugs. Fauna of Delhi *Zool. Surv. India. State Fauna Series*, G: 261–276.

Bergroth, E., 1915. Hemiptera from the Bombay Presidency. *J. Bombay Nat. Hist. Soc.*, 24: 170–179.

Distant, W.L., 1903. The fauna of British India including Ceylon and Burma. *Rhynchota*. 2: 167–191.

Distant, W.L., 1910a. Some underscribed Gerrinae. *Ann. Mag. Nat. Hist.*, 5(8): 140–153.

Distant, W.L., 1910b. The Fauna of British India including Ceylon and Burma. Appendix, 5: 137–166 and 310–353.

Dover, C., 1928 Aquatic Rhynchota in the college of the Agricultural College, Coimbatore, Southern India. *J. Bombay Nat. Hist. Soc.*, 32(3): 614–615.

Ghosh, L.K., Biswas, B., Chakraborty, S.P. and Sen, G.C., 1989. Fauna of Orissa state, Insecta: Hemiptera. State Fauna Series, *Zool. Surv. India*, 1:181–224.

Paiva, C.A., 1919a. Aquatic and semiaquatic Rhynchota from the Satara and Poona Districts. *Rec. Indian Mus.*, 16(1): 152–156.

Paiva, C.A., 1919b. Rhynchota form the Garo Hills, Assam. *Rec. Indian Mus.*, 16(3): 349–377.

Pradhan, K.S., 1950. On the distribution of the Genus *Amemboa* Esaki (Hemiptera: Heteroptera) with the description of a new species. *Rec. Indian Mus.*, 48(3 and 4): 11–16.

Thirumalai, G., 1986. On Gerridae and Notonectidae (Heteroptera : Hemiptera : Insecta) from Silent Valley, Kerala. *Rec. Zool. Surv. India*, 84(14): 9–33.

Thirumalai, G., 1989. Aquatic and semiaquatic Hemiptera (Insecta) from Javadi Hills, Tamil Nadu. Misc. Occ. Pap. *Rec. Zool. Surv. India*, 118: 1–63.

Thirumalia, G., 1994a. Aquatic and semiaquatic Hemiptera (Insecta) of Tamilnadu. I. Dhargmpuri and Pudukkottai Districts, Misc. Ace. Pap. *Rec. Zool. Surv. India*, 165: 1–45.

Thirumalai, G., 1999a. Aquatic and semiaquatic: Heteroptera of India. Indian Association of Aquatic Biologists (IAAB) Publication, 7: 1–74.

Thirumalai, G., 1999b. A checklist of aquatic and semi-aquatic Heteroptera (Insecta) of Tamil Nadu. *Zoos' Print J.*, 1–14(1–10): 132–135.

Thirumalai, G., 2001. Insecta: Aquatic and semiaquatic. Heteroptera Fauna of conservation Area Series, 11. Fauna of Nilgiris Biosphere Reserve, pp. 111–127. Publ. *Zool. Surv. India*, Kolkata.

Thirumalai, G., 2002. A checklist of Gerromorpha (Hemiptera) from India. *Rec. Zool. Surv. India,* 100(1–2): 55–97.

Thirumalai, G. and Krishnan, S., 2000. Diversity of Gerromorpha (Heteroptera: Hemiplera: Insecta) in the Western Ghats States of India. *Rec. Zool. Surv. India,* 98(4): 59–77.

Thirumalai, G. and Sharma, R.M., 2002, Aquatic and semiaquatic Heteroptera (Insecta). Fauna of Ujani, Wetland Ecosystem Series 3: 105–116.

Thirumalai, G., 2002. A checklist of Gerromorpha (Hemiptera) from India. *Rec. Zool. Surv. India,* 100(1–2): 55–97.

Thirumalai, G. and Krishna, S., 2000 Diversity of Gerromorpha (Heteroptera: Hemiptera: Insecta) in the Western Ghats of India. *Rec. Zool. Surv. India,* 98(4): 59–77.

Thirumalai, G. and Sharma, R.M., 2002. Aquatic and semi-aquatic Heteroptra (Insecta). Fauna of Ujani, Wetland Ecosystem Series, 3: 105.

Chapter 6

Biodiversity of Wild Silkmoth Species from Western Maharashtra, India

*T.V. Sathe**

Entomology Division, Department of Zoology, Shivaji University, Kolhapur – 416 004

ABSTRACT

Western Maharashtra is characterised by Western Ghats, hilly terrains, plateaus and heavy rains. Due to its tropical climate, unusual geological stability and evolutionary continuity Western Maharashtra is visualized as the biodiversity rich area in the world. Besides ecological, intrinsic and socioeconomic value, Western Maharashtra has tremendous importance in Tasar and Mulberry sericulture. Hence, present work was aimed to biodiversity, abundance and host food plants of silkworms in the region. In all, 14 species of wild silkmoths have been reported belonging to the genera, *Antheraea* (6), *Actias* (2), *Attactus* (2), *Leopa* (1), *Saturnia* (1) and *Cricula* (1). As Western Maharashtra has a very conducive environment for tasar and mulberry, there is an urgent need to have the greater attention for the development of sericulture.

Keywords: Wild silkmoth, Diversity, Western Ghats, Maharashtra.

---

* E-mail: drtvsathe@rediffmail.com.

Introduction

India is one of the 12 identified centres of mega biodiversity having as many as 18 hot spots of biodiversity of the world including Eastern Himalaya and Western Ghats (Rajamani, 1993). Western Ghats is most important amongst 10 bio-ecographic regions *viz.* Himalayan, Trans Himalayan, Indian Desert, Semiarid, Western Ghats, Deccan Peninsula, Gangetic plains, North-East, Islands and coats. 17500 species of higher plants have been reported from India of which 7500 species have ethnobotanical importance. Likely, out of the world's total, 6.4 per cent diversity, containing 81000 species have been reported from India. Hence, India has a very strong position of biodiversity. But not managed potentially as a manageable resource for mankind. Wild silkmoths can be managed in sericulture business. Hence, for understanding the potential of wild silk moths in Western Maharashtra, biodiversity of wild silkmoths have been studied during the years 2000–2004.

Sericulture is an important source for earning the foreign currency and has tremendous scope hence, deserves a special attention for its advancement. India is the only country in the world producing five kinds of commercially known varieties of silkworms *viz.*, Mulberry, Tasar, Eri Tasar, Oak and Muga. India stands second among the mulberry and tasar silk producing countries (M.S. Publ., 1988). In Maharashtra sericulture is rising with a very rapid rate since every district of state is involved in sericulture. The present work will add relevance in development of Tasar sericulture in Western Maharashtra.

Materials and Methods

Silkworm diversity was studied during the year 2000–2004 by collecting the silkworms and silkmoths from Western Ghats (M.S., Amboli region, Kolhapur Satara and Pune) and other parts of Western Maharahstra, India fortnightly. The cocoons of the silkworms were also collected and reared for adult emergence. The collected material (silkworm larvae, cocoons and silkmoths) were identified by adopting terminology of Hampson (1892) and FAO (1987). The worms were preserved in 4 per cent formalin and the cocoon and moths kept in wooden boxes by pinning them. The specimens have been deposited in the Zoology Department, Shivaji University, Kolhapur.

Results

Attacus atlas Linn.

It characterized by having wing expanse 250 mm in female and 224 mm in male; head, thorax and abdomen red brown; basal segment and abdomen pale, each segment with pale fringe. Lees brown. Forewings with costa brown, basal area brown and red brown edged by red, pale, and black lines. Medial area red-brown; large triangular hyaline spot at the end of cell with a black edge. Hind wings are similar to fore wings, no streak above the hyaline triangular mark.

Larvae

Pale green with brownish speckles.

Cocoon

Pale greyish brown and pyriform.

Distribution

Western Ghats, Amboli, Satara, Pune.

Attacus edwardso

Female

Head, thorax and abdomen brown. The thorax and abdomen darker. Abdominal first segment white, paired dorsal white segmental streaks on third to last segment. Two lateral and two ventral lines present. Fore wing much darker, the antemedial line inwardly black and outwardly white. The triangular hyaline spots edged with yellow-brown and not touching the postmedial line. Hyaline streaks black and short.

Wing Expanse

200 mm.

Actias selene

Female

Head, thorax and abdomen white, legs pinkish. Fore wing bluish-green, 6.5 mm long, 4 mm broad. A dark pink costal fascia present. The outer margin less excised and waved. The yellow markings are less developed. The antemedial line, on a fore wing nearer the base and in hind wing it is absent. The tail is less pink.

Larva

Yellowish green coloured.

Cocoon

Pale brown and oval.

Distribution

Western Ghats (M.S.): Kolhapur, Amboli, Satara.

Actias indica sp.nov.

Female

Similar to *A. selene* except forewing have two marginal lines, forewings longer in length smaller in width.

Wing Expanse

145 million.

Distribution

Western Ghats: Kolhapur.

Antheraea paphia Linn.

Male

Pale brown yellow, costal brown and grey fascia of fore wing reaching the apex, hyaline and ocellated spots much larger, inner lunule and post medial line bright pink. Females 150–190 mm in wing expanse, either pinkish brown or bright yellow form; the hyaline and ocellated spots usually larger than in the males.

Larva

Green with paired dorsal series of yellow humps, lateral purple bordered white lunulate spots on 5th and 6th segments. Spiracles yellow.

Cocoon

Brownish grey, hard, oval with silken peduncle attached to substrate.

Distribution

Western Ghats (M.S.), plains of Western Maharashtra, Marathwada and Vidarbha.

Host Plant

Terminalia tomentosa, T. arjuna and *Z. jujuba.*

A. mylitta Linn.

It is the most yellow. Smaller than *A. paphia.*

Cocoon

Similar to *A. paphia.*

Distribution

Western Maharashtra, Marathwada and Vidarbha etc.

Host Plant

Terminalia tomentosa, T. mjuna and *Z. jujube.*

A. cingaleca

Female

It is dark brownish yellow moth. Wing expanse: 152 mm.

Distribution

Western Ghats: Amboli, Kolhapur, Satara.

Early Record

China, Srilanka, India.

Host Plants

Terminalia tomentosa and *T. arjuna.*

A. koyvetti

Female

Reddish or olive yellow in colour. Reddish yellow colour antemedial line of hind wing. Cotal fascia of fore wing only extending along 2/3 of the costa. Ocelli small with dark lunule on its inner edge, marginal line yellow.

Wing Expanse

156 mm.

Distribution

Western Ghats: Kolhapur, Atllboli.

Host Plants

T. tomentosa, T. arjuna and *Z. jujube*

Old Record

India: Sikkim.

A. andamana

Female

Colour brownish yellow, with yellow patches; two black marginal lines on forewing, dark post medial line prominent. Submarginal line dark with waved lines. Hind wing with three waved lines, two post medial.

Wing Expanse

171 mm.

Distribution

Western Ghats: Kolhapur.

Host Plants

T. tomentosa, T. arjuna, Z. jujube.

Old Record

India: Andamans.

A. heljeri

Male

Differs from *A. koyvetti* in there being a black bloch on the upper side of the ocellus in forewing, female-yellowish.

Wing Expanse

150 mm.

Distribution

Western Ghats: Satara.

Host Plants

Z. jujube, T. tomentosa.

Saturnia anna

Male

Head, thorax and abdomen yellowish brown. Fore and hindwings yellow, with three highly dentate lines. Ocellus large and round, red brown ringed with black and containing a white lunule and black center with a white strick on it, apical patch reddish brown, palpi moderate size, Antennal branches long.

Wing Expanse

112 mm.

Loepa katinka Westw.

Female

90–124 mm wing expanse, bright chrome-yellow; Forewing with costa grey, subbasal more or less with angled pink line; a large round or oval ocellus at the end of cell is pinkish brown containing white and black lunulate marks; a highly waved post medial dark line, double sub marginal lines, inner line pinkish and angled below costa. Hindwing differs from forewing in the first line being further from the base, narrow and dark. The subcostal patch and spot absent.

Larva

Brown hairy, with six pink tubercles on each segment; white sublateral irregular blotches from 4th to 10th segments; claspers pink.

Distribution

Himalayas, Assam, Western Ghats (Amboli region) Maharashtra State.

Cricula trifenestrata

Female

Brownish reddish head, thorax and abdomen. Forewing with waved and antemedial dark line. Three large irregular shaped hyaline spots beyond the cell of forewing often with one or two small once inside them.

Wing Expanse

75 mm.

Larva

Blackish brown.

Distribution

Western Ghats (M.S.).

Discussion and Conclusion

Western Ghats of Maharashtra is a biodiversity rich region of India. 13 species of saturnid moths have been reported from this region. Specially *Antheraea* spp. has great importance in sericulture. In India, about 10 per cent population is from tribal community and dependant on wild silk industry. It is essential to educate the tribal population of Western Maharashtra for undertaking this business scientifically on large scale. The present work has added great relevance in identifying biodiversity of saturnid moths from Western Ghats. There is also need to popularize wild sericulture in Maharashtra so that this natural resource will be utilized for well being of mankind and nature. Some rare species like *Leopa katinka, Saturnia anna, Actias* spp. *Antheraea andamana* be protected and conserved in the region for strengthening the standard of biodiversity of this region.

For protection, conservation and utilization of wild silkmoth biodiversity, an action plan at various levels *viz.* local, regional, state, inter-state, thematic and national level be prepared which may includes.

1. Identification of critical components of biodiversity *i.e.* communities, ecosystem, species, genotypes, etc.
2. Enumeration of the important gene, species, etc. which are endangered and needs protection on priority.
3. To develop a nation-wide data base.
4. Identification of the methods of *in situ* and *ex situ* through a protected area, field gene bank and wild life sanctuaries etc.
5. Documentation of traditional knowledge, innovations, etc. and suitable reward/mechanisms.
6. Funding and monitoring of implementation, sharing benefits out of use of biodiversity.

7. To develop network of all the activities as an integrated approach for conservation and utilization of biodiversity of silk worms.

Acknowledgement

Author is thankful to Shivaji University, Kolhapur (M.S.) for providing facilities and financial assistance.

References

FAO, 1979. *Agri. Seri. Bull.* Manuals on Sericulture, 29: 1–178.

Hampson, G.F., 1892. *Fauna of British India, Moths.* 1: 1–527.

M.S. Publ., 1988. Project Report DSMBC (1988–1992), pp. 1–71.

Sathe, T.V., Mulla, M.K. and Sathe, D.B., 1998. A brief note on silkworm diversity. *Proc. Nat. Man. Cong.*, Imphal.

Chapter 7

Dragonflies (Order: Odonata) of Western Ghats of Sindhudurg District

T.V. Sathe, K.P. Shinde, T.M. Chougle, and P.M. Bhoje

Biosystematics Laboratory, Division of Entomology, Department of Zoology, Shivaji University, Kolhapur, M.S.

ABSTRACT

Dragonflies (Order: Odonata) are exclusively predacious form of class insecta. They feed upon various kinds of insects and supposed to be ideal predaceous insects of the world. They have a potential practical value in biological pest control programmes. Hence, the dragonflies have been studied with respect to taxonomical diversity from Western Ghats of Sindhudurg district, Maharashatra. In all 97 species of dragonflies have been described belonging to 55 genera.

Keywords: Dragonflies, Pest predatory insects, Biodiversity, Western Ghats.

Introduction

Dragonflies (Order: Odonata) are ideal predators in insect world and found darting and dancing actively near ponds, pools, rivers, streams and also marshy places (Parsad and Kulkarni, 2001). Many species of odonates are recorded perching high on trees and shrubs in forest or away from water bodies and forest. They are reported from sea levels to over 3600 ft. from all over the world. Out of 6000 species reported from the world, 502 spp belonging to 139 genera of 17 families have been reported from India (Sathe and Shinde, 2006), Kulkarni and Prasad (2001) reported 71 species from Kerala. Recently Radhkrishnan and Emiliyamma (2003) reported 137 species belonging to 79 genera from Western Ghats of Kerala of the studies are confined to Western Ghats of Kerala and very little attention has been paid on Western Gha:; scattered in Maharashtra.

In spite of rich biodiversity and specialized aquatic habitats of Sindhudurg region of Western Ghats, dragonflies have not been attempted earlier. Therefore, present work was carried out on the biodiversity of dragonflies of Western Ghats of Sindhudurg district. In past, Fraser (1933, 1934, 1930), Lahiri (1987), Prasad (1995, 1996), Prasad and Varshney (1995), Prasad and Kulkarni (2001, 2002), Jafer *et al.* (2002), Kulkarni and Prasad (2002) Radhkrishnan and Emiliyamma (2003), Sathe (2005), Sathe and Shinde (2006 a, b), Sathe *et al.* (2006), Silsby (2001), etc. have worked on biodiversity of Odonata from India.

Materials and Methods

Survey and taxonomical studies of dragonflies were made by visiting selected spots of Western Ghats: Sindhudurg district (Sawantwadi, Amboli, Kudal, Kankawali, Vaibhavwadi and Fonda) at 15 days interval. Spot observations have been followed by photography, morphology and taxonomy of species. After recording characteristics and measurements, the live dragonflies were released in the environment from where they were collected. Dragonflies were identified by consulting Fraser (1933, 1934, 1936) and other workers cited in the literature.

Results and Discussion

Results are recorded in Table 7.1. The taxonomical and distributional details of species found in the Western Ghats of Sindhudurg are given under each species in the text.

Table 7.1: Species Diversity of Dragonflies from Western Ghats of Sindhudurg District

Sl.No.	Genera in India	Number of Species in India	Number of Species in Sindhudurg
1.	*Archibasis* Kirby	1	–
2.	*Ceriagarion* Selys	8	2
3.	*Pseudagrion* Sclys	11	5
4.	*Cercion* Navas	2	–
5.	*Coenagrion* Kirby	1	–
6.	*Himalagrion* Fraser	2	–
7.	*Aciagrion* Selys	8	2
8.	*Pyrrhosoma* Charpentier	1	–
9.	*Enallafama* Charpentier	5	–
10.	*Ischnura* Charpentier	11	3
11.	*Rhodischnura* Laidlaw	1	–
12.	*Awiocnemis* Selys	9	3
13.	*Argiocnemis* Selys	1	–
14.	*Mortonawion* Frase	2	1
15.	*Onycharwia* Selys	2	1
16.	*Calicnemis* Strand	10	–
17.	*Coeliccia* Kirby	12	–
18.	*Indocenemis* Laidlaw	1	–
19.	*Copera* Kirby	6	2
20.	*Plalycnemis* Charpentier	I	–
21.	*Drepanosficta* Laidlaw	3	–
22.	*Plalistica* Selys	I	–
23.	*Protostica* Selys	9	4
24.	*Caconeura* Kirby	5	2
25.	*Esme* Fraser	3	2
26.	*Melanoneura* Fraser	1	1
27.	*Phylloneura* Fraser	1	1
28.	*Elattoneura* Cowely	7	2
29.	*Prodasineura* Cowely	4	I
30.	*Lestes* Leach	16	3

Contd...

Table 7.1–Contd...

Sl.No.	Genera in India	Number of Species in India	Number of Species in Sindhudurg
31.	*Orolestes* Maclachlan	2	–
32.	*Indolestes* Fraser	6	1
33.	*Sumpecma* Burmeister	2	–
34.	*Burmargioletes* Kennedy	1	–
35.	*Megalestes* Selys	6	–
36.	*Philoganga* Kirby	1	–
37.	*Caliphaea* Selys	1	–
38.	*Echo* Selys	2	–
39.	*Matrona* Selys	1	–
40.	*Neurobasis* Selys	1	1
41.	*Vestalis* Selvs	5	2
42.	*Rhinocypha* Rambur	14	1
43.	*Indocypha* Laidlaw	1	–
44.	*Colocypha* Fraser	1	1
45.	*Libellago* Selys	4	1
46.	*Anisopleura* Selys	5	–
47.	*Bayadera* Selys	4	–
48.	*Dysphaea* Selys	2	–
49.	*Epallage* Charpentier	1	–
50.	*Euphaea* Selys	6	2
51.	*Schmidtiphaea* Asahina	1	–
52.	*Epiophlebia* Calvert	1	–
53.	*Anisogomphus* Selys	4	–
54.	*Anormogomphus* Selys	2	–
55.	*Asiagomphus* Asahina	3	–
56.	*Burmagomphus* Williamson	6	2
57.	*Cyclogomphus* Selys	4	–
58.	*Dubitogomphus* Fraser	1	–
59.	*Heliogomphus* Laidlaw	4	1
60.	*Macrogomphus* Selys	5	–
61.	*Merogomphus* Martin	3	1

Contd...

Table 7.1–Contd...

Sl.No.	Genera in India	Number of Species in India	Number of Species in Sindhudurg
62.	*Microgomphus* Selys	4	1
63.	*Phaenandrogomphus* Lieftinck	1	–
64.	*Platyomphus* Selys	1	–
65.	*Davidius* Selys	7	–
66.	*Nepogomphus* Fraser	I	–
67.	*Nihonogomphus* Oguma	1	–
68.	*Stylogomphus* Fraser	1	–
69.	*Acrogomphus* Laidlaw	2	–
70.	*Davidioides* Fraser	1	1
71.	*Megalogomphus* Campioni	5	2
72.	*Onychogomphus* Selys	16	3
73.	*Ophiogomphus* Selys	2	–
74.	*Paragomphus* Cowley	3	–
75.	*Perissogomphus* Laidlaw	1	–
76.	*Gomphidia* Selys	6	–
77.	*Ictinogomphus* Cowley	6	1
78.	*Aeshna* Fabricius	4	–
79.	*Anaciaeschna* Selys	3	–
80.	*Anax* Leach	6	2
81.	*Hemianax* Selys	1	–
82.	*Gynacantha* Rambur	10	1
83.	*Tetracanthaeschna* Selys	1	–
84.	*Austroaeschna* Selys	1	–
85.	*Cephalaeschna* Selys	5	–
86.	*Gynacanthaeschna* Fraser	1	–
87.	*Oligoaeschna* Selys	3	–
88.	*Periaeschna* Martin	4	–
89.	*Petaliaeschna* Fraser	1	–
90.	*Polycanthagyna* Fraser	2	–
91.	*Chlorogomphus* Selys	11	2
92.	*Anotogaster* Selys	3	–

Contd...

Table 7.1–Contd...

Sl.No.	Genera in India	Number of Species in India	Number of Species in Sindhudurg
93.	*Cordulegaster* Leach	3	–
94.	*Neallogaster* Cowley	5	–
95.	*Hemicordulia* Selys	1	1
96.	*Somatochlora* Selys	1	–
97.	*Idionyx* Hagen	14	4
98.	*Macromidia* Martin	1	1
99.	*Epophlhalmia* Burmelster	4	2
100.	*Macromia* Rambur	14	4
101.	*Hylaeothemis* Ris	2	1
102.	*Tetrathemis* Brauer	1	–
103.	*Brachydiplax* Brauer	3	–
104.	*Nannophya* Rambur	2	–
105.	*Argionoptera* Brauer	2	–
106.	*Amphithemis* Selys	2	–
107.	*Cratilla* Kirby	1	1
108.	*Epithemis* Laidlaw	1	1
109.	*Lalhrecista* Kirby	1	1
110.	*Libellula* Linnaeus	1	–
111.	*Nesoxenia* Kirby	1	–
112.	*Lyriothemis* Brauer	3	–
113.	*Orthetrum* Newman	15	5
114.	*Potamareha* Karsch	1	–
115.	*Acisoma* Rambur	1	–
116.	*Brachythemis* Brauer	1	–
117.	*Bradinopyga* Kirby	2	1
118.	*Crocothemis* Brauer	3	1
119.	*Diplacodes* Kirby	3	2
120.	*Indothemis* Ris	2	1
121.	*Neurothemis* Brauer	6	2
122.	*Rhodothemis* Ris	1	–
123.	*Symvetrum* Newman	9	–

Contd...

Table 7.1–Contd...

Sl.No.	Genera in India	Number of Species in India	Number of Species in Sindhudurg
124.	*Trithemis* Brauer	–	4
125.	*Onychothemis* Brauer	1	–
126.	*Palpopleura* Rambur	2	–
127.	*Rhyothemis* Hagen	4	3
128.	*Camacinia* Kirby	1	–
129.	*Hiydrobasileus* Kirby	1	–
130.	*Pantala* Hagen	1	1
131.	*Pseudotramea* Fraser	1	–
132.	*Tramea* Hagen	4	1
133.	*Tholymis* Hagen	1	–
134.	*Zyxomma* Rambur	1	–
135.	*Aethriamanta* Kirby	1	1
136.	*Macrodiplax* Brauer	1	–
137.	*Selysiothemis* Ris	1	–
138.	*Urothemis* Brauer	1	1
139.	*Selysiothemis* Ris	1	–
140.	*Zygonyx* Hagen	6	1
	Total	502	97

Systematic Account

Systematic account of Odonata (Insecta) from Western Maharashtra is described as under:

Suborder–Zygoptera

Superfamily–Coenagrionoidea

Family–Coenasgrionidae

Subfamily–Pseudagrioninae

1. *Ceriagrion cerinorubellum* (Brauer)

Ceriagrion cerinorunellum Sclys, *Bull. Aead. Belg.* (2) Vol. xlii, p. 526 (1876); Selys, *Ann. Mus. Civ. Genova* (2) Vol. vii(xxvii) (1889).

Distribution

Throughout India, Kerala, Maharashtra: Western Ghats (Sindhudurg: Sawantwadi, Amboli, Fonda).

2. *Ceriagrion rubiae* Laidlaw

Ceriagrion rubiae Laid., *Rec. Ind. Mus.* Vol. xii. p. 132–33 (1916); id. ibid. vol. xvi, pp. 188, 190–191 (1919).

Distribution

Orissa and Western Ghats: Kerala, Maharashtra (Sawantwadi, Amboli, Fonda).

3. *Pseudagrion decorum* (Rambur)

Agrion decorum Ramb. *Ins. Nevrop.p* 258 (1842).

Pseudogrion decorum Selys, *Bull Acad. Belg.* (2) Vol. xlii, p. 504 (1876); Kirby, *Cat. Odon.* p. 153 (1890).

Distribution

Western Ghats: Kerala, Maharashtra (Sawantwadi, Amboli and Fonda).

4. *Pseudagrion indicum* Fraser

Pseudagrion indicum Fras., *Rec,. Ind. Mus.* Vol. xxvi, pp. 428, 495–7, Fig. 3, i, ii, iv, v (1924).

Distribution

Tamil Nadu and Western Ghats: Kerala, Maharashtra (Sawantwadi, Amboli, Kudal, Fonda).

5. *Pseudagrion malabaricum* Fraser

Pseudagrion malabaricum Fraser, *Rec. Ind. Mus.* xxvi, pp. 428, 494–5 (1924); id., Ibid., Vol. xxxiii, pp. 448–465 (1931).

Distribution

Bihar, Tamil Nadu, Western Ghats: Kerala, Maharashtra (Fonda, Vaibhavwadi, Amboli).

6. *Pseudogrion microcepllalum* (Rambur)

Agrion microcephalum Ramb., *Ins. Nevrop,* p. 259 (1842).

Pseudagrion microcephalum Selys, *Bull. Acad Belg.* (2) Vol. xlii, p. 504 (1876); Kirby, *Cat. Odon.* p. 135 (1890).

Distribution

Throughout plains of India, Kerala, Western Ghats: Maharashtra (Amboli, Sawantwadi, Fonda, Kankavali).

7. *Pseudogrion rubriceps rubriceps* (Selys)

Pseudogrion rubriceps Selys, *Bull. Acad Belg.* (2) Vol. xlii, p. 510 (1876); Kirby, *Cat. Odon.* p. 153(1890); Selys, *Ann. Mus. Civ. Genova,* (2) Vol. x (xxx), p. 83 (1891).

Distribution

Through out plains of India. Kerala, Western Ghats: Maharashtra (Amboli, Sawantwadi, Fonda, Kudal).

Subfamily–Coenagrioninae

8. *Aciagrion hisopa hisopa* (Selys)

Pseudagrion hispa Selys, *Bull. Acad. Belg.* (2) Vol. x (xxx), pp. 512 (1891); Laid., *Rec. Ind Mus.* Vol. xvi, p. 172 (1919).

Distribution

Karnataka, Kerala, Western Ghats: Maharashtra (Fonda, Amboli, Kudal, Kankavali).

9. *Aciagrion occidentale* Laidlaw

Aciagrionhispa hisopa (Selys)? race occidentalis Laid., *Rec. Ind Mus.* Vol. xvi, p. 186 (1919).

Aciagrion occidentalis Fras. *J. Bombay Nat. Hist. Soc.*, Vol. xxix, p. 749 (1923); id., *Rec. Ind. Mus.* Vol. xxvi, pp. 428, 491, Fig. 3, III, p. 496 (1924).

Distribution

South India, Kerala, Western Ghats: Maharashtra (Amboli, Sawantwadi, Kudal, Vaibhavwadi).

10. *Ischnura aurora aurora* (Braner)

Agrion delicatum Hagen, *Verh. Zoo. bot. ves. Wien,* Vol. viii, p. 497(1858).

Agrion aurora Brauer, *Verh. Zoo. bot. ves. Wien,* Vol. xv, p. 510 (1865).

Ichnura delicata Selys, *Bull. Acad Belg.* (2) Vol. xli, p. 282 (1876); *Mart., Mem. Soc. Zoo.* France, Vol. xix, p. 246 (1901).

Ichnura aurora Ris., Nov. *Caledonia, Zoo.* Vol. ii, 4, p. 67 (1915); Laid., *Rec. Ind. Mus.*Vol. xii, p. 131 (1916).

Distribution

Throughout India, Kerala, Western Ghats: Maharashtra (Amboli, Kudal, Sawantwadi).

11. *Ischnura senegalensis* (Rambur)

Agrion senegalensis Ramb. Ins. Nevrop. p. 276 (1842); Selys, *Rev. Odon.* p. 186 (1850).

Ischnura senegalensis Selys, *Bull. Bel.* (2) Vol. xli, p. 273 (1876); id., *Ann. Soc. Esp. Nat.* Vol. xi, no. 60 (1882).

Distribution

Throughout India, Kerala, Western Ghats: Maharashtra (Amboli, Kudal, Sawantwadi).

12. *Ischnura forcipata* Morton.

Ichnura forcipata Mort., *Trans. Ent. Soc. Lond.* p. 306, p. xxiv, Figs. 1–3 (1907); Laid., *Rec. Ind Mus.* vol. xii, pp. 129–130 (1916); id., ibid. Vol. xvi, pp. 171, 173, 174 (1919).

Ischnura musa Barton. *Rev. Russ. Ent.* Vol. xiii, no. 1, pp. 187–189 (1913).

Ischnura gangetica Laid., *Entomologist*, p. 235 (1913).

Distribution

Western Ghats: Maharashtra (Amboli, Sawantwadi).

Subfamily–Agriocnemidinae

13. *Agriocnemis pieris* Laidlaw

Agriocnemis pieris Laid., *Rec. Ind Mus.* Vol. xvi, pp. 179, 180 (1919); Fras., *J. Bombay. Nat. Hist. Soc.* Vol. xxix, p. 747 (1923); *Id., Rec. Ind. Mus.* Vol. xxvi, p. 490 (1924).

Distribution

South India, Kerala, Western Ghats: Maharashtra (Amboli, Sawantwadi).

14. *Agriocnemis pygmae* (Rambur)

Agriocnemis Pygmae. Ram., *Ins. Nevrop.* p. 278 (1842); Brauer, *Verh. Zool. Bol. Ges. Wilen,* Vol. xiv, p. 161 (1864); id., Reise, *Novora',* p. 103 (1866).

Agriocnemis pygmae Selys, *Syn. Agr. me, Legion:* Agrion (Suite fin), p. 52 (1877); Kirby, *Cat. Odon.* p. 158 (1890).

Distribution

Throughout the Oriental Region, Kerala, Western Ghats: Maharashtra (Fonda, Kudal, Vaibhavwadi).

15. *Agriocnemis splendidissima* Laidlaw

Agriocnemis splendidissima Laid., *Rec. Ond Mus.* Vol. xvi, pp. 180–182 (1919), Fraser, *J. Bombay Nat. Hist. Soc.* Vol. xxix, p. 747 (1923); id., *Rec. Ind. Mus.* vol. xxvi, pp. 470–419 (1924).

Distribution

West Bengal, Madhya Pradesh and South India, Kerala, Western Ghats: Maharashtra (Amboli, Kankavali, Fonda).

16. *Mortongrioll varralli* Fraser

Mortongrion varralli Fras., *J. Bombay Nat. Hist. Soc.* Vol. xxvi, p. 490 (1924); *Rec. Ark. Zool.* Uppsala, Bd. Vol. xxi, A. no. 31, pp. 7: 13–14 (1930); Fras *Rec. Ind. Mus.* Vol. xxiii, pp. 448–464 (1931).

Distribution

North East India and Western Ghats: Kerala, Maharashtra (Amboli, Sawantwadi, Fonda, Kudal, Kankavali).

Subfamily–Argiinae

17. *Onychargia atrocyana* Selys

Onychargia onychargia atrocynan Selys, *Bull. Acad. Belg* (2) Vol. xx, p. 416 (1865); Kirby, *Cat. Odon.* P. 139 (1890); id., *J. Linn. Soc. Lond., Zoo.* Vol. xxiv, p. 563 (1893).

Distribution

Assam, West Bengal, Bihar, Orissa and South India, Western Ghats: Kerala, Maharashtra (Amboli, Sawantwadi, Kudal).

Family–Platycnemididae

Subfamily–Platycnemidinae

18. *Copera marginipes* (Rambur)

Platycnemis marginipes Ramb., Ins. Nevrop. p. 240 (1842).

Copera marginipes Kirby, *Cat. Odon.* P. 129(1890); *id., J. Linn. Soc. Lons., Zoo.* Vol. xxiv, p. 560. (1894); Laid., Fasc. Malay. (Odon), Pt. II, p. 8 (1907); Ris, *Suppl. Ent. nov.* p. 18 (1916).

Distribution

Southern Asia, Western Ghats: Kerala, Maharashtra (Amboli, Fonda, Kudal, Sawantwadi, Vaibhavwadi).

19. *Copera vittata deccanesis* Laidlaw

Psilocnemis vittata Selys, *Bull. Acad Belg.* (2) Vol. xvi, p. 170 (1863); *id. Mem. Cour.* Vol. xxxviii, p. 121 (1886).

Copera vittata Kirby, *Cat. Odon.* p. 129 (1890); Krug., *Stett. Ent. Zeit.* p. 117 (1898).

Copera vittata deccaneusis Laid., *Rec. Ind. Mus.* Vol. xxix, p. 744 (1923); id., *Rec. Ind. Mus.* Vol. xxxiii, p. 448 (1931).

Distribution

Western Ghats: Kerala, Maharashtra (Amboli, Fonda, Kudal, Vaibhavwadi).

Family–Platystictidae

Subfamily–Platystictinae

20. *Protostica antelopoides* Fraser

Protostica antelopoides Fra., *Rec. Ind. Mus.* Vol. xxxiii, pp. 449, 467, 468 (1931).

Distribution

Western Ghats: Kerala, Maharashtra (Amboli, Sawantwadi, Kudal, Fonda, Vaibhavwadi).

21. *Protostica davenporti* Fraser

Protostica davenporti Fra., *J. Bombay Nat. Hist. Soc.* Vol. xxxv, pp. 70, 71, Pl. i. Fig. 9, 10 (1931).

Distribution

Uttar Pradesh, Western Ghats: Kerala, Maharashtra (Amboli, Fonda, Kudal, Sawantwadi).

21. *Protostica gravelyi* Laidlaw

Protostica gravelyi Laid., *Rec. Ind. Mus.* Vol. xi, pp. 389, 390, text. Fig. 2 (1915), id., *ibid.*,Vol. xiii. p. 499 (1924).

Distribution

Maharashtra, Western Ghats: Kerala, Maharashtra (Sawantwadi, Fonda, Vaibhavwadi).

23. *Protostica heaseyi* Fras.

Protostica heaseyi Fras., Rec. Ind. Mus. Vol. xxiv, pp. 6, Pl. i, Figs. 3, 4 (1922); id., *ibid.*,Vol. xxvi, pp. 499 (1924); id, *J. Bombay Nat. Hist. Soc.*, Vol. xxxv, pp. 73, 71, Pl. i. Fig. (1931) *id., Rec. Ind. Mus.* Vol. xxxiii, pp. 449, 466 (1931).

Distribution

Western Ghats: Kerala, Maharashtra (Amboli, Sawantwadi, Kudal, Fonda).

Family–Protoneuridae

Subfamily–Caconeurinae

24. *Caconeura ramburi* (Farser)

Indoneura ramburi Fras. *Rec. Ind. Mus.* Vol. xxiv, pp. 2, 3, Pl. i, (1922); id, *J. Bombay Nat. Hist. Soc.* Vol. xxxv, pp. 71,72, Pl. i. Figs. 5,6 (1931); id., *Rec. Ind. Mus.*, Vol. xxxiii, p. 449, 466 (1931).

Distribution

South India, Western Ghats, Kerala, Maharashtra (Amboli, Sawantwadi, Kankavali, Vaibhavwadi).

25. *Caconeura risi* (Fraser)

Indoneura risi Fraser *Rec. Ind. Mus.* Vol. xxiv, pp. 449, 469, 471, Fig. 6i (1931).

Distribution

Western Ghats: Kerala, Maharashtra (Amboli, Sawantwadi, Kudal, Fonda).

26. *Esme cyaneovitta* Fraser

Esme cyaneovitta Fras. *Mem. Dept. Agric. India.* (*Ent.*) Vol. vii, pp. 45, 46 (1922), id, *Rec. Ind. Mus.* Vol. xxiv, pp. 6, 7, Figs. 5, 6 (1924) id., *ibid.*, Vol. xxvi, pp. 429, 506, Fig. 6, v (1931).

Distribution

Western Ghats: Kerala, Maharashtra (Amboli, Sawantwadi, Kudal, Fonda, Kankavali, Vaibhavwadi).

27. *Esme mudiensis* Fraser

Esme mudiensis Fra. *Rec. Ind. Mus.* Vol. xxiv, pp. 449, 472, 473, Fig. 6, iv (1931).

Distribution

South India, Western Ghats: Kerala, Maharashtra (Fonda, Kudal).

28. *Melanoneura billineata* Fraser

Melanoneura billineata Fra. *Mem. Dept. Agric.* India. (Ent.) Vol. vii, pp. 45, 46 (1922), id, *J. Bombay Nat. Hist. Soc.* Vol. xxxv, pp. 71, 72, Pl. i. Figs. 5, 6 (1931); id., *Rec. Ind. Mus.* Vol. xxxiii, p. 449, 466 (1931).

Distribution

South India, Western Ghats: Kerala, Maharashtra (Amboli, Sawantwadi, Kudal, Vaibhavwadi).

29. *Phylloneura westermanni* (Selys)

Alloneura westermanni Selys, *Bull. Acad. Belg.* (2) Vol. x, p. 447 (1860).

Phylloneura westermanni Fras. *Rec. Ind. Mus.* Vol. xxiv. pp. 3, 4 (1922); id., *J. Bombay Nat. Hist. Soc.* Vol. xxix, p. 743 (1923); id., *Rec. Ind. Mus.* Vol. xxvi, pp. 506–9 (1924).

Distribution

Western Ghats: Kerala, Maharashtra (Amboli, Fonda, Kankavali, Vaibhavwadi).

Sub-family–Disparoneurinae

30. *Elattoneura souteri* (Fraser)

Disparoneura souteri Fras. *Rec. Ind Mus.* Vol. xxxiii, pp. 449, 468 (1931).

Distribution

Western Ghats: Kerala, Maharashtra (Amboli, Fonda, Kankavali, Vaibhavwadi).

31. *Elattoneura tetrica* (Laidlaow)

Disparoneura tetrica Laid. *Rec. Ind. Mus.* Vol. xiii, pp. 323, 345, 346 (1917); Fras., *J. Bombay Nat. Hist. Soc.*, Vol. xxix, p. 743 (1923); id., *Rec. Ind. Mus.* Vol. xxvi, pp. 429, 503 (1924).

Distribution

Western Ghats: Kerala, Maharashtra (Amboli, Kudal, Vaibhavwadi).

32. *Prodasineura verticalis annandaleli* (Fraser)

Caconeura annadalei Fras., *J. Bombay Nat. Hist. Soc.*, Vol. xxix, p. 743 (1923); id., *Rec. Ind. Mus.*, Vol. xxvi, pp. 429, 503 (1924).

Distribution

Western Ghats: Kerala, Maharashtra (Fonda, Kudal, Kankavali, Vaibhavwadi).

Super-family–Lestoidea

Family–Lestidae

Subfamily–Lestinae

33. *Lestes elatus* Hagen

Lestes elata Hagen. Verh. Zool. Bot. Ges Wies., Vol. viii, p. 478 (1858); Selys, *Bull. Acad. Belg.* (1862); Laid., *Rec. Ind. Mus.*, Vol. xix, pp. 146, 153–4, Fig. 2 (1920) *Lestes elatus* Kirby, *Cat Odon.* P. 162 (1890); id; Proc. Zool. Soc. Land. p. 203(1891); id., *J. Linn. Soc. Lond. Zool.*, 27, 565 (1893).

Distribution

Sri Lanka, Peninsular India, Western Ghats: Kerala, Maharashtra (Amboli, Sawantwadi, Kudal, Fonda).

34. *Lestes malaharica* Fraser

Lestes malabarica Fra. *J. Bombay Nat. Hist. Soc.*, Vol. xxiii, pp. 847, 848 (1929); id., *Rec. Ind. Mus.*, Vol. xxiii, pp. 448, 463 (1931).

Distribution

Chandigarh, Andamans Island, South India, Western Ghats: Kerala, Maharashtra (Amboli, Kudal, Fonda).

35. *Lestes praemorsus pramorsus* (Sleys)

Lestes praemorsa Selys, Bull. Acad. Belg. (2) Vol. xii, p.320 (1862); id., *mitt. mus. Dresd.*, Vol. iii. p. 317 (1878); id., *Ann. Soc. Espan. Hist. Nat. Soc.*, Vol. xxix, p. 743 (1923); id., *Rec. Ind. Mus.*, Vol. xxvi, pp. 506–509 (1924).

Distribution

Western Ghats: Kerala, Maharashtra (Amboli, Sawantwadi, Kudal, Fonda, Vaibhavwadi).

Lestes praemorsus Kirby, *Cat Odon*, p. 162 (1890)

Distribution

Western India, Andamans Island, Western Ghats: Kerala Maharashtra (Sawantwadi, Kudal, Amboli, Vaibhavwadi, Fonda).

Subfamily–Sympecmatinae.

36. *Indolestes davenporti* (Fraser)

Lestes gracilic birmanus Ris, nee *Selys, Suppl. Ent.* No. 5, pp. 13, 14 (1916), *Laid. Rec. Ind. Mus.*, Vol. xix, p.158 (1920).

Celyonolestes davenporti Fras. *J. Bombay Nat. Hist. Soc.*, Vol. xxiv, pp. 96, 97 Pl. i. Fig. 7 (1930), id., *Rec. Ind. Mus.* Vol. xxiii, pp. 448, 484, 464 (1931).

Distribution

Western Ghats: Kerala, Maharashtra (Fonda, Amboli, Kudal, Vaibhavwadi).

Superfamily– Calopterygoidea

Family–Calopterygidae

Subfamily–Calopteryginae

37. *Neurobasis chinesis chinensis* (Linnaeus)

Libellula chinensis Linnaeus, *Syst. Nat.*, Vol. i, p. 545 (1758);

Edwards, *Nat. Hist. Bird,* Vol. iii, t. 112 (1750); Donovan, *Ins. China,* t. 46, f:i,I (1798). *Calopteryx chinesis* Rambur, Ins. Nevrop. p. 226 (1842).

Neurohasis chinesis Kirby, *Cat. Odon,* p. 102 (1890). Fraser *Rec. Ind Mus.,* Vol. xxvi, pp. 428, 479, (1924). id., *J. Bombay Nat. Hist. Soc.,* Vol. xxiv, pp. 96, 97, Pl. i. Fig. 7 (1930).

Distribution

Throughout India, Kerala, Western Ghats: Maharashtra (Fonda, Amboli, Kudal).

38. *Vestalis apicalis apicalis* Selys

Vestalis apicalis Selys. *Bull. Acad. Belg.,* (2) Vol. xxxvi, p. 612 (1873); id., *ibid,* (2) Vol. xlvii, p. 362 (1879); Kirby, *Cat. Odon.,* p. 102 (1890); id., *J. Linn. Soc. (Zool.),* Vol. xxiv, pp. 558–559 (1893).

Distribution

Eastern Ghats, N.E. India, Western Ghats (Sawantwadi, Kudal, Amboli, Vaibhavwadi).

39. *Vestalis gracilis gracilis* (Rambur)

Colopteryx gracilis Rambur, *Ins. Nevrop,* p. 224 (1842); Walker, *List Neur. Ins. Brit. Mus.,* IV. p. 611 (1853). *Vestalis gracilis* Sely. *Syn. Cal.,* p. 26 (1853), id. *Mon. Cal.,* p. 84 (1854); *Kirby, Cat. Odon.,* p. 102 (1890); Selys, *Ann. Mus. Civ. Genov,* (2) Vol. x(xxx), p. 487 (1891).

Distribution

Bihar, N.E. India, Andaman and Nicobar Islands, Eastern Ghats, W. Bengal, Western Ghats, Kerala (Amboli, Vaibhavwadi, Kudal, Fonda).

Family–Chlorocyphidae

40. *Rhinocypha (Heliocypha) bisignata* (Selys)

Rhinocypha hisignata Selys. *Syn. Cal.,* 62 (1853); id., *Mon. Cal.,* p. 214 (1855); Kirby, *Cat. Odon,* p. 113 (1890); Laidlaw, *Rec. Ind. Mus.,* Vol. xiii, p. 38 (1917); Fraser, *Mem. Dept. Agric. India (Ent.),* Vol. vii, pp. 80, 81, Pl. x (1922).

Distribution

Southern and Peninsular India, Kerala, Western Ghats: Maharashtra (Sawantwadi, Amboli, Fonda).

41. *Calocypha laidlawi* (Fraser)

Rhinocypha laidlawi Fraser, *Rec. Ind Mus.*, Vol. xxvi, pp. 482–483 (1924).

Calocypha laidlawi Fraser, *J. Bombay Nat. Hist. Soc.*, Vol. xxxiii, pp. 457, 458, Pl. ii, Fig. 3 (1928); id., *Rec. Ind. Mus.*, Vol. xxxiii, p. 448 (1931).

Distribution

Karnataka, Kerala, Western Ghats: Maharashtra (Sawantwadi, Amboli, Fonda, Kudal).

42. *Libellago lineata indica* (Fraser)

Micromerus lineatus Fraser, *Nec. Ind. Mus.*, Vol. xvi, pp. 197, 198 (1919) (larva); *Laidlaw spolia Zeylanica*, Vol. xii, pp. 354–355 (1924).

Micromerus lineatus indica Fraser, *J. Bombay Nat. Hist. Soc.*, Vol. xxxii, pp. 686, 687, Pl. 1, Fig. 5 and Pl. iii, Fig. 2 (1928); id., *Rec. Ind. Mus.*, Vol. xxxiii, pp. 448, 463 (1931).

Distribution

Karnataka, Western Ghats: Kerala, Maharashtra (Amboli, Fonda, Sawantwadi, Kudal).

Family–Euphaeidae

43. *Euphaea cardinalis* Fraser, *Rec. Ind Mus.*, Vol. xxvi, pp. 512, 513 (1924).

Indophaea cardinolis Fraser, *J. Bombay Nat. Hist. Soc.*, Vol. xxxiii, pp. 295, 296 (1929); id., *Rec. Ind. Mus.*, Vol. xxxiii, pp. 448, 463 (1931).

Distribution

South India, Kerala, Western Ghats Maharashtra (Sawantwadi, Fona, Amboli, Kudal).

44. *Euphaea dispar* (Rambur)

Euphaea dispar Rambur, *Ins. Nevrop*, p. 230 (1842); Selys. *Syn. Cal.*, p. 51 (1853); Walker, *List. Neur. Inst. Brit. Mus.*, Vol. iv, p. 640 (1853).

Pseudophaea dispar Kirby, *Cat. Odon.*, p. 109 (1890); Laidlaw *Rec. Ind. Mus.*, Vol. xiii, p. 32 (1917); id., *ibid*, Vol. xix, pp. 25, 27 (1920).

Ihdophaea dispar Fraser, *J. Bombay Nat. Hist. Soc.*, Vol. xxxiii, pp. 294, 295 (1929); id., *Rec. Ind. Mus.*, Vol. xxxiii, pp. 448, 463 (1931).

Distribution

Western Ghats: Kerala, Maharashtra (Amboli, Fonda, Sawantwadi).

Suborder– Anisoptera

Superfamily–Aeshnoidea

Family–Gomphidae

Subfamily–Gomphinae

45. *Burmagomphus pyramidalis* Laidlaw

Burmagomphus pyramidalis Laidlaw, *Rec. Ind. Mus.*, Vol. xxiv, pp. 371, 399, 401, Fig. 17 (1922); Fraser, *J. Bombay Nat. Hist. Soc.*, Vol. xxix, pp. 62, 331 (1923); id., *Rec. Ind. Mus.*, Vol. xxvi, pp. 427, 476 (1924); id. *J. Bombay Nat. Hist. Soc.*, Vol. xxxi, pp. 409–410 (1926).

Distribution

Western India, Southern India, Kerala, Western Ghat: Maharashtra (Sawantwadi, Amboli, Fonda, Vaibhavwadi).

46. *Burmagomphus laidlawi* Fraser

Gomphus sp. Fraser, *Rec. Ind. Mus.*, Vol. xxiv, pp. 419 (1922); id., *J. Bombay Nat. Hist. Soc.*, Vol. xxiv, pp. 62, 330 (1923).

Burmagomphus laidlawi Fraser, *Rec. Ind. Mus.*, Vol. xxvi, pp. 427, 475, 476 (1924); id., *J. Bombay Nat. Hist. Soc.*, Vol. xxxi, pp. 412–413, Pl. 1, Fig. 5 (1926); Laidlaw, *Trans. Ent. Soc. Lond.*, Vol. 1 xxviii, p. 189 (1930); Fraser, *Rec. Ind. Mus.*, Vol. xxxiii, p. 447 (1931); Needham, *Rec. Ind. Mus.*, Vol. xxxiv, pp. 226, 227 (1932).

Distribution

Western Ghats: Maharashtra (Fonda, Vaibhavwadi, Amboli).

47. *Heliogomphus kalarensis* sp.n.

Heliogomphus kalarensis Fraser (nom. nud.), *Ceylon J. Sci.*, B. Vol. xviii, p. 29, Fig. 4d (1933).

Distribution

Nilgiris, South India, Western Ghats (Fonda, Sawantwadi, Amboli).

48. *Merogomphus longistigma longistigma* (Fraser)

Indogomphus longistigma Fraser, *Rec. Ind. Mus.*, Vol. xxiv, pp. 422, 424, Pl. xi, Fig. 8 (1922); id., *J. Bombay Nat. Hist. Soc.*, Vol. xxix, pp. 64, 332 (1923).

Merogomphus longistigma Laidlaw, *Trans. Ent. Soc. Lond.*, Vol. lxxviii, p. 185 (1930); Fraser, *Rec. Ind. Mus.*, Vol. xxxiii, pp. 448, 460 (1931).

Distribution

South India, Kerala, Western Ghats (Fonda, Amboli, Kankavali).

49. *Microgomphus souteri* Fraser

Microgomphus torquatus souteri Fraser, *Rec. Ind. Mus.*, Vol. xxvi, pp. 427, 474 (1924); id., *ibid.* Vol. xxxiii, p. 447 (1931).

Microgomphus souteri Fraser, *J. Bombay Nat. Hist. Soc.*, 30: 853, 854, Pl. 1, Fig. 3 (1925); Laidlaw, *Trans. Ent. Soc. Lond.*, Vol. lxxviii, p. 182 (1930); Needham, *Rec. Ind Mus.*, Vol. xxxiv, p. 221 (1932).

Distribution

Western Ghats, Kerala, Western Ghats (Sawantwadi, Amboli, Fonda, Vaibhavwadi).

Subfamily–Onychogomphidae

50. *Davidioides martin;* Fraser

Davidioides martini Fraser, *Rec. Ind. Mus.*, Vol. xxvi, pp. 427, 472, 473, Fig. 2 (1924); id., *J. Bombay Nat. Hist. Soc.*, Vol. xxvi, pp. 419, 420, Fig. 3 (1926); Laidlaw, *Trans. Ent. Soc. Land.*, Vol. lxxviii, p. 188 (1930); Fraser, *Rec. Ind. Mus.*, Vol. xxxiii, p. 447 (1931); Needham, *Rec. Ind. Mus.*, Vol. xxxiv, p. 226 (1932).

Distribution

Western Ghats (Amboli, Kudal, Kankavali).

51. *Megalogomphus superbus* Fraser

Megalogomphus superbus Fraser, *Rec. Ind. Mus.*, Vol. xxxiii, pp. 448, 460–463, Figs. 3 and 4 (1931); Needham, *Rec. Ind. Mus.*, Vol. xxxiv, p. 222 (1932).

Distribution

South India, Kerala, Western Ghats (Amboli, Sawantwadi, Kankavali, Fonda).

52. *Megalegomphus hannyngton* (Fraser)

Megalegomphus hannyngtoni (Fraser), *J. Bomhay Nat. Hist. Soc.*, Vol. xxix, pp. 674, 676, Fig. 4 (1923).

Megalegomphus hannyngtoni Fraser, *Rec. Ind Mus.*, Vol. xxvi, pp. 428, 478, 479 (1924); id., *Mem. Dept. Agric. India*, Vol., viii, No. 8, pp. 79–81, Figs. 1–3 (1924).

Distribution

South India, Kerala, Western Ghats (Sawantwadi, Amboli, Fonda, Kudal).

53. *Onychogomphus acinaces* (Laidlaw)

Onychogomphus acinaces Laidlaw, *Rec. Ind. Mus.*, Vol. xxiv, pp. 407–408, Fig. 20 (1922).

Onychogomphus acinacesFraser, *J. Bombay Nat. Hist. Soc.*, Vol. xxix, pp. 988–990, Pl. I, Fig. 2 (1924).

Distribution

Western Ghats (Sawantwadi, Kudal, Fonda, Vaibhavwadi).

54. *Onychogomphus nilgiriensis nelgiriensis* (Fraser)

Onychogomphus biforceps nelgiriensis (Fraser), *Rec. Ind. Mus.*, Vol. xxiv, pp. 425, 426, Pl. xi, Fig. 2 (1922).

Lamilligomphus nilgiriensis Fraser, *J. Bombay Nat. Hist. Soc.*, Vol. xxix, pp. 986, 988, Pl. 1, Fig. 1 (1924); id., *Rec. Ind. Mus.*, Vol. xxvi, pp. 427, 477 (1924); Laidlaw, *Trans. Ent. Soc. Land*, Vol. lxxviii, p. 193 (1930).

Distribution

South India; Kerala, Western Ghats (Amboli, Sawantwadi, Kudal, Fonda, Vaibhavwadi, Kankavali).

55. *Onychogomphus malabarensis* (Fraser)

Onychogomphus malabarensis Fraser (nec. Selys), *Rec. Ind. Mus.*, Vol. xxiv, pp. 424, 425, Pl. xi, Fig. 10 (1922).

Distribution

Western Ghats (Fonda, Kudal, Arnboli).

Subfamily–Lindeniinae

56. *Ictinogomphus rapax* (Rambur)

Distatomma rapax Rambur, *Ins. Nevrop*, p. 169 (1842).

Ictinus vorax Rambur (Female) Rambur, *Ins. Nevrop*, p. 171 (1842).

Ictinus rapax Selys, *Bull. Acad. Belg.*, Vol. xxi, Pt. 2, p. 90 (1854); id., *Mon. Gomph.* p. 276 (1857); Kirby, *Cat. Odon*, p. 77 (1890); Martin, Mision Pavie Indo-China, p. 217 (1904).

Distribution

Throughout India, Western Ghats: Maharashtra (Amboli, Sawantwadi).

Subfamily–Aeshninae

57. *Anax guttatus* (Burmeister)

Aeshna guttata Brumeister, *Hanib. Ent.*, Vol. ii, p. 840 (1839).

Anax magnus Rambur, *Ins. Nevrop*, p. 188 (1842); Breuer, *Reise Novara, Neur.* p. 62 (1866).

Anax guttatus Hagen, *Verh. Zool. Bot. Ges. Wien*, Vol. xvi, p. 39 (1867); Kirby *Cat. Odon*, p. 84 (1890); Martin, *Cat. Coil.* Selys, Fase, Vol. xviii, p. 23, Fig. 17 (1908).

Distribution

Throughout India, Kerala, Western Ghats (Sawantwadi, Arnboli, Fonda, Vaibhavwadi).

58. *Anax immaculifrons* Rambur

Anax immaculifrons Rambur, *Ins. Nevrop.*, p. 189 (1842); Brauer, *Reised-Novora Neur*, p. 60 (1866); Kirby *Cat. Odon*, p. 84 (1890);

Martin, *Cat. Coli.* Selys (Aesehnines), Fase. xviii, pp. 18, 19, Fig. 12 (1908).

Distribution

Bihar, Chandigarh, West Bengal, Kerala, Western Ghats: Maharashtra (Fonda, Arnholi, Sawantwadi, Kudal).

59. *Gynacantha dravida* Lieftinck

Gynacantha hyaline Selys; *Ann. Soc. Espan. Hist. Nat.*, Vol. xi, p. 19 (1882); Karsch, *Ent. Nachr*, Vol. xvii, p. 228 (1891); Kruger, *Stell. Ent. Zeit.*, Vol. lix, p. 275 (1898).

Distribution

Japan, Burma, Sri Lanka, China, India (Assam, Meghalaya, Bihar, Kerala, West Bengal, Western Ghats (Fonda, Kankavali, Sawantwadi, Amboli).

Superfamily–Cordulegasteroidea

Family–Cordulegasteridae

Subfamily–Chlorogomphinae

60. *Chlorogomphus campioni* (Fraser)

Orogomphus campioni Fraser, *Rec. Ind. Mus.*, Vol. xxvi, pp. 427, 467–469 (1924); id., *ibid.* Vol. xxvii, pp. 423–429 (1925).

Chlorogomphus campioni Fraser, *Rec. Ind. Mus.*, Vol. xxxiii, pp. 447, 457 (1931).

Distribution

Western Ghats (Amboli, Fonda, Sawantwadi, Vaibhavwadi), Kerala.

61. *Chlorogomphus xanthoptera* (Fraser)

Chlorogomphus xanthoptera Fraser, *J. Bombay Nat. Hist. Soc.*, Vol. xxvi, pp. 874, 875, Fig. 19 (9).

Chlorogomphus xanthoptera Fraser, *Rec. Ind. Mus.*, Vol. xxvi, pp. 427, 469 (1924); id., *ibid.* Vol. xxxiii, pp. 447–456, Figs. 2, a,b (1931); id., *Mem. Ind. Mus.*, Vol. ix, pp. 456, 457 (1933).

Distribution

Western Ghats (Amboli, Sawantwadi, Kudal, Kankavali, Vaibhavwadi).

Superfamily–Libelluloidea

Family–Cordullidac

Subfamily–Cordullinae

62. *Hemicordulia asiatica* Selys. *Bull. Acad. Belg.*, (2) Vol. xiv, p. 186 (1878); Kirby. *Cat. Odon*, p. 47 (1890); Martin. *Cat. Coll.* Selys (Cordulinos), fase. xvii, p. 13 (1906).

Distribution

Assam, Meghalaya, Kerala, Maharashtra, Western Ghats (Fonda, Sawantwadi, Amboli, Kankavali, Kudal).

Subfamily–Idionychinae

63. *Idionyx burliyarensis* Fraser

Idionyx corona race *nilgiriensis* Fraser, *Mem. Depl. Agric. India.* (ENT), Vol. vii, No. 7, pp. 65, 66 (1922).

Idionyx corona burliyarensis Fraser, *Rec. Ind. Mus.*, Vol. xxvi, pp. 427, 461–462 (1924).

Distribution

Western Ghats (Amboli, Sawantwadi, Fonda), Kerala.

64. *Idionyx minima* Fraser

Idionyx minima Fraser, *Rec. Ind. Mus.*, Vol. xxxiii, pp. 447, 453–455 (1931); id., *J. Bombay Nat. Hist. Soc.*, Vol. xxxvii, pp. 558, 562 Pl. I, Figs. 1, 5 (1934).

Distribution

Western Ghats (Sawantwadi, Fonda, Kudal, Kankavali, Vaibhavwadi), Kerala.

65. *Idionyx saffronata* Fraser

Idionyx saffronata Fraser, *Rec. Ind. Mus.*, Vol. xxvi, pp. 427, 458–460 (1924).

Distribution

Western Ghats, Kerala. Maharashtra (Amboli, Sawantwadi).

66. *Idionyx travancorensis* Fraser

Rec. Ind. Mus., Vol. xxxiii, pp. 447, 455, 456 (1931).

Distribution

Western Ghats: Maharashtra (Amboli, Sawantwadi, Fonda, Kudal), Kerala.

67. *Macromidia donaldi* (Fraser)

Indomacromia donaldi Fraser. *Rec. Ind. Mus.*, Vol. xxvi, pp. 515, 516, text Figs. 4 and 7c,d (1924).

Distribution

Western Ghats: Kerala, Maharashtra (Fonda, Vaibhavwadi, Kudal, Amboli).

Subfamily–Macromiinae

68. *Epophthalmia frontalis binocellata* (Fraser)

Macromia binocellata Fraser, *Rec. Ind. Mus.*, Vol. xxvi, pp. 451, 452 (1924).

Distribution

Western Ghats: Maharashtra (Amboli, Sawantwadi, Vaibhavwadi, Kudal, Fonda), Kerala.

69. *Epophthalmia vittata vittata* Burmeister

Epophthalmia vittata Burmeister, *Handb. Ent.*, Vol. ii, p. 845 (1839).

Distribution

Uttar Pradesh, West Bengal, Peninsular India, Kerala, Western Ghats: Maharashtra (Amboli, Fonda, Kankavali, Vaibhavwadi).

70. *Macromia annaimallaiensis* Fraser

Macromia annaimallaiensis Fraser, *Rec. Ind. Mus.*, Vol. xxxiii, p. 447, 452, 453 (1931).

Distribution

Western Ghats: Maharashtra (Fonda, Vaibhavwadi, Amboli), Kerala.

71. *Macromia flavocolorafa* Fraser

Macromia flavocolorata Fraser (female). *J. Bombay Nat. Hist. Soc.*, Vol. xxviii, p. 702, Fig. 2 (1922).

Distribution

North Bengal, Western Ghats (Amboli, Fonda, Sawantwadi), Kerala.

72. *Macromia indica* Fraser

Macromia indica Fraser, *Rec. Ind. Mus.*, Vol. xxvi, pp. 427, 447–449, pi. xxv, Fig. 5 (1924); *id.*, Vol. xxxiii, pp. 447, 453 (1931); Needham, *Rec. Ind. Mus.*, vol. xxxiv, pp. 211, 212 (1932).

Distribution

Western Ghats: Kerala, Maharashtra (Vaibhavwadi, Kudal, Amboli).

73. *Macromia irata* Fraser

Macromia irata Fraser, *Rec. Ind. Mus.*, Vol. xxvi, pp. 427, 454–455, pl. xxv, Fig. 6 (1924).

Distribution

Western Ghats: Kerala, Maharashtra (Sawantwadi, Amboli)

Family–Libellulidae

Subfamily–Tetrathemistinae

74. *Hytaeothemis fruhstorferi* (Karsch)

Tetrathemis fruhstorferi Karsch. *Ent. Nachr.*, Vol. xv, p. 3211 (1889); Kirby, *Cat. Odon.* p. 44 (1890).

Distribution

Western India, Western Ghats: Maharashtra (Sawantwadi, Amboli, Fonda), Kerala.

Subfamily–Libellulinae

75. *Cratilla lineata calverti* Foerster

Orthemis lineata Brauer, Sitzungsber, Akad, Wien., Vol. lxxvii, p. 9 (1770).

Distribution

Orissa, Uttar Pradesh, Western Ghats: Maharahstra (Amboli, Sawantwadi, Kudal), Kerala.

76. *Epithemis mariae* Laidlaw

Amphithemis mariae Laidlaw, *Rec. Ind. Mus.*, Vol. xi, pp. 337, 339 (1915).

Distribution

Western Ghats: Maharashtra (Fonda, Amboli), Kerala.

77. *Lathrecista asiatica asiatica* (Fabricius)

Libellula asiatics Fabricius, *Ent. Syst. Suppl.*, p. 283 (1798).

Lathrecista asiatica asiatica Ris, *Cat. Coli. Selys. Fasc.*, Vol. ix, pp. 129–132, (1909); Fraser, *J. Bombay Nat. Hist. Soc.*, Vol. xxvi, pp. 147–149, Figs. 20–22 (1918).

Distribution

South Andaman Islands, West Bengal, Western India, Kerala, Western Ghats: Maharashtra (Sawantwadi, Amboli, Fonda).

78. *Orthetrum chrycea* (Selys)

Libella testacea race? Chrysis Selys, *Ann. Mus. Civ. Genova*, Vol. xxx, p. 462 (1891), *Orthetrum Chrysis*, Ris, Archivtar *Natur. Bd.*, I, p. 186, Pl. ix, Fig. 2 (1900).

Distribution

Himachal Pradesh, Orissa, Andaman Islands, Western Ghats: Maharashtra (Amboli, Kudal, Fonda), Kerala.

79. *Orthetrum galaucum* (Brauer)

Libellula glauca Brauer, *Verh. Zool. Bot. Ges. Mien*, Vol. xv, p. 1012 (1865).

Orthetrum glaucum Kirby, *Cat. Odon*, p. 39 (1890); id., *J. Linn. Soc., Zool.*, Vol. xxiv, p. 555 (1893).

Distribution

Throughout India, Kerala, Western Ghats: Maharashtra (Sawantwadi, Fonda, Kankavali).

80. *Orthetrum luzonicum* (Brauer)

Libella luzonica (Brauer), *Verh. Zool. Bot. Ges. Mien*, Vol. xviii, pp. 169, 732 (1868).

Orthetrum luzonicum Kirby, *Cat. Odon*, p. 38 (1890).

Distribution

Throughout India, Kerala, Western Ghats: Maharashtra (Fonda, Amboli, Kudal).

81. *Orthetrum sabina* (Drury)

Libellula sabina Drury, III *Exot. Ins.*, Vol. 1, Pl. xlviii, Fig. 4, pp. 114, 115 (1770).

Orthetrum sabina Kirby, *Trans. Zool. Soc. Lond.*, Vol. xii, p. 302, Pl. iv, Fig. 5 (1889).

Distribution

Throughout India, Kerala, Western Ghats: Maharashtra (Amboli, Kankavali).

82. *Orthetrum triangulare triangulare* (Selys).

Libella triangularis Selys. *Mitth. Mus. Dresden*, p. 314 (1878); id., *Ann. Mus. Civ.* Genova, Vol. xxx, p. 461 (1891).

Orthetrum triangulare triangulare Ris, *Cat. Coil. Selys.* Fase., Vol. ix, pp. 181, 243–244 (1909).

Distribution

Arunachal Pradesh, Haryana, Meghalaya, Mizoram, South India, Kerala, Western Ghats: Maharashtra (Amboli, Fonda, Vaibhavwadi).

83. *Bradinopyga geminata* (Rambur)

Libellula geminata Rambur, *Ins. Nevrop.*, p. 90 (1842).

Bradinopyga geminata Ris, *Cat. Coil. Selys, Fass.*, Vol. xiii, pp. 545, 548, Fig. 324 (1911).

Distribution

Bihar, West Bengal, Orissa, Uttar Pradesh, Central India, Haryana, Rajasthan, Peninsular India, Kerala, Western Ghats: Maharashtra (Fonda, Kudal, Vaibhavwadi).

84. *Crocothemis servilia servilia* (Drury)

Libellula servilia Drury, III Ins., Vol. 1, Pl. xlviii, Fig. 6, pp. 112, 113 (1770); Rambur, *Ins. Nevrop*, p. 80 (1842).

Distribution

Throughout India, Kerala, Western Ghats: Maharashtra (Amboli, Sawantwadi, Fonda).

85. *Diplacodes trivialis* (Rambur)

Libellula trivialis Rambur, *Ins. Nevrop.*, p. 115 (1842).

Diplacodes trivialis Karsch, *Ent. Maehr.*, Vol. xvii, p. 246 (1891).

Distribution

Throughout India, Kerala, Western Ghats (Fonda, Vaibhavwadi, Kudal).

86. *Diplacodes nebulosa* (Fabricius)

Libellula nebulosa Fabricius, *Ent. Syst.*, Vol. ii, p. 379 (1793).

Diplaeodes nebulosa Kirby, *Trans. Zool. Soc. Lond.*, Vol. xii, p. 308 (1889).

Distribution

Throughout India, Kerala, Western Ghats: Maharashtra (Fonda, Amboli, Kankavali).

87. *Indothemis carnatica* Fabricius

Libellula caesia Rambur, *Ins. Nevrop.*, p. 95 (1842).

Indothemis caesia Ris, *Cat. Coli.* Selys, Fosc., Vol. xiii, pp. 529–531 (1911).

Distribution

West Bengal, Western Ghats: Maharashtra (Sawantwadi, Amboli), Kerala.

88. *Neurothemis intermedia intermedia* (Rambur)

Libellula intermedia Rambur, *Ins. Nevrop.*, p. 91 (1842).

Neurothemis intermedia intermedia Ris, *Cat. Coil.* Selys. Fasc., Vol. xiii, pp. 551, 563–564 (1911).

Distribution

Throughout India, Western Ghats: Kerala, Maharashtra (Sawantwadi, Kudal, Vaibhavwadi).

89. *Neurothemis tullis tullia* (Drury)

Libellula tullia Drury, III *Exot. Ins.*, Vol. ii, Pl. xlvi, Fig. 3 p. 85 (1773).

Neurothemis tullia Kirby, *Cat. Odon.*, p. 8 (1890).

Distribution

Uttar Pradesh, Himachal Pradesh, Bihar, West Bengal, Orissa, South India, Western Ghats: Kerala, Maharashtra (Fonda, Kankavali, Kudal).

90. *Rhyothemis variegata variegata* (Linnaeus)

Rhyothemis variegata Linnaeus, Amacnitates Acad., Vol. vi, p. 412 (1763); Linnaeus, *Syst. Nat.*, ed. Vol. xii, Vol. ii, p. 904 (1766); Rambur, Ins. Nevrop., p. 44 (1842); Hagen, Stett. Ent. Zeit., Vol. vi, p. 156 (1845); id., *Verh. Zool. Bot. Ges. Wien*, Vol. viii, p. 480 (1958).

Rhyothemis phyllis phyllis Fraser

Rec. Ind. Mus., Vol. xxvi, pp. 426, 443 (1924).

Distribution

Western Ghats: Maharashtra (Amboli, Fonda, Vaibhavwadi), Sri Lanka, Myanmar.

91. *Rhyothemis triangularis* Kirby

Rhyothemis triangularis Kirby *Trans. Zool. Soc. Lond.*, Vol. xii, p. 319, (1889); id., *Cat. Odon.*, p. 6 (1890); Fraser, *Rec. Ind. Mus.*, Vol. xxxiii, p. 446 (1931); Liefhrick, Treubia, Vol. xiv, p. 425 (1934).

Distribution

Assam, Western Ghats: Maharashtra (Sawantwadi, Amboli), Java, Sumatra, Myanmar, Sri Lanka, Malaysia.

92. *Rhyothemis yeshwanti* Sathe and Shinde

J. Natcon., 18(2) 421–424 (2006).

Distribution

Western Ghats: Maharashtra (Amboli, Fonda).

93. *Pantala flavescens* (Fabricius)

Libellula flavescens (Fabricius), *Ent. Syst. Suppl.*, p. 285 (1778); Hagen, Stett. Ent. Zeit., Vol. xvii, pp. 366, 369, 370 (1856); Hist. Cuba, p. 442 (1857).

Libelula viridula Palirot de Beauvais, *Ins. Aft. ef Amer.* p. 69, Pl. III, Fig. 4 (1805); Rambur. *Ins. Nevrop.*, p. 38 (1842); Selys-*Hogen; Rev. Odon.* p. 322 (1850); Fraser. *Rec. Ind. Mus.*, Vol. xxxiii, p. 446 (1931); Liettneck, Treubia, Vol. xiv, pp. 424, 425 (1934).

Distribution

Western Ghats: Maharashtra (Fonda, Sawautwadi), Sri Lanka, Tibet, Myanmar.

94. *Tramea limbata sililata* (Rambur)

Libellula limbata Desjardius, *Rapport Soc. Maurice,* 1 (1832).

Tramea limbata Kirby, *Trans Zool. Soc. Lond.*, Vol. xii, p. 318 (1889).

Tramea limbata race *Similata* Fraser, *J. Bombay Nat. Hist. Soc.*, Vol. xxvii, pp. 53, 54 (1920).

Distribution

South India, Western Ghats: Kerala, Maharashtra (Amboli, Fonda, Kudal).

95. *Aethriamanta brevipennis* (Rambur)

Libellula brevipennis Rambur. *Ins. Mevrop.*, p. 114 (1842).

Aethriamanta brevipennis Kirby, *Trans. Zool. Soc. Lond.*, Vol. xii, p. 283, Pl. lili, Fig. 3 (1869).

Distribution

Bihar, West Bengal, Western Ghats: Kerala, Maharashtra (Sawantwadi, Fonda, Amboli).

96. *Urithemis signata signata* (Rambur)

Libellula souguinea Burmeister, *Handb. Ent.*, Vol. ii, p. 858 (1839).

Urothemis signata signata Ris. *Cat. Call. Selys. Tasc.*, Vol. xvi, pp. 1023, 1024 (1913).

Distribution

Throughout Peninsular India, Western Ghats: Kerala, Maharashtra (Vaibhavwadi, Kudal, Amboli).

97. *Zygonyx metallicu* Fraser

Zygonyx iris metallica Fraser, *Rec. Ind. Mus.*, Vol. xxxiii, pp. 446, 450, 451 (1931).

Distribution

Western Ghats: Kerala, Maharashtra (Amboli, Sawantwadi).

Acknowledgement

Authors are thankful to Shivaji University for providing facilities.

References

Emiliyamma, K.G. and Radhakrishnan, C., 2000. Odonata (Insecta) of Parambikulam Wildlife Sanctuary, Kerala, India. *Rec. Zool. Sur. India*, 98(1): 157–167.

Emiliyamma, K.G. and Radhakrishnan, C., 2002. Addition to Odonata (Insecta) of Thiruvananthapuram district, Kerala. *Zoo Print Journal*, 17(10): 914–917.

Fraser, F.C., 1931. Addition to the Survey of the Odonata (Dragontly), Fauna of Western India with descriptions of nine new species. *Rec. Ind Mus.*, 33: 443–474.

Fraser, F.C., 1933. The Fauna of British India, including Ceylon and Burma. *Odonata, Vol. 1.* Taylor and Francis Ltd., London, p. 1–423.

Fraser, F.C., 1934. The Fauna of British India, including Ceylon and Burma. *Odonata, Vol. 2.* Taylor and Francis Ltd., London, p. 1–398.

Fraser, F.C., 1936. The Fauna of British India, including Ceylon and Burma. *Odonata, Vol. 3.* Taylor and Francis Ltd., London, p. 1–461.

Kulkarni, P.P. and Prasad, 2002. Insecta: Odonata. *Zool. Surv., India*, Wetland Ecosystem Series No. 3, Fauna of Ujani, p. 91–104.

Lahiri, A.R., 1987. Studies on Odonata fauna of Meghalaya. *Rec. Zool. Surv. India*, Misc. Publi., Paper, 90, 402 pp.

Mathavan, S. and Miller, M.L., 1989. A collection of dragonflies (Odonata) Made in Periyar National Park, Kerala, S. India. *Soc. Int. Odonalol. Rapid. Comm.* (*Suppl.*), 10: 1–10.

Mitra, T.R., 2003. Ecology and biography of Odonata with special reference to Indian fauna. *Rec. Zool. Sur. India, Occ., Pap.*, 202: 1–41.

Oriental Ins., 29: 385–428.

Palot, Jafer, Cheruvat, M. Dinesan, Emiliyamm, K.G. and Radhakrishanan, C., 2002. Dragonfly Manace at the National Fish Seed Farm, Malampuzha, Kerala. *Fishing Chimes*, 22(5): 56 and 60.

Peters, Gunther, 1981. Trockenzeit-Libellen aus dem indischen Tietland. *Dtsh. Ent. Z (N.F) 28 Helft I–II, Seite*, p. 93–108.

Prasad, M. and Kulkarni, P.P., 2002. Insecta. *Zool. Surv. India:* Fauna of Eravikulam National Park. Conservation Area Series, 13: 7–9.

Prasad, M. and Varshney, R.K., 1995. A check list of Odonata India including data on larval studies.

Prasad, M., 1995. On a collection of Odonata from Goa, India. *Fraseria* (M.S.), 2(12): 7–8.

Prasad, M., 1996. An account of the Odonata of Maharashtra State, India. *Rec. Zool. Surv. India*, 95(3–14): 305–327.

Radhkrishnan, C., 1997. Ecology and Conservation Status of Entomofauna of Malabar "2005". *Print* 11: 2–5.

Ramchandra Rao, K. and Lahiri, A.R., 1982. First records of Odonates (Arthropoda: Insecta) from the Silent Valley and New Amarambalam Reserved Forests. *J. Bombay Nat. Hist. Soc.*, 79(3): 557–562.

Sathe, T.V. and Shinde, K.P., 2006a. On a new species of the genus *Rhyothemis* Hagen (Order: Odonata) from India. *J. Nat. Con.*, 18(2): 421–424.

Sathe, T.V. and Shinde, K.P., 2006b. On a new species of the genus *Agriocnemis* Selys (Order: Odonata) from India. *J. Curr. Sci.*, 9(2): 799–802.

Sathe, T.V., 2005. Impact of dragonflies population suppression of paddy pests in agroecosystem of Kolhapur district, India. In: *Proc. 4th Inf. Nat. Sym. Odonatology*, Pontevedra, Spain, pp. 56.

Sathe, T.V., Shinde, K.P., Jadhav, B.V. and Chogule, T.M., 2006. Role of dragonflies for suppression of pests in mulberry ecosystem in Kolhapur, India. *Proc. Asia Pacific Cong. Seri. and Insect Biotech.*, Sangju, Korea, 52: 159.

Sathe, T.V., Shinde, Kiran, Bhoje, P.M. and Patil, R.G. 2005. Biodiversity of dragonflies (Odonata) from Western Ghats of Maharashtra, India. In: *Proc.* 4th *Inf. Nat. Sym. Odonataology,* Pontevedra, Spain, pp. (18)70.

Silsby, J., 2001. *Dragonflies of the World: Natural History.* Museum and Plymbridge Distributors Ltd., p. 1–124.

Chapter 8

Conservation of Tasar Fauna and Flora: Perspectives and Limitations

Ramakrishna Naika[1], *B. Sannappa*[2]
and S. Basavarajappa[3]
[1]*Sericulture College, University of Agricultural Sciences, Chintamani – 563 125*
[2]*Department of Sericulture,* [3]*Department of Zoology, University of Mysore, Mansagangotri, Mysore – 570 006*

ABSTRACT

Sericulture has attracted special attention of India since independence due to rising internal demand and its export potential for silk. Non-mulberry sericulture in general and tasar culture in particular is the finest facets of India's rich tribal culture. The culture is totally a forest based besides providing gainful utilization of vast natural resources, with huge employment generation in rural areas; it also checks destruction of forests. Owing to a prolific wealth of food plants, abundant manpower and ideal climatic conditions, survey was carried out during the year 2004–05, to identify the tasar flora in hilly areas of southern districts of Karnataka. The details regarding tasar flora, opportunities for tasar culture, tribal involved in tasar culture were collected. The conservation of flora, fauna and their economic perspectives and limitations were discussed.

Conservation of tasar fauna and flora is of great importance since majority of the activities of tasar culture are forests based and eco-friendly.

Keywords: *Tasar flora, Fauna, Conservation, Southern Karnataka.*

Introduction

Silk has mesmerized the mankind transcending all man made barriers ever since its invention. It is the natural fibre that spells splendour, luster and elegance and has been an inseparable part of Indian culture and tradition over many years. Silk is the main product of sericulture activity and this activity has attracted special attention of India since independence due to rising internal demand and its export potential. India has the unique distinction of being the only country in the world bestowed by nature with all the four commercially exploited natural silks *viz.* mulberry, tasar, eri and muga. Non-mulberry sericulture in general and tasar culture in particular is the finest facets of India's rich tribal culture and commercial potential. Tasar silkworm rearing is an indispensable part of tribal culture. It is estimated that the country has 11.168 million ha of forests having different primary and secondary food plants. In India, 284 MT of tasar silk is being producing annually (Anonymous, 2003).

Tasar culture is an age-old practice in the hands of aboriginals and is a forest based occupation. It is one of the most labour intensive sectors of the Indian economy, that provides means of livelihood to a large section of the population, which mainly belong to the economically weaker sections of the society *i.e.* the tribals.

Tasar culture today is being viewed as the right medium of transformation of the rural economy as it focuses on three basic objectives: (*i*) socio-economic upliftment of the rural, (*ii*) preservation of forest ecology and (*iii*) nurturing the traditional culture associated with it. The challenge before the sericulturists is to utilize this potential source systematically and scientifically so as to bring about a balanced development of the silk industry in general and the rural economy of the developing nation like India in particular. In addition to providing gainful utilization of vast natural resources, with huge employment generation in rural areas, tasar culture also checks destruction of forests. One of the main constraints affecting the expansion of tasar culture is fast depletion of forests encompassing the host plants. The production of tasar silk decreased sharply, despite its potential, mainly due to non-availability of sufficient food plants in the forests and associated problems. Conservation is the most efficient and beneficial utilization of natural resources. Hence, conservation of tasar flora and fauna needs attention since majority of the activities in tasar culture are forest based and eco-friendly. Reports on conservation of tasar flora and fauna are meager in this region. Hence, an attempt was made in this communication.

Methodology Followed

To explore and examine the projected issues, the published and experimental research data were gathered from different sources and from the available information. An attempt has been made to empirically analyze and discuss the prospects and problems of conservation of tasar flora and fauna as per Munirajappa and Jayaramaiah (1989).

Results and Discussion

Tasar Flora

Tasar culture is a subsidiary small scale forest based activity, which provides remunerative income to the tribals. The silkworm employed in tasar culture is tropical, tasar silkworm, *Antheraea mylitta* Drury is polyphagous insect having primary and secondary food plants. It thrives well on three primary food plants *viz.*, Asan (*Terminalia tomentosa*), Arjun (*Terminalia arjuna*) both belonging to the family Combretaceae and Sal (*Shorea robusta*) belonging to the family Dipterocarpaceae. Distribution of primary and secondary food plants of tasar silkworm are given in Tables 8.1 and 8.2. The other primary host plants *viz.*, *Lagerstroemia* sp., *Zizyphus mauritiana* and *Hardwickia binata*. The secondary host plants comprise of *Terminalia chebula*, *T. belerica*, *T. catappa*, *T. paniculata*, *Zizyphus jujuba*, *Tectona grandis*, and *Ficus* spp., are also very much useful for the commercial exploitation of tasar.

Indian traditional tasar occurs mostly in the tract of the country bounded by the Ganges in the north, the Godavari in the south, Orissa coast in the south-east and the Narmada river and the Kaimur mountains in the north-east. Oak tasar on the other hand occurs in

Table 8.1: Distribution of Primary Food Plants of Tropical Tasar Silkworm

Sl.No.	*Food Plants*	*Distribution*
1.	*Terminalia* sp.	India, Nepal, Burma, Laos, Sri Lanka, Indonesia, Malaysia, New Guinea, Australia and Africa
2.	*T. lomenlosa*	North India up to Nepal and Central India
3.	*T. arjuna*	Central and Eastern India
4.	*Shorea robusta*	Throughout the tropics and sub-tropics
5.	*Lagerstroemia* sp.	India, Sri Lanka, China, Vietnam, Malaysia, Indonesia and Australia
6.	*Hardwickia binata*	Tropics particularly Africa and Western Peninsular India
7.	*Ziziphus maurtiana*	India, Afghanistan, Malaysia, China and Australia

Table 8.2: Tasar Food Plants Found in Karnataka and Maharashtra

Sl.No.	*Local Name*	*Botanical Name*	*Family*
1.	Saja	*Terminalia tomentosa* W.	Combretaceae
2.	Arjun	*T. arjuna* Bedd.	Combretaceae
3.	Harra	*T. chebula* Retz	Combretaceae
4.	Bahera	*T. belerica* Roxb.	Combretaceae
5.	Dhaora	*Anogeissus latifolio* Wall.	Combretaceae
6.	Ber	*Zizyphys mauritiana* Lamk.	Rhamnaceae
7.	Dhobin	*Dalbergio paniculata* Roxb.	Papilionaceae
8.	Shishan	*D. latifolia* Roxb.	Papilionaceae
9.	Amt	*Bauhinia malabaricum* Roxb.	Caesalpinaceae
10.	Jamun	*Syzigium cumini* Linn.	Myrtaceae
11.	*Kumbhi*	*Careya arhorea* Roxb.	Myrtaceae
12.	Lendia	*Lagerstroemio parviflora* Roxb.	Latyraceae
13.	Mahua	*Madhuca indica* Gmel.	Sapotaceae
14.	Sag	*Tectona grandis* Linn.	Verbenaceae

the Sub-Himalayan belt stretching from Jammu and Kashmir in the north-west to Manipur in the north-east. Tasar food plants grow luxuriantly at lower altitudes of upto 600 m AMSL and extending between 231/2°N and 231/2°S latitudes. However, the other primary host plants species like *Lagerstroemia* have a distribution from base of western Himalayan to the south of Indo-Gangetic plain and extending upto central India. *Hardwickia binata* is more prevalent in western and peninsular India. Most of these host plants are also found in other tropical countries like Nepal, Burma, Sri Lanka, Malaysia, Australia, Africa, Vietnam and Afghanistan. The tasar host plants growing areas of Karnataka are enlisted in Table 8.3. Table 8.4 shows the forest area and its availability for tasar culture in south India.

Table 8.3: Tasar Host Plants Growing Areas of Karnataka

Sl.No.	*Locality*	*District*	*Type of Vegetation*
1.	Bannerghatta and Savandurga	Bangalore	Scrubby
2.	Devarayanadurga	Tumkur	Semi-evergreen
3.	Beligirirangana hills	Mysore	Semi-evergreen, scrubby
4.	Sagar and Agumbe	Shimoga	Semi-evergreen
5.	Gorur, Ballupet, Sakaleshpur	Hassan	Scrubby, semi-evergreen
6.	Sangameshwarpet	Chickmagalore	Semi-evergreen
7.	Khanapur	Belgaun	Dry deciduous

Table 8.4: Forests Area and its Availability for Tasar Culture in South India

Sl.No.	*State*	*Total Forests Area (ha)*	*Area Under Tasar Food Plants (ha)*
1.	Andhra Pradesh	43,29,000	13,02,000
2.	Karnataka	32,40.300	5,21,000
3.	Maharashtra	46,14,300	5,21,000
	Total	1,21,83,600	28,27,000

Need for Conservation

Since time immemorial tasar culture had been practiced by the tribals on nature grown plants of Arjun, Asan and Sal. This had maintained healthy environment in the natural habitat of the tribals through protection of the ecosystem. However, on account of indiscriminate felling and merciless lopping and pollarding of trees in natural forest the plant wealth is being lost at an alarming rate (Kapila, 1990). Man's intervention into the forest for wood and other forest utilities has enormously influenced the forest cover in general and tasar flora in particular. The host plants are also affected by both insect and non-insect pests and diseases, which poses limitations for availability of food to tasar silkworms. Several control measures were tried in their abode, but becomes impracticable due to its wild nature (Table 8.5). Varied climates, habitat loss, industrialization, pollution, etc., are important causes for variations in the food plants of tasar flora. Hence, a conscious and responsible utilization of forests is the need of the hour and conservation of tasar flora needs attention.

Table 8.5: Important Pests and Diseases of Host Plants of Tasar Silkworm

Sl.No.	*Pest/Disease*	*Host Plants Attacked*
		Pests
1.	Stem borers: *Sphenoptera* sp.	*Terminalia arjuna* and *Terminalia tomentosa*
2.	Gall fly: *Phylloplecta hirusta* and *Triaza* sp.	*Terminalia tomentosa*
3.	Termites: *Microtermes, Odontotermes*, etc.	All the food plants
4	May-June beetle: *Anamola* sp.	All the food plants
		Diseases
1.	Stem canker	*Terminalia arjuna* and *Terminalia tomentosa*
2.	Leaf curl	*Terminalia tomentosa*
3.	Root rot	*Shorea robusta* and *Terminalia tomentosa*

Tasar Eco-races and Distribution

Different tasar fauna and their common associates inspite of different functions have occupied the same habitat and therefore, occur together in one community, but their population structure varies. The adaptation of different species in a community differs due to their ecological niche. The distribution of 34 tasar ecoraces extends over a large area from West Bengal to Karnataka (Alam *et al.*, 1998). Further, about 32 nature grown *Antheraea mylitta* ecoraces have been reported from different parts of India and some of the important ecoraces are listed in Table 8.6. Of all, Andhra local, Laria, Sarihan, Modal, Raily, Bhandara and Tira are economically important (Rao *et al.*, 1999), occupying different geographical regions with specific host relationship. Furthermore, 13 tasar silkworm lines have been evolved (Aron Ramanathan, 1997) to suit to *Terminalia arjuna* and *T. tomentosa* with appropriate disease control measures. The tasar hybrids *viz.*, Raily, Daba TV and Sukinda are extensively used in Bastar (Madhya Pradesh), Bihar and Orissa state respectively. These eco-races show variability with regard to their voltinism and economic parameters. The majority of eco-races exist in Bihar, Orissa and Madhya Pradesh and few in other states (Table 8.6). Among these eco-races some are very popular due to their high productivity and profitability. These eco-races provide a large base of variability and gene pool to be exploited in the synthesis and fixation of new breeds. Use of hybrids developed by interracial hybridization of few eco-races has proved to be beneficial. The cross between Daba x Raily and its reciprocal showed heterosis to the extent of 50 per cent over mid-parental value for shell weight (Jolly *et al.*, 1979). Similarly, F_1 crosses between Raily x Daba and Raily x Sukinda has resulted better performance than their parents (Rao *et al.*, 1999) with 28 per cent enhancement in effective rate of rearing.

Moreoever, *Antheraea proylei* is widely reared along the sub-Himalayan belt to harness abundant oak (*Quercus* spp.) flora available there in the form of natural forests. About 58 species of *Quercus*, distributed all along the western sub-Himalayan range at 1200 to 2100 m AMSL (Pankaj *et al.*, 1992). Of all, *Q. incana*, *Q. serrata*, *Q. semicarpifolia* and *Q. himalayana* are more commonly exploited for *Antheraea proylei* culture at farmers level.

Threat to Tasar Eco-races

The tasar culture is subject to the vagaries of nature like extreme climate, heavy rainfall, storms, predators, pests and pestilence. Due to enormous exploitation by man, depletion of forests and natural calamities, the population reveals decreasing trend day by day. The major cause of concern over conservation of tasar silkworms are (*i*) Collection of tasar cocoons by the tribals without leaving sufficient population to multiply in nature. (*ii*) Natural enemies of tasar silkworms (Table 8.7 and 8.8). (*iii*) Indiscriminate deforestation leading to depletion of tasar host plants. Though each species adopted unique niche, several species are becoming extinct due to human activities such as development pressures, encroachment, exploitation, human induced disasters and uncontrolled use of natural resources and political as well as policy issues. Further, loss due to infectious diseases, *i.e.* viral, bacterial and fungal is about 42–45 per cent (Harish Chandra, 2002). The problem is challenging and directly affecting the tasar industry.

Table 8.6: Some Eco-races of Tasar Silkworm, *A. mylitta*

Sl.No.	*Eco-race*	*Food Plant*	*Availability*
1.	*Daba*	*Terminalia* sp.	Bihar
2.	Moonga	*Terminalia* sp.	Bihar
3.	Mungia	*Shorea* sp.	Bihar
4.	Barhawra	*Shorea* sp.	Bihar
5.	Modia	*Shorea* sp.	Bihar
6.	Laria	*Shorea* sp.	Bihar
7.	Laria	*Terminalia* sp.	Madhya Pradesh
8.	Raily	*Shorea* sp.	Chattisgarh
9.	Nalia	*Shorea* sp.	Orissa
10.	Modal	*Shorea* sp.	Orissa
11.	Sukinda	*Terminalia* sp.	Orissa
12.	Tira	*Lagerstroemia* sp.	West Bengal
13.	Bhandara	*Terminalia* sp.	Maharashtra
14.	Andhra Local	*Terminalia* sp.	Andhra Pradesh
15.	Belgaun Local	*Hardwickia binata*	Karnataka

Conservation Techniques

In India the methods implemented to conserve the living organisms which are decreasing alarmingly are species preservation, habitat preservation, habitat improvement, breeding the species in protected areas and releasing the new born offspring's in natural area, setting of research projects, legal protection, etc. The tropical forests are considered as the lungs of the earth and termed as life support systems. Non-availability of host plants of tasar silkworms due to fast depletion of forest is one of the main reasons for decreased production. Though developmental agencies have taken a number of initiatives for systematic development of host plants through various projects that have given encouraging results, still a large number of tribals depend on the forest flora for their livelihood. Tasar culture being carried out by tribals to a greater extent, tasar biodiversity needs to be saved to nourish the various sections of the people.

Table 8.7: Natural Enemies of Tasar Silkworm

Sl.No.	*Natural Enemy*	*Stage of Attack*
1.	*Blepharipha zebina*	4th and 5th instar
2.	*Pimpla punctata*	Spinning stage
3.	*Canthecona furcellata*	1st, 2nd and early 3rd instar
4.	*Sycanus collaris*	1st to late 4th instar
5.	*Hierodulla bipapilla*	3rd and 4th instar
6.	*Oecophylla smaragdina*	1st and 2nd instar
7.	*Vespa orientalis*	2nd and early 3rd instar
8.	Rats	All stages
9.	Snakes	All stages
10.	Birds	All stages

Table 8.8: Important Parasites and Predators of *Antherea mylilta*

Sl.No.	*Parasite/Predator*	*Family*	*Incidence (per cent)*	*Season*
		Parasite		
1.	*Blepharipa zebina* (Uzi fly)	Tachinidae	20–25	Oct.–Nov.
2.	*Pimpla punctuator* (Ichneuman fly)	Icbneumonidae	10–15	Oct.–Nov.
		Predator		
3.	*Canthecona furcellata*	Pentatomidae	10–15	July–Nov.
4.	*Sycanus collaris*	Reduviidae	7–10	July–Nov.
5.	*Hierodulla bipapilla*	Mantidae	5–7	Sept.
6.	*Vespa orientalis*	Vespidae	10–15	July–Nov.
7.	*Polistes hehraeus*	Vespidae	10–15	July–Nov.
8.	*Oecophylla smaragdina*	Formicidae	5–7	July–Nov.

Conserving the diversity plays a key role in proper utilization of the tasar species for silk production and sustainable use. The host plantation of tropical tasar silkworm can be raised in the scrub and semi-evergreen forests, water-shed areas, panchayat lands, tank bunds with minimum expenditure. Fallow degraded lands with poor fertility and acidic in nature can also be well exploited for raising of tasar food plantation (Munirajappa and Jayaramaiah, 1989). In this regard, the Forest Department, Government of Karnataka has initiated to cultivate different woody taxa at *Kallkere arboretum* near Bannerghatta (Bangalore district), Lakkunda reserve forest area near Agumbe (Shimoga district), Lingadahalli, Kalla Koppa, Ulloor Reserve Forests near Sagar. It was found suitable for raising *Terminalia tomentosa, Anogeissus latifolia* in these areas of Karnataka. This could be made possible by pooling different eco-races and conserving them *in-situ i.e.*, in their own ecosystem specific to ecotypes (Ajit Kumar *et al.*, 1993). Some of the key steps followed for conserving the tasar fauna and flora are (*i*) Exploitation and collection (*ii*) Characterization and evaluation (*iii*) Utilization and conservation. To prevent the role of genotype and environmental interactions, efforts could be made to maintain the various eco-races *in situ* in their respective habitats. Several models have been worked out and suggested to conserve the various eco-races of tasar silkworm. In order to arrive at an *in situ* conservation, five methods were adopted to release the raily eco-race into the forest patch isolated from adjacent forests and devoid of natural population. The methods included (*i*) Release of seed cocoons (*ii*) Release of male and female moths (*iii*) Release of gravid moths (*iv*) Release of eggs in leaf cups of *Shorea* (*v*) Release of chawki worms. The release of eggs in leaf cups found to be economically viable and having maximum profitable index in addition to maintaining original commercial characters. This model has been adopted successfully by the State Government of Chattisgarh (Rao *et al.*, 1998). Another model has been suggested to collect the cocoons from the periphery of the protected area and put it back in the central population (Alam *et al.*, 1998).

The polyphagous nature of tasar silkworm calls for extensive collection, maintenance and documentation of germplasm of a wide range of its food plants. The role of plant genetic resources in improvement of cultivated plants has been well recognized. The germplasm collection of these host plants need to be much more extensively maintained and documented so that breeders biotechnologists can readily obtain what they need for future and also to counteract the genetic vulnerability arising due to rapid deforestation (Kumar *et al.*, 2003). Further, it is suggested that the conservation models should be developed suiting into different ecological conditions for conserving the need based ecotypes. Adopting stringent measures to prevent plucking and marketing of wild cocoons indiscriminately may be one of the tools to conserve the same. Research based conservation projects needs to be developed and implemented.

Acknowledgements

Authors are thankful to the Director of Instruction (Seri.), Sericulture College, UAS(B), Chintamani and the Chairman, DOS in Sericultural Science and Zoology, University of Mysore, Mysore for the encouragement.

References

Alam, M.O., Sinha, B.R.R.P. and Sinha, S.S., 1998. Conservation and utilization of wild tasar eco-race "Modal" of *Antheraea mylitta*. In: *Third International Conference on Wild Silkmoths*, 11–14th November, Orissa, Abstract No. 126.

Anonymous, 2003. *Sericulture and Silk Industry Statistics*. Central Silk Board, Bangalore, p. 52.

Chandra, Harish, 2002. Viral diseases in oak tasar silkworm and its management. *Indian Silk*, 41(2): 17–20.

Jolly, M.S., Sen, S.K., Sonwalkar, T.N. and Prasad, G.K., 1979. *Non-Mulberry Silks*. FAO Bulletin, Rome, p. 178.

Kapila, M.L., 1990. Tasar silkworm rearings can help environmental improvement and social forestry in backward areas. *Indian Silk*, 30(12): 6–9.

Kumar, Ajit, Bajpai, S. and Thangavelu, K., 1993. Genetic resources of tropical tasar silkworm: Its exploitation and utilization. *Indian Silk*, 32(9): 47–49.

Kumar, R., Joshi, M.C., Beck, S. and Gangppadhyaya, A., 2003. Tasar host plants germplasm: Establishment, management and cataloguing. *Indian Silk*, 42(3): 1920.

Munirajappa and Jayaramaiah, V.C., 1989. Tasar culture in Karnataka: Its prospects. *Myforest*, 25(4): 365–368.

Rajadurai, S. and Thangavelu, K., 1996. Development of Tasar in Bhandara: Problems and prospects. *Indian Silk*, 35(6 and 7): 19–20.

Ramanathan, Arun, 1997. Magic of silk. *Indian Silk*, 36(4): 6–10.

Rao, K.V.S., Ramkumar, Mahobia, G.P., Pande, V.K. and Thangavelu, K., 1999. Viability of tasar hybrids: A perspective. *Indian Silk*, 38(7): 15 –17.

Rao, K.V.S., Ramkumar, Mahobia, G.P., Pande, V.K., Ade, A.L. and Sinha, B.R.R.P., 1998. Studies on *in-situ* conservation of "Raily" eco-race of tasar insect *Antheraea mylilta* D. (Lepidoptera: Saturniidae). In: *Third International Conference on Wild Silkmoths*, 11–14th November, Orissa, Abstract No. 120.

Singh, R.N. and Thangavelu, K., 1991. Parasites and predators of tasar silkworm. *Indian Silk*, 29(12): 32–34.

Chapter 9

Study of Pesticides Use in Croplands Fields of Maidan Area, Karnataka, India

S. Basavarajappa[1], *B. Sannappa*[2] *and R. Naika*[3]

[1]*Department of Zoology,* [2]*Department of Sericulture, Manasagangotri, University of Mysore, Mysore – 570 006*

[3]*Department of Sericulture, Chinthamani College, Chinthamani, Kolar District*

ABSTRACT

Different chemical substances *viz.* pesticide, insecticide, fungicide and herbicides are widely used under irrigated agro-ecosystems of Maidan area of Karnataka. The noxious chemicals use is ruthless. Most of the agriculturists are dependent on varieties of chemical compounds for maintaining cash crops and to control pests and other enemies in cropland. Due to regular application of these chemicals, impact on the living beings, including man is threatened. Hence, the present study aims to reveal the details in this regard.

Keywords: Pesticides, Irrigated croplands, Maidan area, Karnataka.

Introduction

Agriculture is the way of life for millions of families in rural India. It is a potential economic activity to most of the farmers, who are putting sincere efforts to enhance yield potential in different croplands. Accordingly, high tech agricultural implements are used at various levels; intensive cultivation practices, application of more synthetic chemical fertilizers and pesticides, which are being employed ruthlessly in croplands. In farming environments, controlling pests, enemies and diseases is of utmost important. Because, they claim considerable amount of crop loss and results in severe economic loss to the farmers. The consequences due to indiscriminate use of chemical pesticides in cropland fields have altered the delicate ecological balance and may envenom the farming environments.

Pest is any microorganism (*i.e.* bacteria and virus) or terrestrial or aquatic plant (*i.e.* weed, fungus) or animal (*i.e.* nematode, insect, mites, rodent) that harms crops and their products. However, as per farmers, pest means the crops damaging agent's *viz.* insect, mites, snails, slugs, rodents and plant disease causing microbe's *viz.* fungi, bacteria, viruses and nematodes. Owing to wide diversity of pest population in the crop field (Table 9.1 and 9.2) various pesticides (chemicals that kill pest species) has been synthesized, formulated and tested on target pest species (Khanna *et al.*, 2003).

Table 9.1: Commonly Occurring Pests in the Study Area

Sl.No.	*Name of Crop*	*Major Pests*
1.	Paddy	*Hieroglyphus banian* (Padhka grasshopper), *Leptocorisa acuta* (Gundhi bug), *Oxya* spp., *Tryporyza incertulas* (Stem borer), *Spodoptera mauritia* (Swarming caterpillar)
2.	Cotton	*Earias vittella* (Spotted bolloworm), *Pectinophora gossypiella* (Pink bolloworm), *Amrasca* spp. (Cotton Jassid)
3.	Sugarcane	*Pyrilla perpusilla* (Pyrilla leaf hopper), *Scirpophaga novella* (Top shoot borer)
4.	Maize	*Chilo* spp. (Stem borer)
5.	Jawar	*Heliothis armigera* (Gram pod borer), *Agrotis* spp. (Cut worm)
6.	Ragi	–do–

Note: Insect pests are identified as per Hill (1994), Gillott (1995) and Mani (1982).

Table 9.2: Commonly Occurring Pests in Horticultural Croplands

Sl.No.	Name of Crop	Major Pests
1.	Coconut	*Oryctes rhinoceros* (Rhinoceros beetle), *Nephantis serinopa* (Black-headed caterpillar)
2.	Arecanut	*Aspidiotis destructor*, *Elymnias* spp., *Oryctes rhinoceros* (Rhinoceros beetle)
3.	Mango	*Sternochetus* spp. (Mango stone weevil), *Drosicha* spp. (Mango mealy bug), *Idioscopus* spp. (Mango-hoppers)

Note: Insect pests are identified as per Hill (1994), Gillott (1995) and Mani (1982).

India, being a tropical country, diversified with various types of agri-horticultural and forest ecosystems. Under various agronomic conditions, farmers rely heavily upon various types of pesticides to control pests of agricultural crops (Tables 9.1 and 9.2). More than 155 pesticides have been registered under Insecticide Act 1968, which includes 57 insecticides, 44 fungicides, 33 herbicides, 4 fumigants and 3 each of acaricides, nematicides, molluscicides and rodenticides. These chemical compounds are applied on various agri-horticultural crops under irrigated conditions. Thus, India becomes the third largest consumer of pesticides in the world and highest among the South Asian countries (Khanna *et al.*, 2003).

Of all the crops grown, the Paddy (*Oryza sativa* L.) plantation has attained a good status as an important cash crop under irrigated conditions in different districts of Karnataka. Being an important cereal crop, paddy is infested by more than eighty insect pests during different stages of its growth (Panwar, 2002). However, despite best efforts the overall returns from the paddy crop is marginal. The farmers are getting on an average yield of 20–25 quintals of paddy per crop. By and large, the paddy growers use different types of synthetic pesticides *viz.* insecticides, fungicides, and herbicides from the day of paddy plantation until harvest. It is due to the damage caused by various insect pests and diseases, which are more common during different seasons and results in severe crop loss. Therefore, it is essential to control the pests before it reaches the level of economic injury (Fenemore and Prakash, 1992; Manikandan, 1999). Thus, agriculturists are regularly applying these synthetic chemical pesticides to control pests and other enemies in their croplands. Hence, the pesticides use is ruthless. Widespread use of pesticides in agriculture poses a health hazard to animals and human beings who are exposed to it (Gayathri *et al.*, 1999) and pollute the aquatic environment (Geetha *et al.*, 1999). The toxicity of some of the pesticides (*e.g.* Organophosphorus insecticides) has received a tremendous amount of attention since three decades (Chawla, 1996). Therefore, present study was made to collect scientific information on various types of synthetic pesticides used to conduct further research and to remedy the envenoming situations in cropland fields of irrigated ecosystems.

Materials and Methods

A survey was conducted in Channageri taluk of Davangere district, Karnataka State, India, during 2003–04. The data on application of different synthetic pesticides, mode of their application and other agriculturally related details were collected from potential paddy growers. Farmers were randomly chosen irrespective of their land holding and they were met personally during different months *i.e.*, from the day of plantation of paddy crop until harvest using a pre-tested questionnaire. The questionnaire included:

1. Type of pesticide *viz.*, insecticide/fungicide/herbicide/rodenticide.
2. Chemical class and trade name of pesticide.
3. Commonly used pesticides.
4. Physical state of pesticides.
5. Effect of pesticides on livestock (if any).
6. Common symptoms recorded among the field workers during the application of pesticides and other details.

Further, the information regarding different types of insecticides and their marketing details were also collected from nearby pesticide dealers. Collected data was analyzed by following standard methods as described by Saha (1992).

As the greater attention was paid to collect data on the application of synthetic pesticide in paddy cropland during the survey, it was thought to be more meaningful to collect information about impact of these chemicals on beneficial animals including man. Consequently, emphasis was given to collect data on incidence of toxicity on animals in cropland fields so as to create awareness among the farmers about the health hazards due to the ruthless application of pesticides.

Further, the dead animals were recorded in the premises of cropland after critical symptomatological observations. The laborers who are assigned to spray and spread of pesticides in cropland and the agriculturists who are in the field during pesticides spray were met to collect the toxicity related information. Pests were collected as per the description of Fenemore and Prakash (1992) by using trap nets. Collected insect pests were isolated group wise and preserved separately in 70 per cent ethyl alcohol. They were identified with the help of field guides and Entomology Text Books (Mani, 1982; Fenemore and Prakash, 1992; Hill, 1994; Gillott, 1995, Van Emden and Service, 2004) up to classification of families.

Results and Discussions

Different types of synthetic chemical pesticides, their brand/ trade name, chemical name and class and other details are enlisted in Table 9.3. Fifty three pesticides were chosen by the farmers for application on different crops. Of all, insecticides claims 81.2 per cent followed by both fungicides (7.5 per cent) and herbicides (7.5 per cent). However, both the rodenticide and nematicide application was 1.9 each (Table 9.4). Thus, the farmers rely more upon wide range of pesticides to control pests in their croplands. The specific chemical class of insecticide and their present use depicted in Figure 9.1.

The organophosphate insectides were applied more (28.3 per cent) in the study area. The commonly used organophosphate compounds were: Methyl Parathion with different concentrations in the brand/trade name of Metacid-50 and Folidol. The Dichlorvos (DDVP 76 per cent EC) used in the brand/trade name Nukem 776, Dash and DDVP. Whereas, the Fenobucarb (50 per cent EC) was used in the name of Merlin-BPMC. The Malathion, Ethyl Parathion and Fenitrothion compounds were used in the trade name Cythion, Parathion and Sumithion respectively. However, the Chlorpyriphos (20 per cent EC) was used as Clearout™, Classic-20 and Noorani 505 (Chlorpyrifos 50 per cent + Lypermethrin 5 per cent EC). The Fortex 10G and Monocrotophos 36 per cent SL (Luphos-36) are also commonly used organophosphates in this area (Table 9.3).

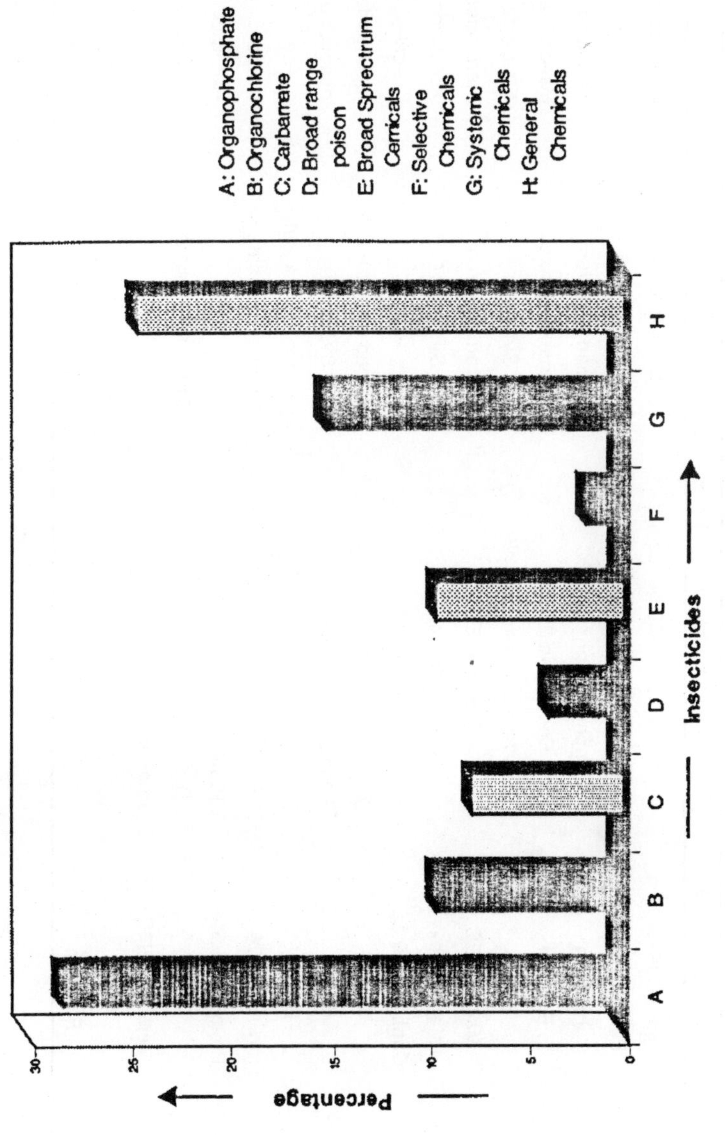

Figure 9.1: Chemical Class of Pesticides

Table 9.3: Use of Different Synthetic Chemical Pes'icides in Croplands

Sl.No.	*Brand/Trade Name*	*Chemical Name*	*Chemical Class*	*Type*
1.	Metacid–50	Methyl Parathion–50 per cent	Organophosphate	Insecticide
2.	Folidol (Dust)	Methyl Patrathion–2 per cent	–do–	–do–
3.	Hinosan	Edinphos 50 per cent EC	Broad spectrum	Fungicide
4.	Endocel	Endosulfan 35EC	Organochlorine	Insecticide
5.	Beam	Tricyclazole 75 per cent WP	Systemic	Fungicide
6.	Machete EC	Mechete EC	Broad spectrum	Herbicide
7.	Metasystox	Oxydemelon methyl	Systemic	Insecticide
8.	Canfidor 200 SL	Imidacloprid	–do–	–do–
9.	Ratol	Zinc Phosphide 80 per cent	Broad spectrum	Rodenticide
10.	Bipvin	Fenobucarb 50 per cent EC	Systemic	Insecticide
11.	BPMC	Fenobucarb 50 per cent EC	Systemic (Knock out)	–do–
12.	Nukem 776	DDVP 76 per cent EC– Dichlorvos	Organophosphate	–do–
13.	Pretty herbicide	Pretilachlor 50 per cent EC	–	Herbicide
14.	Dash	Dichlorvos 76 per cent EC	Organophosphate	Insecticide
15.	Merlin – BPMC	Fenobucarb 50 per cent EC	–do–	–do–
16.	Aldrin	Aldrin	Organochlorine	–do–
17.	Furadan	Carbofuran	Carbamate	–do–

Contd...

Table 9.3–Contd...

Sl.No.	Brand/Trade Name	Chemical Name	Chemical Class	Type
18.	Rogor	Dimethoate	Organophosphate	Insecticide
19.	DDVP	Dichlorovos	–do–	–do–
20.	Cythion	Malathion	–do–	–do–
21.	Lannate	Methomyl	Carbamate	–do–
22.	Sevin	Carbaryl	–do–	–do–
23.	Heptachlor	Heptachlor	Organochlorine	–do–
24.	Gammexane (BHC)	Benzene hexachloride	–do–	–do–
25.	Dantop	Cithianidin 50 per cent	–	–do–
26.	Acephate	Sodi. Dioactyl sulfosuccinate 0.50 per cent	–	–do–
27.	Thimet	Phorate	–	–do–
28.	Methyl Bromide	Methyl bromide	–	–do–
29.	Phosphamidon	Phosphamidon	–	–do–
30.	Parathion	Ethyl Parathion	Organophosphate	–do–
31.	Fenvalerate	Fenvalerate	–	–do–
32.	Sumithion	Fenitrothion	Organophosphate	–do–

Contd...

Table 9.3–Contd...

Sl.No.	*Brand/Trade Name*	*Chemical Name*	*Chemical Class*	*Type*
33.	Kabaddi™	Lambda Cyhalothrin 2.5 per cent EC	Broad range Poison	Insecticide
34.	Mahan	2,4–D amine salt 58 per cent SL	Selective	Herbicide
35.	Clearout™	Chlorpyriphos 20 per cent EC	Organophosphate	Insecticide
36.	Fortex	Fortex–10 G	Organophosphate	Insecticide
37.	Kitazin	Kitazin 48 per cent EC	Systemic	Fungicide
38.	Diafuran	Diafuran	Broad spectrum	Insecticide
39.	Classic–20	Chlorpyriphos 20 per cent EC	Organophosphate	–do–
40.	Carbofuran	Carbofuran 3 per cent CG	Broad spectrum	Nematicide
41.	Luphos–36	Monocrotophos 36 per cent SL	Organophosphate	Insecticide
42.	Noorani 505	Chlorpyrifos 50 per cent + Cypermethrin 5 per cent EC	Organophosphate	–do–
43.	Comfort	Hexaconazole 5 per cent EC	Systemic	Fungicide
44.	Fastest Herbicide	Butachlor 50 per cent EW	–	Herbicide
45.	Bravo™	Triazophos 40 per cent EC	–	Insecticide
46.	Wrestler™	Lambda Cyhalothrin 5 per cent EC	Poison	–do–
47.	Anumida	Imidaclorpid 17.8 per cent SL	Systemic	–do–
48.	Phosmite	Phosmite 50C	–	–

Contd...

Table 9.3–Contd...

Sl.No.	Brand/Trade Name	Chemical Name	Chemical Class	Type
49.	Divop	Divop 100	–	–
50.	DDT	DDT	Organochlorine	Insecticide
51.	Demacron	Demacron	Carbamate	–do–
52.	Ecalax	Ecalax	–	–
53.	Navon	Navon	–	–

Note: Data is based on 20 field observations.

However, the organochlorine compounds application was only 9.4 per cent. Endosulfan-35EC (Endocel), Aldrin, Heptachlor, Benzene hexachloride (Gammexane BHC) and DDT chemicals were the commonly used organochlorine insecticides in this area (Table 9.3). Moreover, the commonly used carbamate compounds were Carbofuran (Furadan), Methomyl (Lannate), Carbaryl (Sevin), Demacron and their use was only 7.6 per cent (Table 9.3 and Figure 9.1). In comparison to the organochlorine and carbamates, organophosphates were selected by and large with different brand names. Thus, there was a statistical significance ($X^2 = 12.37; P>0.05$) between the organophosphate, organochlorine and carbamate insecticides applied in the field.

The chemical compounds *viz.* Cithianidin 50 per cent (Dantop), Sodi. Dioactyl sulfosuccinate 0.5 per cent (Acephate), Phorate (Thimet), Methyl Bromide, Phosphamidon, Fenvalerate, Butachlor 50 per cent EW, Triazophos 40 per cent EC (Bravo™), Phosmite 50C, Divop 100, Ecalax and Navon (Table 9.3) are also applied more often as insecticides in this area. However, all these chemical compounds are not used specifically against particular group of target pests. Hence, these insecticides (24.5 per cent) were considered as general chemical compounds used to control pest population (Figure 9.1).

Further, the fungicides *viz.* Beam (Tricyclazole 75 per cent WP), Kitazin (Kitazin 48 per cent EC), Comfort (Hexaconazole 5 per cent EC) and insecticides *viz.*, Metasystox (Oxydemelon methyl), Canfidor 200SL (Imidacloprid), Fenobucarb 50 per cent EC (Bipvin and BPMC), Anumida (Imidaclorpid 17.8 per cent EC) were the only compounds applied as systemic chemicals against various pests (Figure 9.1).

The Fungicide–Hinosan (Edinphos 50 per cent EC), Herbicide-Mechete, Rodenticide–Ratol (Zinc phosphide 80 per cent), Insecticides–Kabaddi™ (Lamda Cythalothrin 2.5 per cent EC) and Diafuran, Nematicide–Carbofuran (Carbofuran 3 per cent CG) were used as broad spectrum chemical compounds, applied on all types of pests which could inflict crop loss regularly in the field (Table 9.5 and Figure 9.1). Furthermore, the Herbicide–Mahan (2,4-D amine salt 58 per cent SL) was the only selective chemical compound used on weeds in the field. Thus, among all the insecticides applied, 24.5 per cent of the compounds were of general type followed by systemic (15.1 per cent), broad spectrum (9.4 per cent), systemic and broad range poison (3.8 per cent) and selective (1.9 per cent) types (Table 9.5 and Figure 9.1). The analysis of variance of the data revealed that, there was a significant variation (F = 114.05; P>0.05) existing among the different types of chemicals used in the field.

Table 9.4: Different Group of Pesticides Used in the Study Area

Sl.No.	*Name*	*Per cent*
1.	Insecticide	81.2
2.	Fungicide	7.5
3.	Herbicide	7.5
4.	Rodenticide	1.9
5.	Nematicide	1.9
	Total	100.0

Table 9.5: Specific Nature of Insecticides Used in the Field

Sl.No.	*Name of Insecticide*	*Per cent*
1.	General	24.5
2.	Broad spectrum	9.4
3.	Systemic	15.1
4.	Selective	1.9
5.	Broad range poison	3.8
	Total	54.7

Table 9.6 shows the physical state of different pesticides applied in the cropland. The liquid (47.3 per cent), powder (18.2 per cent) and granulated (5.4 per cent) formulations were used along with additives *viz.* canal water, paddy husk and sand particles respectively.

Effect of Pesticides

The farmers and field workers are regularly exposing themselves to pesticides through various means. Moreover, increased application of pesticides had an adverse effect on the natural resources *viz.* air, soil and water (Basavarajappa, 2003), every pesticide (which contain noxious chemical substance or poison) put into the environment would ultimately reach to man through food chain. Further, the direct exposure to pesticides would occur through vector control, residues in food and water (Khanna *et al.*, 2003) tends to have low or larger dose, chronic or persistent exposure to pesticides. Certain persistent pesticides move through air, soil and water finding their way into living tissues, where they accumulate or undergo biological magnification (Khanna *et al.*, 2003). The residues of pesticides when entered into living system are capable of altering the functions of the immune system (Khanna *et al.*, 2003), and cause a variety of problems or disorders (Table 9.7) among the field workers.

Table 9.6: Physical State of Pesticides

Sl.No.	*Physical State*	*Per cent*	*Additives Used*
1.	Solid:		
	A. Granules	5.4	Sand particles
	B. Powder	18.2	Paddy husk
2.	Liquid	47.3	Canal water
3.	Solid/Liquid	29.1	Mixed with other pesticides

Table 9.7: Symptoms Recorded Among the Field Workers

Sl.No.	*Toxicity*	*Symptoms*
1.	Acute Toxicity:	Nausea, abdominal cramps, headache, sweating,
A.	General:	Vomiting, diarrhoea, blurred vision, breathing difficulties, excessive salivation
B.	Severe (Due to very high doses)	Unconsciousness, Giddiness
2.	Chronic toxicity	An influenza like condition with headache, muscle ache, nausea, loss of appetite, weakness

Note: Symptoms were recorded directly from the field workers during the pesticides spray and after the spray in the field.

During this investigation, various symptoms were recorded among the field workers and farmers. The recorded symptoms are enlisted in Table 9.7. The symptoms are: Nausea, abdominal cramps, headache, sweating, unconsciousness, vomiting, diarrhoea, blurred vision and excessive salivation.

There was a common trend in the field that, the workers who come for pesticides application are allowed to apply continuously for 4 to 5 hours with 5 to 10 minutes break. They were not allowed to eat any thing and even not allowed to drink the water also. Most of the time, field workers will not use any mask to face or protective clothing's to hands or body. After completion of pesticides application, they were allowed to wash their hands and asked them to bath by applying detergents. However, very few field workers do follow strictly the safety instructions given by the pesticides dealer/ manufacturers and land owners, before and after the application of pesticides. In spite of it, majority of the field workers are not serious and not taking any safety measures while applying the pesticides. Most of the field workers are illiterates and many of the times, land owners has to watch them keenly until completion of pesticide spray. Even then, some times unconsciousness and giddiness symptoms were observed among pesticide applicators. The exact reason for this type of symptoms was not known. Under such circumstances, land owners have to take them to nearby hospitals for medical treatment. However, such sick persons were met personally and asked for the cause in their houses. They said that, it is because of continuous involvement in pesticides application in the field. When only pesticides application starts in the crop field, they could get headache, giddiness, vomiting and other symptoms. Interestingly, these workers experience no such problems during other days

Further, during severe incidence of pests and the outbreak of plant diseases, farmers forcibly apply pesticides 1–2 times per week. Under such circumstances, field workers (labourers) are exposing themselves continuously to the pesticide residues through air, water and food. Few workers and the owners of the cropland have shown influenza like condition with headache, muscle ache and loss of appetite and general weakness. These symptoms were noticed commonly among many farmers who were met personally. All these problems may lead to organ toxicity if proper care is not taken.

Several workers (Abott and Goulding, 1968; Abou-Donia, 1981; Basheer *et al.*, 1999; Bidstrup *et al.*, 1953; Chang *et al.*, 1977; Hardell, 1981; Hartage and Devesa, 1992; Hoar, 1986; Nexo, 1995; Repetto and Baliga, 1996; Rosenstock *et al.*, 1991; Khanna *et al.*, 2003; Tiwari and Bhatt, 1996; Wecker *et al.*, 1978; Zahm and Blair, 1992) have reported on the organ toxicity due to pesticides in animals including man. The detail of organ toxicity due to insecticides is depicted in Table 9.8. The organophosphate compounds act on nervous system through cholinesterase inhibition, which affect nervous system. Whereas, the organochlorines are carcinogenic, while carbamate compounds affect other organs system in animal body. Thus, the problems caused by various pesticides have long-term effect on various organs in our body. Therefore, to allay the public concern, to protect environment regulatory restrictions are to be imposed on these groups of pesticides (Khanna *et al.*, 2003). Moreover, proper care must be taken, while applying the toxic pesticides on target pest population in the field.

Table 9.8: Organ Toxicity Due to Insecticides in Animals Including Man

Sl.No.	*Insecticides*	*Organ Toxicity*
1.	Dichlorvos	Nervous system through cholinesterase inhibition, blockage of an enzyme required for proper nerve functioning
2.	Malathion	Central nervous system, immune system, adrenal glands, liver blood.
3.	Parathion	Liver, Brain, nervous system by inhibiting cholinesterase.
4.	Dimethoate	Testis, kidney, liver and spleen.
5.	Fenitrothion	Lungs, liver.
6.	Chlorpyrifos	Nervous system.
7.	Monocrotophos	Central system by inhibiting cholinesterase.
8.	Phorate	Nervous system.
9.	Carbaryl	Affects the lungs, kidney and liver. Affect immune system in animals and insects.
10.	Carbofuran	Cholinesterase inhibition in both humans and animals, affecting nervous system.
11.	DDT	Carcinogenic.
12.	Endosulfan	Kidney, liver, blood parathyroid gland.
13.	Methomyl	Lungs, skin, eyes, gastro-intestinal tract, kidney, spleen, blood-forming organs.

Ref: http://extoxent.orst.edu/pips/insecticides.htm.

Further to note that, many of the time, farmers use general chemical compounds (24.5 per cent) and broad range poison (3.8

per cent) which were not used specifically against particular target pests and plant diseases (Table 9.3). The majority of the cases, insecticides were applied on the crop canopy level. Many insect species spend their pupa stage in the soil. Pupa is an important development stage, which is susceptible to insecticides irrespective of its origin. It appears as a sedentary stage of life cycle but within the pupation all the vital activities are carried out at full pace. The pupa before emerging into adult and start laying eggs on the surface of plant, should be killed at this stage itself. This type of pest control would avoid indiscriminate use of broad-spectrum pesticides, which becomes toxic to human being (Dwivedi and Bajaj, 2000). Because, the pesticides residues if present above the normal levels in the mature crop, could be harmful to the consumer (Gupta and Bhatnagar, 1998). Therefore, field experiments are to be conducted to study the pesticides persistence in soil, standing crop and food grains *i.e.* in kernels, paddy straw, polished and unpolished rice grains.

Bhatnagar *et al.* (2001) have shown the residues of Malathion, Phosphamidon, Monocrotophos, Dimethoate, Fenvalerate, Endosulfan, Methyl demeton and Quinalphos on Tomato crop. The samples showed insecticide residues above Maximum Residue Limit (MRL). Salgare (2001) has reported on the sensitivity of plant Gawar, *Cyamopsis tetragonoloba* to herbicide in the field. Monocrotophos pesticide affects the metabolic system in crab, *Barytelphusa guerini* (Venkateshwarlu and Sunita, 1996). Further, optimum density and proper balance of different microbes in soil is essential for good soil fertility (Singh, and Gautam, 1999). The pesticides have very strong inhibitory effect on soil bacteria (Singh and Singh, 1997; Singh and Gautam, 1999). This kind of application may envenom the cropland and in turn contaminate the soil. Hence, there is a need to explore the synthetic insecticides and their indiscriminate use under irrigated croplands. It may help farmers to take proper care while selecting the specific pesticides, which could be of more useful to combat pest species. This would help reduce the killing of beneficial insects and microorganisms in the field.

Incidentally, some of the dead animals that belong to various animal groups were found in the premises of paddy field and in the water sample during the present investigation (Table 9.9). The reason for the death of these animals was not known. However, it is presumed that, the pesticide application might have poisoned their food *viz.* microorganisms and other animals on which they depend in this ecosystem. These incidences are commonly observed during the period of pesticides application. Surprisingly, no animal death was noticed due to pesticide poison during other days of the year in paddy fields. Therefore, it needs further in depth study, to identify scientific reasons and to collect detailed data to justify these incidences. Perhaps, this kind of study may create awareness among concerned, that would help to chose selective, less harmful but more effective pesticides to control the pests. Further, it may also help reduce the killing of useful living beings that are part of the food chain and food web of many animals under irrigated ecosystems.

Table 9.9: Dead Animals and their Carcasses Observed in Pesticides Applied Crop Field

Sl.No.	*Animal Group*	*Affected/Dead Animals Observed*
		Vertebrates
1.	Mammals (livestock)	Food poisoning effect was noticed in Dog (*Canis* sp), Buffalo, Sheep and Goat. Dead and the carcasses of Squirrel (*Funambulus* sp.) and Fox were recorded.
2.	Birds	House sparrow (*Passer domesticus*), Crow (*Corvus* sp.), Koel (*Eudynamys* sp.), Peacock (*Pavo cristatus*), Crow-pheasant, Fowl
3.	Reptiles	Snakes
4.	Amphibians	Frog (*Rana* sp.), Toad (*Bufo* sp.)
5.	Fishes	
		Invertebrates
6.	Protozoans	Euglena, Amoeba, Paramecium
7.	Coelentarates	Hydra
8.	Annelida	Earthworm
9.	Arthropoda	Decapoda: Crab Larval forms: Nauplius, Megalopa, Zoea Insects: Butterflies, moths Others: Daphnia, Cyclops Spiders, Giant water bug (*Belostoma* sp.), Water scorpion (*Nepa* sp.).
10.	Molluscs	Snails, Garden snail (*Helix* sp.), Gray slug [(*Limax* sp.), Apple snail (*Pila* sp.)]

Acknowledgements

First author is thankful to Prof. S.N. Hegde, the Chairman, Department of Zoology, University of Mysore, Mysore for the necessary facility and encouragement.

References

Abott, D.C. and Goulding, R., 1968. Organochlorine pesticide residues in human fat in Great Britain. *Br. Med. J.*, 3: 146–149.

Abou-Donia, M.B., 1981. Organophosphorus ester-induced delayed neurotoxicity. *Ann. Rev. Pharmacol. Toxicol.*, 21: 511–548.

Basavarajappa, S., 2003. A holistic biotechnological approach in farming environments for quality agriproducts in the 21st century. In: *Recent Advances in Environmental Science*, (Ed.) K.G. Hiremath. Discovery Publishing House, New Delhi, pp. 86–106.

Basavarajappa, S., Vijayan, V.A. and Vijayakumar, S., 2006. Impact of pesticides and other noxious chemicals used in cropland fields of Maidan area of Central Karnataka, India. *INCQZ–2006.*

Bhatnagar, A., Parihar, N.S., Gupta, A. and Singh, B., 2001. Pesticide residues in farmgate simples of tomato. *Ad. Bios.*, 20(4): 29–34.

Bidstrup, P.L., Bonner, J.A. and Beckett, A.G., 1953. Paralysis following poisoning by a new organophosphorus insecticide (Mipafox). *Br. Med. J.*, 1: 1068–1072.

Chang, L.W., Reuhl, K.R. and Lee, G.W., 1977. Degenerative changes in the developing nervous system as a result of uterus exposure to methyl mercury. *Environ. Res.*, 14: 414–423.

Chawla, R., 1996. Acute oral toxicity studies with K-othrine (A synthetic pyrethroid insecticide) in Swiss Albino Mice. *Uttar Pradesh J. Zool.*, 16(1): 67–68.

Dhar, G., Barat, S. and Dhar, K.M., 2004. Influence of the organophosphorus insecticide Phosphamidon on lentic water. *J. Environ. Biol.*, 25(3): 359–363.

Dwivedi, S.C. and Bajaj, M., 2000. Efficacy of *Acacia nilotica* leaf extract as pupicidal against *Trigoderma granarium* (Everts). *J. Exp. Zool. India*, 3(2): 153–155.

Fenemore, P.G. and Prakash, A., 1992. *Applied Entomology*. New Age International (P) Ltd., Publishers, New Delhi.

Gayathri, M.R., Murthy, P.B.K, Pillai, KS., Kandasarny, R., Ramamoorthy, S., Banu Priya, C.A.Y. and Jayashri, K., 1999. Sublethal toxicity of chlorpyrifos 48 per cent EC to Wistar Rats. *Ad. Bios.*, 18(11): 25–34.

Geetha, N., Manavalaramanujam, R., Ramesh, M. and Chezhian, A., 1999. Influence of Methomyl, a Carbamate pesticide on biochemical components of a freshwater fish *Catla catla. Ad. Bios.*, 18(11): 1–6.

Gillott, C., 1995. *Entomology*, 2nd edn. Plenum Press. New York.

Gupta, A. and Bhatnagar, A., 1998. A study on the dissipation of Quinalphos in groundnut and loamy sand soil. *Ad. Bios.*, 17(11): 67–74.

Hardell, I., 1981. Malignant lymphoma and exposure to chemicals, especially organic solvents chlorophenols and phenoxy acids. *British J. of Cancer*, 43: 169–176.

Hartage, P. and Devesa, S.S., 1992. Quantification of the impact of known risk factors on time trends in Non-Hodgkin's Lymphoma incidence. *Cancer Res.*, World Resources Institute, Washington, DC.

Hoar, S.K., 1986. Agricultural herbicides use and risk of Lymphoma and soft tissue Sarcoma. *J. Amer. Med. Assn.*, 54: 214–231.

http://extoxnet.orst.edu/pips/carbamates.htm.

http://extoxnet.orst.edu/pips/organochlorines.htm.

http://extoxnet.orst.edu/pips/pesticides.htm.

Hill, D.S., 1994. *Agricultural Entomology*. Timber Press Inc. Orgen, USA.

Khanna, S., Srivastava, M.M. and Srivastava, S., 2003. Eco-friendly pesticides: An emerging trend. In: *Recent Advances in Environmental Science*, (Ed.) K.G. Hiremath. Discovery Publishing House, New Delhi.

Mani, M.S., 1982. *General Entomology*. Oxford and IBH Publishing Co., New Delhi.

Manikandan, P., 1999. Effect of insecticides on eggs of *Helicoverpa armigera* (Hubner) (Lepidoptera : Noctuidae). *J. Exp. Zool., India*, 2(2): 111–113.

Nexo, B.A., 1995. Risk assessment methodologies for carcinogenic compounds in indoor air, in scan journal of work. *Envt. and Health*, 21(5): 3776–3381.

Panwar, V.P.S., 2002. *Agricultural Insect Pests of Crops and their Control.* Kalyani Publishers, Ludhiana.

Reptto, R. and Baliga, S., 1996. Pesticides and the immune system: The public health risks. World Resources Institute, Washington DC.

Rosenstock, L., Keifer, M. and Daniell, W.E., 1991. Chronic central system effects of acute organophosphate pesticide intoxication. *Lancet*, 338: 223–227.

Saha, T.K., 1992. *Statistics in Theory and Practices.* Emkay Publication, Delhi.

Salgare, S.S., 2001. Monitoring of herbicide (MH) toxicity by using pollen as indicators: Pollen of Gawar. *Ad. Bios.*, 20(1): 75–82.

Singh, B.K. and Singh, A.K., 1997. Adverse effect on Coumaphos on soil dehydrogenase activity. *J. Environ. and Pollution*, 4(3).

Singh, B.K. and Gautam, M., 1999. Effect of Coumaphos on soil microbial population. *Ad. Bios.*, 18(11): 111–116.

Tiwari, S.K. and Bhatt, R.S., 1996. Methoxychlor and dimethoate induced changes in biochemical components of the haemolymph and fat body of the larva of rice moth, *Corcyra cephalonica* Staint (Lepidoptera : Pyralidae). *Uttar Pradesh J. Zool.*, 16(1): 1–8.

Van Emden, H.P. and Service, M.W., 2004. *Pest and Vector Control.* Cambridge University Press. United Kingdom.

Venkateshwarlu, M. and Sunita, A., 1996. Changes in the phosphorylase enzyme activity levels in the tissues of the *Barytelphusa guerini* (Milne Edwards) due to monocrotophos stress. *Uttar Pradesh J. Zool.*, 16(1): 54–56.

Wecker, L., Kiauta, T. and Dettbarn, W.D., 1978. Relationship between acetylcholinesterase inhibition and the development of a myopathy. *J. Pharmacol. Exp. Ther.*, 206: 97–104.

Zahm, S.H. and Blair, A., 1992. Pesticides and Non Hodgikn's Lymphoma. *Cancer Res.*, World Resources Institute, Washington DC.

Chapter 10

Culture of *Moina macrocopa* under Maintained and Static Conditions

Satinder Kour
Zoology Department, GCW, Gandhi Nagar, Jammu, J&K State

ABSTRACT

Cladocera are found in majority of fresh water systems. These organisms are mostly filter feeders and sustain on bacteria and algae as well as artificial food supplements like cow dung and rice bran. Being important fish food organisms these needs to be grown in sustainable cultures. The accumulation of metabolic products creates unfavourable conditions in static cultures. The maintained cultures give better results as compared to static cultures.

Keywords: *Growth and reproduction of Moina macrocopa, Sustainable culture, Static and Maintained culture.*

Introduction

Cladocera are found in majority of fresh water systems. These organisms are mostly filter feeders and sustain on bacteria and algae as well as artificial food supplements like cow dung extract and rice bran. Being important fish food organisms these need to be grown in sustainable cultures. The accumulation of metabolic products create unfavourable conditions in static cultures. The maintained cultures give better results compared to static cultures. Zooplankton especially cladocerans are primary food of fish during its early stages. These are preferred live feeds for cultivable fish. The cladocerans are natural feeds for larvae and juvenile of many finfish and crustaceans. Consequently there is considerable interest in the use of cladocerans feed sources for small marine larval fish. Majority of diets serve as supplement than complete substitutions (Kumlu and Jones, 1995). Success is subject to creation of diet that possesses less than the ideal characteristics. The highest level of success has been predominantly realized with herbivorous forms of crustacean larvae (Bautista *et al.*, 1989; Koshio *et al.*, 1989).

Materials and Methods

The culture was raised under laboratory conditions (Room temperature 19.3–27°C). During experimentation, four plastic containers (each of 50 litres capacity and 1 foot deep) were used for culture purposes. The sets were run in duplicate. In 1st set throughout the experiments, water was changed (maintained conditions) once a week. In 2nd set (static set) the water level was neither maintained nor changed. The experiments were carried out in duplicate and the arithmetic mean was used in the results.

The inoculation of *M. macrocopa* was done at the rate of 3 ind./lit. The nutrient sources used were a combination of finely ground rice bran + cow-dung extract at the rate of 1 g/lit. each in the ratio of 1 : 1. The water was changed using a plastic pipe of 5 mm diameter. One end was dipped in the culture container and other connected to the sieve (mesh size =18 meshes/cm) to avoid even least loss of organism. The observations were recorded twice a week. For each reading 250 cc of water was taken from different depths and sides, so as to make the final quantity up to 1 litre. The above readings were repeated five times. The arithmetic mean was taken for the final recording. The animals were examined and their mean length, brood size and population were recorded. The readings were done as per the method used in experiment 1st.

Results and Discussion

Culture of *Moina macrocopa* in Sets with Maintained and Static Conditions

Perusal of Table 10.1, Figures 10.1 and 10.4 revealed that the population of *M. macrocopa* in maintained culture set increased from 105 inds/35 lit. (initial inoculation) to 23985 inds/35 lit. after 15 days. This set recorded a suddèn decrease in population till 20th day. From here onward there is an increase in the population up to 30th day reaching as high as 63610 inds/35 lit. There is a gradual

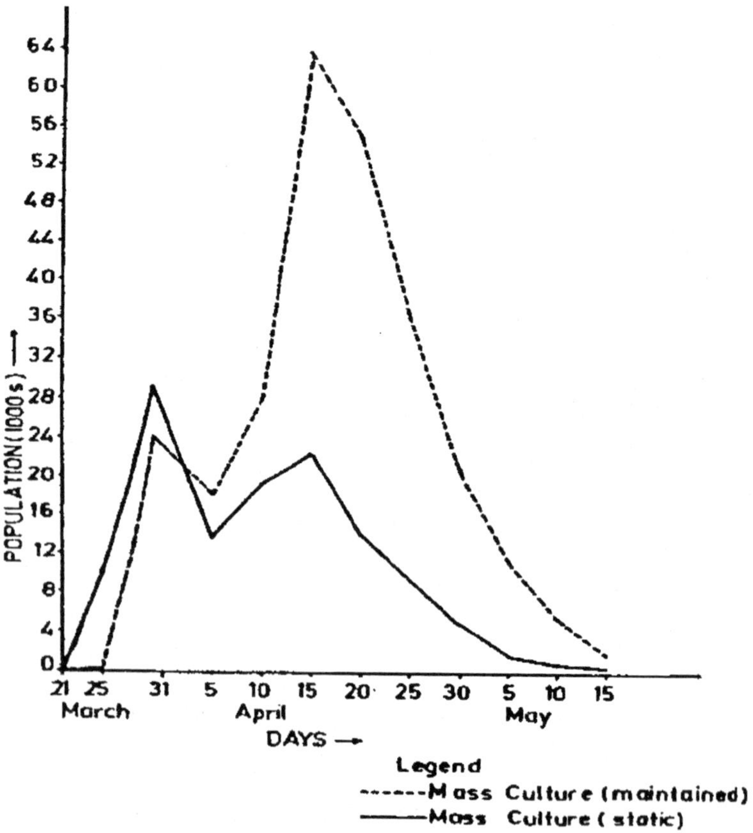

Figure 10.1: Graph Showing Variation in Population in with No. of Days in the Culture Studies of *Moina macrocopa* Under Maintained and Static Culture

Table 10.1: Showing Population, Average Length, Average Breadth and Average Brood Size Under Maintained and Static Culture Conditions of *M. macrocopa*

Days	Population		Length (mm)		Breadth (mm)		Brood	
	Maintained Set	Static Set	Maintained Set	Static Set	Maintained Set	Static Set	Maintained Set	Static Set
1	105	105	0.79	0.82	0.29	0.31	6	7
5	245	286	0.84	0.98	0.31	0.86	12	10
10	12530	10000	1.02	1.03	0.62	0.90	10	12
15	23985	20385	0.92	1.00	0.80	0.82	17	12
20	18130	29045	0.98	0.96	0.62	0.87	14	14
25	28070	13 548	1.04	1.00	0.89	0.84	19	15
30	63610	19389	1.10	1.02	0.98	0.76	5	4
35	54940	22461	1.07	0.99	0.93	0.72	8	3
40	36535	14150	1.01	0.96	0.92	0.62	6	3
45	29575	9329	1.00	0.93	0.89	0.62	–	–
50	11191	4807	0.99	0.96	0.88	0.60	5	–
55	5388	1680	0.93	0.99	0.81	0.06	–	–
60	1808	786	0.92	0.90	0.80	0.64	–	–
Mean	21316.30	11276.53	0.97	0.96	0.72	0.69	7.86	6.38

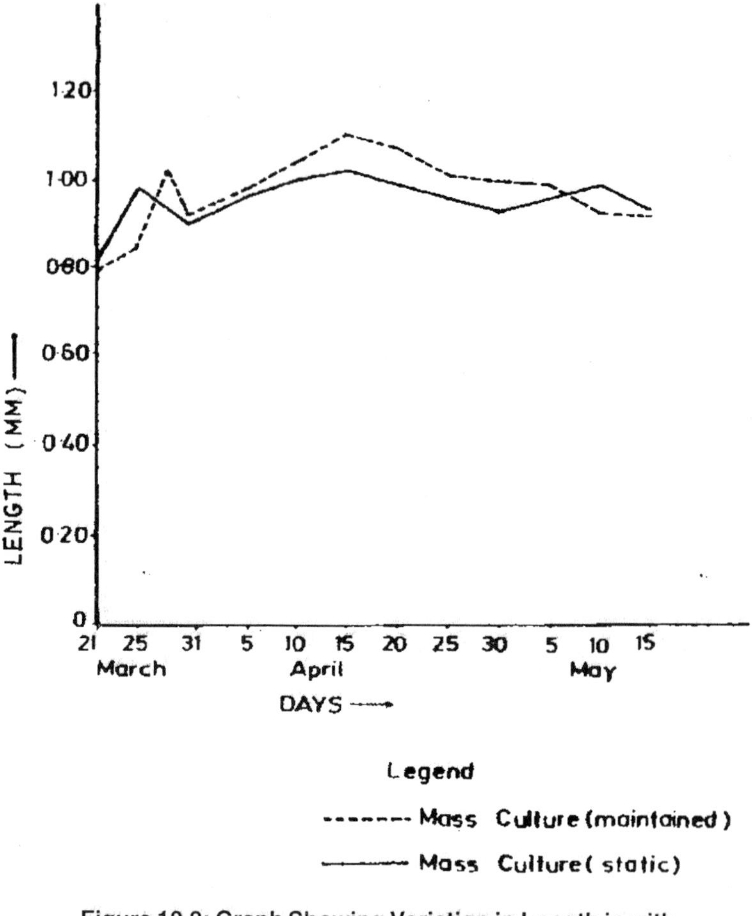

Figure 10.2: Graph Showing Variation in Length in with No. of Days in the Culture Studies of *Moina macrocopa* Under Maintained and Static Culture

decrease (after this peak) to 1808 inds/35 lits. after 60 days. Such variations in the population during present studies thus indicates that in *M. macrocopa* harvesting should start around 30th days, after the initial inoculation in maintained culture.

In the static set it has been observed (Table 10.1 and Figure 10.1) that the population of *M. macrocopa* reached a peak (from 1.05 inds/ 35 lits. of initial inoculation to 29045 inds135 lit.) after only 20 days. This set recorded sudden decreases in population up to 25th day

(13548 inds/35 lit.) then a gradual increase till 35th day reaching 22461 inds/35 lits. This increment in population is followed by gradual decreases till termination of experiment (786 inds/35 lit after 60 days). From the present observations it is indicated that the population fluctuations in static set does not follow any steady predictable pattern and it is thus suggested that harvesting in this set should start after 20 days of initial inoculation.

In both maintained and static culture sets of *M. macrocopa*, a sudden increase in their population is followed by a decrease. This decrease in both experimental sets may probably be due to an inverse relationship of population growth with density of organisms. Similar relationship of population with density has also been reported by Jana and Pal (1985b) in *M. micrura.*

Length

There is a gradual increase in length of *M. macrocopa* reared in maintained set (0.79 mm–1.02 mm) from 1st to 10th day of experiment. It is followed by sudden decrease till 20th day. A second peak in growing population of *M. macrocopa* is again recorded on 30th day (Table 10.1 and Figures 10.1, 10.2 and 10.5). From here onward there is gradual decline in average length till the termination of experiment (60 days).

In static culture sets also, a gradual increase in the length (0.082 mm–1.03 mm) from 1st to 10th day has been observed which is followed by a sudden decrease in the average length till 20th day. This decline in average length is again followed by an increase after 25th day up to 30th days (Table 10.1 and Figure 10.2) where after the average length recorded a gradual decrease till the end of experiment.

The observations made during present studies on maintained and static culture of *M. macrocopa* reveal that the decrease in average length of *M. macrocopa* is the outcome of the appearance of young ones (increase in population) in the culture sets. Similar decline in average length of *M. micrura,* with addition of young ones has also been reported by Jana and Pal (1985b). Since presently a decline in average body length of *M. macrocopa* is followed by an increase in average length. It may obviously indicate that such fluctuation (in average length presently investigated) may be attributable to the appearance and disappearance of different stages of *M. macrocopa* (Table 10.1 and Figures 10.1–10.4) in the static and maintained sets.

Brood Size

A look at the Table 10.1; Figure 10.3 reveal that in the maintained culture set of *M. macrocopa*. It has been observed that the average brood size recorded increase in the culture up to 15th day, followed by a decrease up to 21st day and a further increase till 26th day. This is followed by a gradual decrease till the end of the experiment. It has also been observed that the average breadth of *M. macrocopa* shows a direct relationship with the average brood size (Table 10.1 and Figure 10.3).

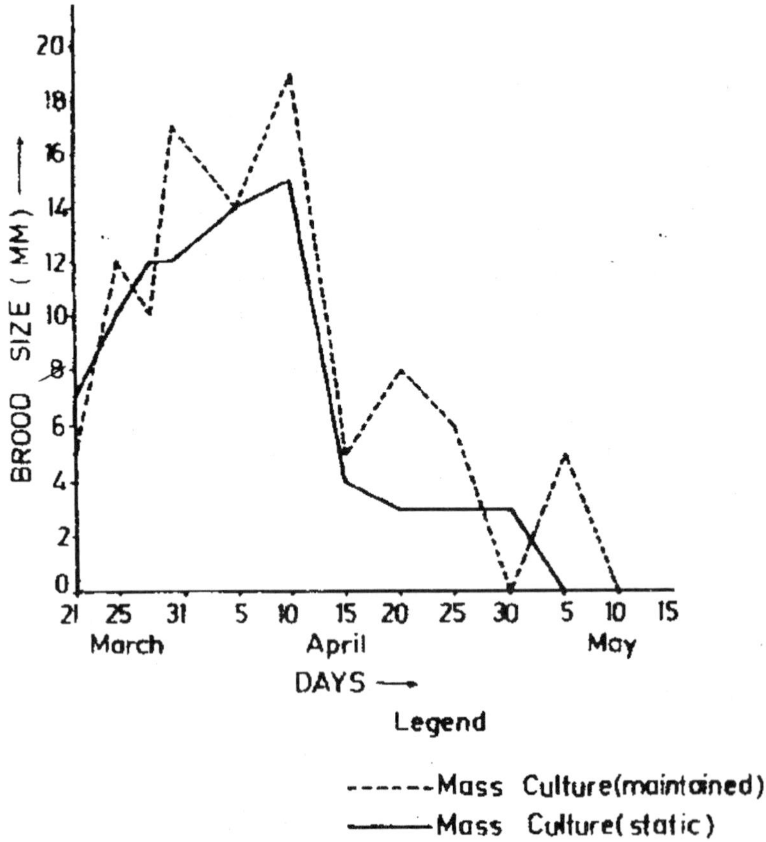

Figure 10.3: Graph Showing Variation in Brood Size in *Moina macrocopa* Under Maintained and Static Culture

In the static culture of *M. macrocopa* the average brood size shows only one peak from 21st to 26th day, followed by a marked decline after 26th day. The breadth of *M. macrocopa* in this set also registered a direct relationship with the brood size (Table 10.1 and Figures 10.3 and 10.4).

As evident from Table 10.1, there is a marked decrease in the brood size with increase in population density. Such depletion in egg production has been attributed to the scarcity of food and

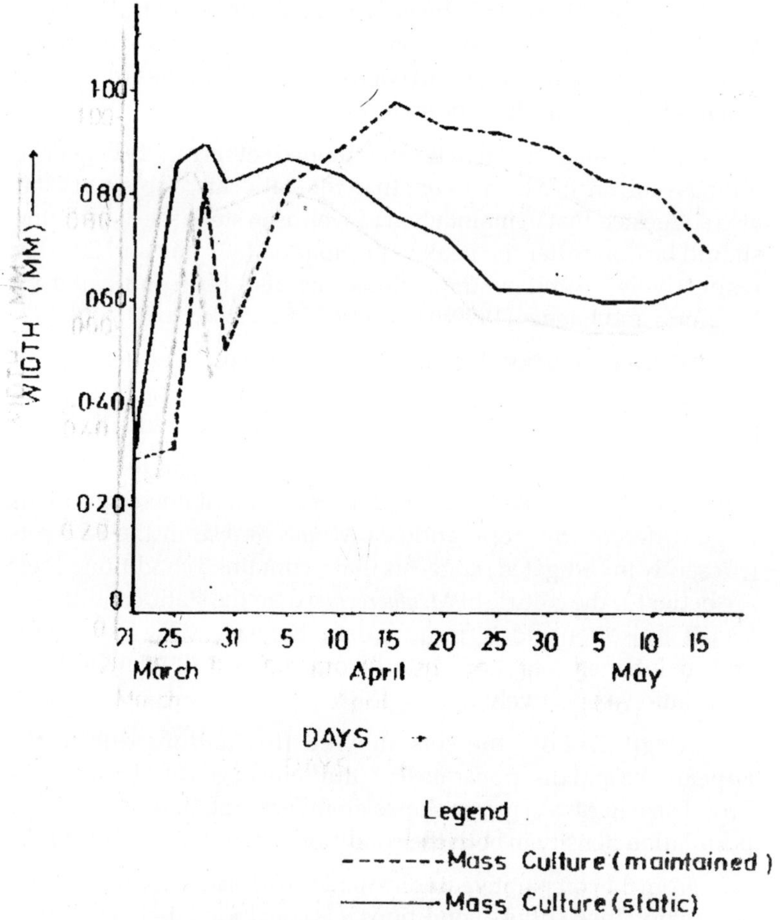

Figure 10.4: Graph Showing Variation in Width (Size) with No. of Days in *Moina macrocopa* Under Maintained and Static Culture

accumulated metabolic wastes by Jana and Pal (1985b) in case of *M. micrura*. and Remirez *et al.* (2004) in static cultures of *M. macrocopa*. A perusal of Table 10.1, further indicates that brood size is dependent on residual density of organisms since the number of eggs/brood during present studies is higher when the population is less. Such views have also been substantiated earlier by Pratt (1943), wherein he stated that both the egg production and death rate are strongly density dependent in *D. magna*. Frank (1952) also reported similar results in *D. pulcaria* and *Simocephalus vetulus*.

The suggestions made by Jana and Pal (1985b) of transferring neonates of *M. micrura* (Y oungones) accelerated egg production by about 45 per cent investigation) was more useful for mass culture of fish food organisms than the static one.

It is thus suggested that the maintained sets are more beneficial for mass culture of *M. macrocopa* than the static sets (Table 10.1). It is also suggested that in maintained as well as in static sets harvesting should be done after the peak in population (30th day and 20th day respectively). Similar suggestions has also been indicated by Malhotra and Langer (1990b) in case of *M. macrocopa*.

An inverse relationship has been observed between density and population in both the culture sets of *M. macrocopa*. A peak in population has been observed to be followed by a sudden decline.

In the static sets maximum number of individuals recorded are 29045 inds/35 lit. (after 20 days) and in the maintained conditions large difference in population of *M. macrocopa*, in the two sets (presently investigated) suggests that maintained conditioned sets are better for the culture of *M. macrocopa* than the static conditioned sets. It is also concluded from the data that harvesting in the static and maintained sets should start from 20th and 30th day of initial inoculation respectively.

Length in both the sets showed fluctuations due to the appearance and disappearance of different stages of *M. macrocopa*. Brood size in *M. macrocopa* shows an inverse relationship with the population density in both the conditions (maintained and static).

From the comparative studies made on *M. macrocopa* under static and maintained culture conditions it is concluded that maintained sets are more beneficial for culture of *M. macrocopa* than the static sets, which causes depletion in population along with the other abnormalities in growth and reproduction.

References

Bautista, M.N., Millamena, O.M. and Kanazawa, A., 1989. Use of kappa-carrageenan microbound diet (CMBD) as feed for *Penaeus japonicus* larvae. *Marine Biology*, 103: 169–173.

Benider, A., *et al.*, 2002. Growth of *Moina macrocopa* (Straus 1820) (Gustacea, Cladocera): Influence of trophic conditions, population density and temperature. *Hydrobiologia*, 1–3: 468.

Frank, P.W., 1952. A laboratory study of intra-species and interspecies competition in *Daphnia pulicaria* (Forbes) and *Simnocephalus vetulus* O.F. Muller. *Physiol. Zool.*, 25: 178–204.

Jana and Pal, 1985b. Effects of inoculum density on growth, reproductive potential and population size in *Moina micrura* (Kurz.). *Limnologica* (Berlin), 16(2): 315–324.

Koshio, S., Kanazawa, A., Teshima, S. and Castell, J.D., 1989. Nutritive evaluation of crab protein for larval *Penaeus japonicus* fed microparticulate diets. *Aquaculture*, 81: 145–154.

Kumlu, M. and Jones, D.A., 1995. The effect of live and artificial diets on growth, survival, and trypsin activity in larvae of *Penaeus indicus*. *Journal of the World Aquaculture Society*, 26(4): 406–415.

Malhotra, Y.R. and Langer, S., 1990b. Biological aspects of *Moina macrocopa* in relation to temperature variations. *J. Freshwater Biol.*, 2(2): 111–115.

Pratt, D.M., 1943. Analysis of population development in *Daphnia* in different temperatures. *Biol. Bull.*, 85: 116–140.

Ramirez, E.M., Sarma, S.S.S. and Nandini, S., 2004. Recovery patterns of *Moina macrocopa* exposed previously to different concentrations of Cadmium and Methyl Parathion. Life table Demography and Population Growth Studies. *Hydrobiologea*, 1: 526.

Chapter 11

Biochemical Mobilization in a Freshwater Crab, *Oziotelphusa senex senex* Due to Aqueous Extract of Endocrine Tissues of *Emerita asiatica*

P. Indumathy[1], *M. Sangeetha*[1], *P. Lakshmi Devi*[1], *J. Jayanthi*[2] *and M.G. Ragunathan*[2]

[1]*Department of Zoology, Queen Marys College, Chennai – 600 004*

[2]*Department of Advanced Zoology and Biotechnology, Guru Nanak College, Chennai – 600 042*

ABSTRACT

The quantitative analysis for protein, carbohydrates and lipids were analysed in a fresh water crab, *Oziotelphusa senex senex*. Three groups of crabs were injected with aqueous extract of brain, thoracic ganglion and eyestalks of the matured female crab, *Emerita asiatica*. The mobilization of organic contents from hepatopancreas to ovary and spermatheca were highly significant in thoracic ganglion treated crab followed by brain, the eyestalks extract treated group showed significant decrease in the organic contents.

Keywords: *Quantitative analysis, Biochemical mobilization, Organic contents, Crabs, Emerita asiatica, Oziotelphusa senex senex.*

Introduction

Reproduction in crustaceans is one of the energy demanding processes (Gomez and Nayar, 1965). Several authors have reported the mobilization of organic reserves, during gonadal development. Like wise the presence of gonad stimulating hormone in brain and thoracic ganglion has been reported by Otsu (1963), Gomez (1965) and the gonad inhibiting hormone in eyestalks by Panouse (1944). According to Van Herp (1992) the reproductive processes in crustacean require interaction between factors from nervous system and endocrine glands. The main regulatory function were fulfilled by biologically active compounds such as neuro hormones and neuro transmitters (Mathieu, 1997).

The use of extracts from the brain or from the thoracic ganglion containing reproductive hormones has to be explored to initiate gonadal maturation (Diwan, 2005). Extracts prepared from the thoracic ganglion of female fiddler crab, *Uca pugilator* induced precocious ovarian maturation in intact and eyestalk crabs (Eastman-Reks and Fingerman, 1984). Induction of ovarian maturation in *Penaeus vannamei* has been accomplished by injecting lobster brain extract and by injecting thoracic ganglion extracts of *P. semisulcatus*. From the survey of literature reports on the effect of aqueous extracts of brain, thoracic ganglion and eyestalks from a mature female crab, *E. asiatica* has not been reported, hence the present investigation has been undertaken.

Materials and Methods

The crab *O. senex senex*, of stage I *i.e.* early stages of vitellogenesis were brought from the Vandalur Lake, 40 kms away from Chennai city and were acclimatized in the lab and were divided into five groups of 10 each. Group A is the initial control and were sacrificed on the first day of the experiment and the tissues such as haemolymph, hepatopancreas, ovary and spermatheca were taken for biochemical analysis. Group B was injected with physiological saline of 25 µl/crab and the Groups C, D and E were treated with aqueous brain extract, thoracic ganglion extract and eyestalks extract of 25 µl/crab respectively on 1st, 5th and 10th days of the experiment. On the 15th day the crabs were dissected out and the tissues were taken for biochemical analysis as said earlier.

The aqueous extract of brain, thoracic ganglion and eyestalks were prepared from the mature female crab, *E. asiatica* collected from the Marina beach. About 10 crabs were dissected out and their brain, thoracic ganglion and eyestalks were separately homogenized with 1 ml of physiological saline and were centrifuged at 5000 rpm for 5 minutes and the supernatant was used for injection.

The biochemical analysis were done by following standard methods. Lowry *et al.* (1951) for protein, Roe (1955) method for carbohydrates and Barnes and Blackstock (1973) method for lipids. The results were statistically analyzed and presented.

Results and Discussion

The results obtained were neatly presented as shown in the tables. All the organic compounds in the hepatopancreas has significantly decreased and significantly increased in the haemolymph, ovary and spermatheca. The increase in the organic contents is more in the thoracic ganglion extract treated groups followed by brain treated and there is significant decrease in eyestalks extract treated group when compared with their control. The trend of depletion of protein, carbohydrates and lipids from hepatopancreas to ovary and spermatheca were similar in all the groups, *i.e.* B, C, D and group E (Tables 11.1–11.3).

Many researchers have studied the biochemical composition in different body parts in relation to reproductive cycles in crustaceans such as *Uca annulipes, Portunus pelagicus* and *Metapenaeus affinis,* (Pillay and Nair, 1973) and in *Machrobrachium kistenesis,* (Mirajker *et al.*, 1983). Kulkarni *et al.* (2002a) have studied the mobilization of biochemical constituents to ovary ttom the hepatopancreas during post reproductive activity. Further, in (2002b) they have reported the biochemical variation in different body tissues in a freshwater prawn, *M.lamarrei lamarrei.* Susan *et al.* (2006) have reported the biochemical mobilization during testicular maturation in a fresh water crab, *Paratelphusa hydrodromous.* Like wise, Ragunathan *et al.* (2007) have reported the biochemical mobilization in a fresh water crab, *O. senex senex* due to Flouxetine.

The results in the present investigation suggest that the organic contents have depleted from hepatopancreas and deposited in the ovary and spermatheca by mediating through haemolymph and the highly significant results in the thoracic ganglion and brain suggests

Table 11.1: Total Protein Levels in Different Tissues of *Oziotelphusa senex senex*

Experiments	*Haemolymph*	*Hepatopancreas*	*Ovary*	*Spermatheca*
Group A (Normal untreated)	45.33±0.815	131.15±1.271	64.20±1.351	56.3±1.193
Group B (Saline treated)	*50.77±0.972	*120.79±0.995	*75.75±0.806	*69.43±9.386
Group C (Brain treated)	**52.82±0.764	*109.76±0.926	*79.62±0.792	*76.93±0.655
Group D (Thoracic ganglion treated)	*55.89±0.634	*115.46±11.886	*80.46±1.122	*78.51±0.650
Group E (Eyestalks treated)	***50.3±18.363	*128.62±1.469	*64.28±0.770	*55.37±1.031

Each value is mean±SEM of 10 samples expressed as mg/gm of wet tissues and mg/ml haemolymph.

Group A Vs B; B Vs C; B Vs D; B Vs E. *: $P < 0.001$; **: $P < 0.05$; ***: Insignificant.

Table 11.2: Total Carbohydrate Levels in Different Tissues of *Oziotelphusa senex senex*

Experiments	*Haemolymph*	*Hepatopancreas*	*Ovary*	*Spermatheca*
Group A (Normal untreated)	16.47±0.497	49.47±0.545	13.25±0.599	15.45±0.963
Group B (Saline treated)	**19.64±0.681	*42.4±0.794	*16.47±0.56	*20.59±0.688
Group C (Brain treated)	**22.57±0.699	*38.47±0.585	*20.55±0.682	**22.06±0.730
Group D (Thoracic ganglion treated)	*25.28±0.535	*35.60±0.933	**18.08±0.862	*24.64±1.086
Group E (Eyestalks treated)	**16.72±0.611	*50.82±1.l07	*12.47±0.563	**14.49±0.931

Each value is mean±SEM of 10 samples expressed as mg/gm of wet tissues and mg/ml haemolymph.

Group A Vs B; B Vs C; B Vs D; B Vs E. *: $P < 0.001$; **: $P < 0.05$.

Table 11.3: Total Lipid Levels in Different Tissues of *Oziotelphusa senex senex*

Experiments	*Haemolymph*	*Hepatopancreas*	*Ovary*	*Spermatheca*
Group A (Normal untreated)	25.49±0.698	104.93±1.105	32.19±0.633	28.4±0.801
Group B (Saline treated)	*29.52±0.639	**100.56±0.744	*36.41±0.607	*33.41±0.941
Group C (Brain treated)	*33.4±10.645	*95.83±0.957	**39.195±0.746	**37.24±0.393
Group D (Thoracic ganglion treated)	*37.63±0.813	*93.38±0.861	*42.4±0.579	*40.53±9.560
Group E (Eyestalks treated)	*25.48±0.821	*105.31±0.888	*32.06±0.451	*27.53±0.671

Each value is mean±SEM of 10 samples expressed as mg/gm of wet tissues and mg/ml haemolymph.

Group A Vs B; B Vs C; B Vs D; B Vs E. *: $P < 0.001$; **: $P < 0.05$.

that thoracic ganglion has more gonad stimulating property followed by the brain. The decreased trend in the eyestalks extract treated group suggests that the hormones present in the eyestalks would have inhibited gonadal maturation there by suppressing the biochemical mobilization.

References

Barnes, H. and Blackstock, Z.J., 1973. Estimation of lipids in marine animals and tissues. Detailed investigation of the Phosphovanilin method for total lipids. *J. Exp. Mar. Bid. Ecol.*, 12: 103–118.

Diwan, A.D., 2005. Current progress in shrimp endocrinology: A review. *Ind. J. Exp. Biol.*, 43: 209–223.

Eastman-Reks, S. and Fingerman, M., 1984. Effect of neuroendocrine tissue and cyclic AMP on ovarian growth *in vivo* and *in vitro* in the fiddler crab, *Uca pugilator. Comp. Biochem. Physiol.*, 79A: 679–684.

Gomez, R., 1965. Acceleration of development of gonads by implantation of brain in the crab, *Paratelphusa hydrodromous. Naturwissenschaften*, 56: 216.

Gomez, R. and Nayar, K.K., 1965. Certain endocrine influence in the reproduction of the crab, *Paratelphusa hydrodromous. Zool. Jb (Abt. Physiol)*, 71: 649–701.

Kulkarni, R.R., Sarojini, R. and Nagabhushanam, R., 2002a. Transfer of energy from hepatopancreas to ovary during reproductive cycle of freshwater prawn, *Macrobrachium lamarrei. J. Aqua. Biol.*, 17(1): 40–42.

Kulkarni, R.R., Sarojini, R. and Nagabhushanam, R., 2002b. Biochemical variations in the different body tissues of the fresh water prawn, *Macrobrachium lamarrei* during ovarian development. *J. Aqua. Biol.*, 17(1): 37–39.

Lowry, O.H., Rosebrough N.K., Farr, A.L. and Randall, R.J., 1951. Protein measurement with Folin-phenol reagent. *J. Biol. Chem.*, 193: 265–275.

Mathieu, M., 1997. Neuroendocrine control of growth, reproduction and carbohydrate metabolism in molluscs. In: *Recent Advances in Marine Biotechnology, Vol. 1: Endocrinology and Reproduction*,

(Eds.) M. Fingerman, R. Nagabhushanam and M.F. Thompson. Oxford and IBH Publishing Co. Pvt. Ltd., New Delhi, pp. 93–108.

Mirajkar, M.S., Sarojini, R. and Nagabhushanam, R., 1983. Biochemical changes associated with the reproductive cycle in the male fresh water prawn, *Macrobrachium kistnensis. Biol. Bull., India,* 5(1): 37–41.

Otsu, T., 1963. Bihormonal control of sexual cycle in the fresh water crab, *Potamon dehaani. Embryologica,* 8: 1–20.

Panouse, J.B., 1943. Influence de l'ablation du peduncle Sur la croissance de l'ovaire chez la crecette *Leander serratus. C.R. Acad. Sci. Paris,* 217: 553–555.

Pillay, K.K. and Nair, N.B., 1973. Observations on the biochemical changes in gonads and other organs of *Uca annulipes, Portunus pelagicus* and *Metapenaeus affinis* (Decapoda : Crustacea) during the reproductive cycle. *Mar. Biol.,* 18: 167–198.

Ragunathan, M.G., Meera, N. and Jayanthi, J., 2007. Prolonged light and fluoxetine stimulates deposition of organic reserves in the gonads of a female brackish water crab, *Uca lactea annulipes. J. Aqua. Biol.,* 22(1): 166–168.

Roe, J.R., 1955. The determination of sugar in blood and spinal fluid with anthrone reagent. *J. Biol. Chem.,* 20: 335–343.

Susan, T., Eswaralakshmi, R., Jayanthi, J. and Ragunathan, M.G., 2006. Quantitative changes in organic contents due to certain neurotransmitters in a female freshwater crab, *Paratelphusa hydrodromous. Uttar Pradesh Journal of Zoology,* 26(1): 101–103.

Van Herp, F., 1992. Inhibition and stimulating neuropeptides controlling reproduction in crustacea. *Invert. Reprod. and Devel.,* 22: 21–30.

Chapter 12

Combined Effects of Endocrine Tissue Extracts of *Emerita asiatica* and Continuous Light on the Organic Contents of a Freshwater Crab, *Oziotelphusa senex senex*

P. Lakshmi Devi[1]*, P. Indumathy*[1]*, M. Sangeetha*[1]*,*
J. Jayanthi[2] *and M.G. Ragunathan*[2]

[1]*Department of Zoology, Queen Marys College, Chennai – 600 004*
[2]*Department of Advanced Zoology and Biotechnology,*
Guru Nanak College, Chennai – 600 042

ABSTRACT

Aqueous extract of brain, thoracic ganglion and eyestalk taken from matured female sand crab, *Emerita asiatica* were injected in early vitellogenic females of freshwater crab, *Oziotelphusa senex senex*. The result shows drastic mobilization of organic contents from hepatopancreas to ovary and spermatheca through haemolymph. The results are highly significant in thoracic ganglion treated groups and the eyestalk-injected group shows significant decrease in the organic contents.

Keywords: *Light, Protein, Carbohydrates, Lipids, Brain, Thoracic ganglion, Organic content mobilization.*

Introduction

The reproduction in crustaceans is effected by exogenous factors like rainfall, temperature, pH, dissolved oxygen, photoperiod etc. Apart from endogenous hormonal factors the brain and thoracic ganglion of the crustacean have been reported to contain gonad stimulating factors (Otsu, 1963; Gomez, 1965) and the eyestalk contained gonadal inhibiting factors. The impact of photoperiodic regimes on reproduction was reported by several authors (Wurts and Stickney, 1984; Reddy, 1986; Kulkarni *et al.*, 2003 a,b; Ragunathan *et al.*, 2007). Apart from the hormones and external factors, there are reports on the injection of brain extracts and thoracic ganglion extracts on the gonadal maturation of crustaceans. The induction of ovarian maturation in *Penaeus vannamei* has been reported by injecting lobster brain extracts and by injecting thoracic ganglion extracts of *P.semisulcatus* (Yano, 1993) has suggested that brain and thoracic ganglia from maturing female shrimp to induce ovarian maturation in different crustaceans of high economic value. From the literature survey a paucity still exists on the biochemical mobilization during ovarian maturation due to endocrine extracts of the sand crab, *E. asiatica* in a female freshwater crab, *O. senex senex*, hence the present investigation has been undertaken.

Materials and Methods

The crab, *O. senex senex*, which were at the stage I *i.e.*, early stages of vitellogenesis were brought from the vandalur Lake 40 kms away from Chennai City and were acclimatized in the lab and were divided into six groups of ten each. Group A is initial control and was sacrificed on the first day of the experiment and the tissues such as haemolymph, hepatopancreas, ovary and spermatheca were taken for the biochemical analysis. Crabs of the groups A, B, C, D and E were subjected to continuous light of 1100 Lux intensity. Group B was injected with physiological saline of 25 µl/crab and the group C, D and E were treated with aqueous brain extracts, thoracic ganglion extracts and eyestalk extracts of 25 µl/crab respectively on the 1st, 5th and 10th days of the experiment. On the 15th day, the crabs were dissected out and tissues were taken for biochemical analysis as said earlier.

The aqueous extracts of brain, thoracic ganglion and eyestalks were prepared from the matured male crab, *E. asiatica* collected from

Table 12.1: Total Protein Levels in Different Tissues of Light Treated *Oziotelphusa senex senex*

Experiments	*Haemolymph*	*Hepatopancreas*	*Ovary*	*Spermatheca*
Group A (Normal untreated)	72.51±1.213	115.63±1.236	85.33±0.762	80.45±1.679
Group B (Saline treated)	*75.33±1.424	*106.46±0.750	*91.22±0.447	*85. 28±0.798
Group C (Brain extract treated)	*81.239±1.207	*98.26±1.436	*96.48±0.544	*91.17±0.719
Group D (Thoracic ganglion extract treated)	*85.51±1.264	90.38±1. 225	*100.38±0.618	*98.00±0.740
Group E (Eyestalks extract treated)	**72.18±0.766	*115.74±2.064	*86.37±1.427	**81.35±0.538

Each value is mean±SEM of 10 samples expressed as mg/g of wet tissues and mg/ml haemolymph.

Group A Vs B; B Vs C; B Vs D; B Vs E. *: $P < 0.001$; **: $P < 0.05$.

Table 12.2: Total Carbohydrate Levels in Different Tissues of Light Treated *Oziotelphusa senex senex*

Experiments	*Haemolymph*	*Hepatopancreas*	*Ovary*	*Spermatheca*
Group A (Normal untreated)	27.53±0.822	36.64±0.861	16.21±0.651	20.7 4±0.794
Group B (Saline treated)	**30.21±0.781	*31.13±0.900	*20.26±0.509	**23.69±1.054
Group C (Brain extract treated)	**34.35±34.493	*26.43±0.469	*26.25±0.906	*28.64±0.773
Group D (Thoracic ganglion extract treated)	*38.43±0.528	*26.43±0.469	*30.36±0. 853	*32.47±0.524
Group E (Eyestalks extract treated)	**27.32±0.651	*37.16±0.957	*16.22±0.610	**20.87±0.272

Each value is mean±SEM of 10 samples expressed as mg/g of wet tissues and mg/ml haemolymph.

Group A Vs B; B Vs C; B Vs D; B Vs E. *: $P < 0.001$; **: $P < 0.05$.

Table 12.3: Total Lipids Levels in Different Tissues of Light Treated *Oziotelphusa senex senex*

Experiments	*Haemolymph*	*Hepatopancreas*	*Ovary*	*Spermatheca*
Group A (Normal untreated)	33.64±0.637	107.45±0.849	85.88±0.994	57.49±0.838
Group B (Saline treated)	**36.37±0.990	*100.32±0.828	*93.37±0.766	*63.23±0.606
Group C (Brain extract treated)	**40.48±1.264	*95.34±0.498	*98.39±0.550	*70.6±0.966
Group D (Thoracic ganglion extract treated)	**44.58±2.965	*85.35±1.660	*102.4±1.043	*73.71±0.955
Group E (Eyestalks extract treated)	**34.30±0.553	*106.45±1.367	*86.4±0.790	*58.27±0.894

Each value is mean±SEM of 10 samples expressed as mg/g of wet tissues and mg/ml haemolymph.

Group A Vs B; B Vs C; B Vs D; B Vs E. *: $P < 0.001$; **: $P < 0.05$.

the marina beach. About ten crabs were dissected and their brain, thoracic ganglion and eyestalk were separately homogenized with 1 ml of physiological saline and were centrifuged at 5000 rpm for 5 minutes and the supernatant were used for injection.

The biochemical analyses were done by the following standard methods Lowry *et al.* (1951) for protein, Roe (1955) method for carbohydrates and Barnes and Black stock, (1973) for lipids. The results were statistically analysed and presented.

Results and Discussion

The results obtained in the present investigation reveals interesting information. The mobilization of organic contents from the hepatopancreas to the ovary and spermatheca through haemolymph are pronounced in thoracic ganglion extract treated group followed by brain extract treated group. The eyestalks extract treated group shows significant decrease in the organic in contents when compared with their respective controls.

The effects of light on the gonadal development has been supported by many earlier workers (Wurt and Stickney, 1984; Nadarajalingam and Subramoniam, 1987; Kulkarni *et al.*, 2003 a,b; Maitra *et al.*, 2007; Ragunathan *et al.*, 2007). Likewise, there are reports on the endocrine aqueous extract treatment on the gonadal maturation of crustaceans (Takayanagi *et al.*, 1986; Yano, 1993; Jayanthi, 2007). Similarly, the biochemical mobilization during gonadal development has been reported by several authors (Soroka *et al.*, 1999; Gunamalai and Subramoniarn, 2002; Arcos *et al.*, 2003; Meera *et al.*, 2006 a,b; Ragunathan *et al.*, 2007). From the present investigation the highly significant deposition of organic content in the ovary and spermatheca from the hepatopancreas is mainly due to the combined action of continuous light and aqueous extract of brain and thoracic ganglion. Likewise, the significant decrease in the organic contents in the eyestalks treated group is due to the gonad-inhibiting factor present in the eyestalks. These reports gain supporting evidences from the above-mentioned workers.

References

Arcos, G.F., Ibarra, A.M., Boucard, C.V., Palacics, E. and Racotta, I.S., 2003. Haemolymph metabolic variables in relation to eyestalks ablation and gonad development of pacific white shrimp,

Litopenaeus vannamei Buone. *Aquaculture Research*, 34: 749–755.

Barnes, H. and Blackstock, Z.J., 1973. Estimation of lipids in marine animals and tissues. Detailed investigation of the Phosphovanilin method for total lipids. *J. Exp. Mar. Bid. Ecol.*, 12: 103–118.

Gomez, R., 1965. Acceleration of development of gonads by implantation of brain in the crab, *Paratelphusa hydrodromous. Naturwissenschaften*, 56: 216.

Gunamalai, V. and Subramoniam, T., 2002. Role of hemolymph lipoproteins in moulting and reproduction of mole crab, *Emerita asiatica*. In: *Abstract Nat. Symp. on Physiology and Biochemistry of Cultivable Crustaceans*, University of Madras.

Jayanthi, J., 2007. Effect of methyl farnesoate on the gonadal development of a freshwater crab, *Paratelphusa hydrodromous.* Report of the UGC minor project submitted to University Grants Commission, South Eastern Regional Office, Hyderabad.

Kulkarni, R.R., Sarojini, R. and Nagabhushanam, R., 2003a. Impact of photoperiod on gonadal maturation of freshwater prawn, *Macrobrachium lamarrei. J. Aqua. Biol.*, 18(2): 111–113.

Kulkarni, R.R., Sarojini, R. and Nagabhushanam, R., 2003b. The influence of extrinsic factors on the spawning patterns of *Macrobrachjum lamarrei. J. Aqua. Biol.*, 18(2): 115–118.

Lowry, O.H., Rosebrough, N.K., Farr, A.L. and Randall, R.J., 1951. Protein measurement with folin-phenol reagent. *J. Biol. Chem.*, 193: 265–275.

Maitra, S., Chattoraj, A. and Seth, M., 2007. Oocyte maturation in Indian carp, *Catla catla* in the context of photoperiods, photoreception and rhythmicity of pineal hormone melatonin. In: *Abstracts 25th National Symposium on Rep. Biol. and Comp. Endocrinol.*, Thiruvananthapuram, p. 44.

Meera, N., Jayanthi, J. and Ragunathan, M.G., 2006a. Stimulation of ovarian maturation due to fluoxetine in a brackish water crab, *Uca lactea annulipes:* A biochemical study. *J. Aqua. Biol.*, 21(2): 178–179.

Meera, N., Jayanthi, J. and Ragunathan, M.G., 2006b. Nutrient deposition in gonads during their maturation due to fluoxetine in an eyestalk ablated fiddler crab, *Uca lacteal annulipes*. *J. Aqua. Biol.*, 21(2): 180–181.

Nadarajalingarn, K. and Subramoniam, T., 1987. Influence of light on endocrine system and ovarian activity in the Ocypodid crabs, *Ocypoda platytarsis* and *Ocypoda macrocera*. *Marine Ecol. Prog. Ser.*, 36: 43–53.

Otsu, T., 1963. Bihormonal control of sexual cycle in the freshwater crab, *Potamon dehaani*. *Embryologica*, 8: 1–20.

Ragunathan, M.G., Meera, N. and Jayanthi, J., 2007. Prolonged light and fluoxetine stimulates deposition of organic reserves in the gonads of a female brackish water crab, *Uca lactea annulipes*. *J. Aqua. Biol.*, 22(1): 166–168.

Reddy, G.M., 1986. Reproductive endocrinology of the West Coast Xanthid crab, *Ozius rugulosus*. *Ph.D. Thesis*, Marathwada University, Aurangabad, India.

Roe, J.R., 1955. The determination of sugar in blood and spinal fluid with anthrone reagent. *J. Biol. Chem.*, 20: 335–343.

Soroka, Y., Milner, Y. and Sagi, A., 2000. The hepatopancreas as a site of yolk protein synthesis in the prawn *Macrobrachium rosenbergii*. *Invertebrate Reproduction and Development*, 37(1): 61–68.

Takayanagi, H., Yamamato, Y. and Takeda, N., 1986. An ovary-stimulating factor in the shrimp, *Paratya compressa*. *J. Expt. Zool.*, 240(2): 203–209.

Wurt, W.A. and Stickney, R.R., 1984. An hypothesis on the light requirements for spawning penaeid shrimp, with emphasis on *Penaeus setiferus*. *Aquaculture*, 41: 93–98.

Yano, I., 1993. Ultra intensive culture and maturation in captivity of penaeid shrimp. In: *Handbook of Mariculture: Crustacean Aquaculture*, (Ed.) J.P. McVey. CRC Press, Boca Raton, FL, USA, pp. 289–313.

Chapter 13

Influence of Endocrine Tissue Aqueous Extracts of *Emerita asiatica* on the Biochemical Contents of an Eyestalk Ablated Female Freshwater Crab, *Oziotelphusa senex senex*

M. Sangeetha[1], *P. Lakshmi Devi*[1], *P. Indumathy*[1],
J. Jayanthi[2] *and M.G. Ragunathan*[2]
[1]*Department of Zoology, Queen Marys College, Chennai – 600 004*
[2]*Department of Advanced Zoology and Biotechnology, Guru Nanak College, Chennai – 600 042*

ABSTRACT

The brain, thoracic ganglion and eyestalks were removed from matured sand crab, *Emerita asiatica* and aqueous extracts were prepared from them. The mobilization of organic contents due to eyestalk ablation is highly significant than their controls, further it has significantly increased in the thoracic ganglion aqueous extract treated crabs, followed by brain extract treated. The eyestalk extract treated groups showed significant decrease in the organic contents.

Keywords: *Eyestalk ablation, Unilateral ablation, Bilateral ablation, Organic contents, Brain, Thoracic ganglion, Aqueous extract mobilization.*

Introduction

The study on the influence of endocrines on the reproduction in crustaceans got impetus after the eyestalk ablation experiments by Panouse, (1943) reporting that it results in precocious gonadal development in the prawn, *Palaemon serratus*. Ferrero *et al.* (1989) have suggested that eyestalk ablation can be useful for increasing recruitment of matured females in natural and experimental populations and also increased the vitellogenesis and ovarian maturation in *Squilla mantis*. The brain and thoracic ganglion of crustaceans has been reported to contain gonad stimulating hormone by Otsu (1963); Gomez (1965). It has been further supported by successive researchers that the extracts of thoracic ganglion of female fiddler crab, *Uca pugilator* induced precocious ovarian maturation in intact and eyestalkless crabs (Eastman-Reks and Fingerman, 1984). The injection of lobster brain extract or thoracic ganglion extracts induced ovarian maturation in *Penaeus vannamei* (Yano, 1993). The biochemical mobilization due to eyestalk ablation has been well established by several workers. (Adiyodi, 1968; Anilkumar and Adiyodi, 1985; Ferrero *et al.*, 1989; Meera *et al.*, 2006). The survey of literature shows that still a paucity exists on the effects of endocrine tissue extracts in a freshwater crab, hence the present investigation has been undertaken.

Materials and Methods

The crab *O. senex senex* which were at the Stage I *i.e.* early stages of vitellogenesis were brought from the Vandalur Lake, 40 kms away from the Chennai City were acclimatized in the lab and were divided into six groups of ten each; Group A is initial control and were sacrificed on the first day of the experiment and the tissues such as haemolymph, hepatopancreas, ovary and spermatheca were taken for biochemical analysis, crabs of the group B, C, D and E were subjected to eyestalk ablation and both the eyestalks were ablated and the group B was injected with physiological saline of 25 µl/crab and the groups C, D and E were treated with aqueous brain extract, thoracic ganglion extract and eyestalk extract of 25 µl/crab respectively on the 1st, 5th and 10th days of the experiment. On the 15th day the crabs were dissected out and tissues were taken for biochemical analysis as said earlier.

The aqueous extract of brain, thoracic ganglion, and eyestalk were prepared from the matured female mole crab, *E. asiatica* collected from the marina beach. About ten crabs were dissected out and their brain, thoracic ganglion and eyestalks were separately homogenized with 1 ml of physiological saline and were centrifuged at 5000 rpm for 5 minutes and the supernatant were used for injection.

The biochemical analysis were done by following the standard methods, Lowry *et al.* (1951) for proteins, Roe (1953) method for carbohydrate and Barnes and Blackstock method (1973) for lipids. The results were statistically analysed and presented.

Results and Discussion

The results obtained are very interesting. The organic contents show significant deposition in the ovary and spermatheca from the hepatopancreas via haemolymph in the eyestalk normal untreated crabs, the same trend further increased in the brain extract treated and further increased in the thoracic ganglion treated crabs. The eyestalk extract treated group A showed significant decreasing results. The total content of proteins, total carbohydrates and the total lipids are well presented in Tables 13.1–13.3.

The mobilization of organic contents due to eyestalk ablation has been well supported by other workers. Predominant alteration in the biochemical levels of various tissues take place during ovarian cycle, gonads, the sites for gametogenesis are the organs of dynamic functions with complex morphology and the centres of considerable metabolic activity (Kulkarni *et al.*, 2002). Hepatopancreas is a labile organ whose function and size varies with metabolic demands and is compared to insects fat body and vertebrate liver as it secretes digestive enzymes and act as a chief center for the intermediary metabolism of carbohydrates and fat (Johnsten and Davis, 1972). Takayanagi *et al.* (1986) reported that both the brain and thoracic ganglion extracts stimulated vitellogenesis in oocytes suggested that the ovary stimulating factor in the central nervous system would have acted directly on ovarian vitellogenesis. Thus, in the present investigation highly significant deposition of organic contents may be due to eyestalk ablation and the gonadal stimulating factor in the brain and thoracic ganglion.

Table 13.1: Total Protein Levels in Different Tissues of Eyestalks Ablated *Oziotelphusa senex senex*

Experiments	*Haemolymph*	*Hepatopancreas*	*Ovary*	*Spermatheca*
Group A (Normal Untreated)	75.43±0.915	116.0±0.939	89.32±0.833	80.06±0571
Group B (Saline treated)	*80.32±0.955	*110.3±0.865	*93.44±0.799	*86.28±0.853
Group C (Brain extract treated)	*85.63±0.709	*105.393±1.834	**96.12±1.152	**90.35±0.929
Group D (Thoracic ganglion extract treated)	*89.45±0.679	*100.33±1.586	*99.17±0.794	*93.58±0.735
Group E (Eyestalks extract treated)	*7451±0.587	*116.31±0.601	*89.75±0526	*79.49±0.749

Each value is mean±SEM of 10 samples expressed as mg/g of wet tissues and mg/ml haemolymph.

Group A Vs B; B Vs C; B Vs D; B Vs E. *: $P < 0.001$; **: $P < 0.05$.

Table 13.2: Total Carbohydrate Levels in Different Tissues of Eyestalk Ablated *Oziotelphusa senex senex*

Experiments	*Haemolymph*	*Hepatopancreas*	*Ovary*	*Spermatheca*
Group A (Normal Untreated)	23.30±0.709	39.83±1.101	14.36±0.454	24.86±0.480
Group B (Saline treated)	*28.29±0.627	**36.50±0.659	*18.37±0.534	**26.69±1.555
Group C (Brain extract treated)	**32.21±0.613	*32±0.517	**21.28±0.542	*30.6±0.591
Group D (Thoracic ganglion extract treated)	*36.53±1.085	*28.52±0.592	*26.13±0.689	*32.4±0.992
Group E (Eyestalks extract treated)	**25.33±0.750	**40.79±0.605	*14.31±1.869	**24.22±0.894

Each value is mean±SEM of 10 samples expressed as mg/g of wet tissues and mg/ml haemolymph.

Group A Vs B; B Vs C; B Vs D; B Vs E. *: $P < 0.001$; **: $P < 0.05$.

Table 13.3: Total Lipid Levels in Different Tissues of Eyestalk Ablated *Oziotelphusa senex senex*

Experiments	*Haemolymph*	*Hepatopancreas*	*Ovary*	*Spermatheca*
Group A (Normal Untreated)	34.39±0.740	105.5±1.173	81.11±1.564	49.35±0.590
Group B (Saline treated)	**37.36±0.568	*98.54±0.486	*86.61±0.449	*53.98±0.421
Group C (Brain extract treated)	**40.51±0.554	*90.31±0.741	**90.30±0.986	**57.24±0.558
Group D (Thoracic ganglion extract treated)	*43.40±1.515	*86.50±0.562	*96.30±1.087	*62.44±0.358
Group E (Eyestalks extract treated)	**33.23±0.568	*106.28±0.877	*81.90±0.872	**50.89±0.366

Each value is mean±SEM of 10 samples expressed as mg/g of wet tissues and mg/ml haemolymph.

Group A Vs B; B Vs C; B Vs D; B Vs E. *: $P < 0.001$; **: $P < 0.05$.

References

Adiyodi, R.G., 1968. On reproduction and moulting in the crab, *Paratelphusa hydrodromous. Physiol. Zool.*, 41: 204–209.

Anil Kumar, G. and Adiyodi, K.G., 1985. The role of eyestalk hormones in vitellogenesis during the breeding season in the crab, *Paratelphusa hydrodromous* (Herbst). *Biol. Bull.*, 169: 686–695.

Barnes, H. and Blackstock, Z.J., 1973. Estimation of lipids in marine animals and tissues. Detailed investigation of the phosphovanilin method for total lipids. *J. Exp. Mar. Biol. Ecol.*, 12: 103–118.

Eastman-Reks, S. and Fingerman, M., 1984. Effect of neuroendocrine tissue and cyclic AMP on ovarian growth *in vivo* and *in vitro* in the fiddler crab, *Uca pugilator. Comp. Biochem. Physiol.*, 79A: 679–684.

Ferrero, E.A., Marzari, R., Mosco, A. and Riggio, D., 1989. Effects of eyestalk ablation on vitellogenesis and molting in *Squilla mantis* (Crustacea, Stomatopoda). In: *Aquaculture: A Biotechnology in Progress*, (Eds.) N. De. Pauw, E. Jaspers, H. Ackefors, N. Wilkins. Euro. Aqua. Soc., Bredene, Belgium, p. 517–523.

Gomez, R., 1965. Acceleration of development of gonads by implantation of brain in the crab, *Paratelphusa hydrodromous. Naturwissenschaften*, 56: 216.

Johnsten, M.A. and Davis, P.S., 1972. Carbohydrates of the hepatopancreas and blood tissues of *Carcinus. Comp. Biochem. Physiol.*, 41(B): 433–443.

Kulkarni, R.R., Sarojini, R. and Nagabhushanam, R., 2002b. Biochemical variations in the different body tissues of the freshwater prawn, *Macrobrachium lamarrei* during ovarian development. *J. Aqua Biol.*, 17(1): 37–39.

Lowry, O.H., Rosebrough N.K., Farr, A.L. and Randall, R.J., 1951. Protein measurement with folin-phenol reagent. *J. Biol. Chem.*, 193: 265–275.

Meera, N., Jayanthi, J. and Ragunathan, M.G., 2006b. Nutrient deposition in gonads during their maturation due to fluoxetine in an eyestalk ablated fiddler crab, *Uca lactea annulipes. J. Aqua. Biol.*, 21(2): 180–181.

Otsu, T., 1963. Bihormonal control of sexual cycle in the fresh water crab, *Potamon dehaani. Embryologica*, 8: 1–20.

Panouse, J.B., 1943. Influence de l'ablation du peduncle Sur la croissance de l'ovaire chez la crecette *Leander serratus. C.R. Acad. Sci., Paris*, 217: 553–555.

Raghunathan, M.G., Meera, N. and Jayanthi, J., 2007. Prolonged light and fluxetin stimulates deposition of organic reserves in the gonads of a female brackish water crab, *Ucalactea annulipis. J. Aqua. Biol.*, 22(1): 166–168.

Roe, J.R., 1955. The determination of sugar in blood and spinal fluid with anthrone reagent. *J. Biol. Chem.*, 20: 335–343.

Takayanagi, H., Yamamato, Y. and Takeda, N., 1986. An ovary-stimulating factor in the shrimp, *Paratya compressa. J. Expt. Zool.*, 240(2): 203–209.

Yano, I., 1993. Ultra intensive culture and maturation in captivity of penaeid shrimp. In: *Handbook of Mariculture: Crustacean Aquaculture*, (Ed.) J.P. McVey. CRC Press, Boca Raton, FL, USA, pp. 289–313.

Chapter 14

Haematological and Biochemical Alterations in *Clarias batrachus* due to Experimental *Procamallanus* Infection

Shashi Ruhela[1], *A.K. Pandey*[2] *and A.K. Khare*[1]

[1]*Department of Zoology, Meerut College, Meerut – 250 001*

[2]*National Bureau of Fish Genetic Resources (ICAR), Canal Ring Road, Telibagh, P O Dilkusha, Lucknow – 226 002*

ABSTRACT

Comparative haematological and biochemical changes were studied in *Clarias batrachus* experimentally infected with *Procamallanus*. Blood parameters of the catfish such as total erythrocyte count (TEC), haemoglobin (Hb), total leucocyte count (TLC), haematocrit percentage (PCV), mean corpuscular volume (MCV), mean corpuscular haemoglobin (MCH) and mean corpuscular haemoglobin concentration (MCHC) were determined and calculated on day 15, 30, 45 and 60 post-infection. TEC, Hb and PCV recorded a decline whereas TLC registered an increase during the post-infection period. While MCV and MCH recorded an increase on day 15 and 30 days (except on day 45) post-infection, MCRC displayed a significant decrease (except on day 30) in *Clarias batrachus* during the post-infection period. The catfish showed significant alteration in the serum glucose, total protein, cholesterol, urea, SGOT and

SGPT levels during the post-infection. Serum glucose, total protein and cholesterol levels decreased in the catfish during infection. Hypoglycemia was attributed to the disturbances in carbohydrate metabolism while hypoproteinaemia was due to alterations in protein metabolism leading to malabsorption of the nutrients. Hypocholesterolaemia may probably be due to increased RBC destruction, the major cause of anaemia. Serum urea, SGPT and SGOT levels increased during the post-infection period. The elevation in urea level may be the cause of the fraction of non-protein nitrogen component of the blood. The observed increase in SGPT and SGOT levels may be due to the damage to various tissues and/or release of toxins (endotoxins/exotoxins) by the nematode parasites harbouring intestine of the catfish.

Keywords: *Procamallanus, TEC, Hb, TLC, PCV, MCV, MH, MCHC, Macrocytic anaemia, Clarias batrachus.*

Introduction

The usefulness of fish as a cheap source of animal protein in the diet has been emphasized during the recent years (Pandian, 2001; Anon, 2006). *Procamallanus* (Order Spiruride; Family Camallanidae) is one of the most important nematode parasite in the stomach and intestine of fishes inhabiting freshwater, brackishwater and marine ecosystems of the world (Ali, 1960; Yamaguti, 1961; Agarwal, 1966; Sinha, 1988a,b; Zaman and Leong, 1988; Chandra, 1994; Martens and Moens, 1995; Bijukumar, 1996; Bharatha Lakshmi and Sudha, 1999; Gonzalez-Solis *et al.*, 2002; Katoch, 2002; Moravec *et al.*, 2003, 2004a,b,c). Though the intermediate host (vector) of the parasites is copepods (Moravec, 1975; De *et al.*, 1986; Morvec *et al.*, 1993; Chandra and Modak, 1995; De, 1995; De and Maity, 1999, 2000), there exist reports that both adult as well as larval 1995; De, 1995; De and Maity, 1999, 2000), there exist reports that both adult as well as larval stages of *Procamallanus* are pathogenic to fish (Moravec and Vargas-Vazquez, 1996; De *et al.*, 1986; De and Maity, 2000; Ruhela *et al.*, 2006a,b, 2008). Heavy infestations *of Procamallanus* have been recorded in the commercially important catfishes like *Heteropneustes fossilis* and *Clarias batrachus* (Furtado and Low, 1973; Basirullah and Hafizuddin, 1974; Sinha, 1988b; Sinha and Sinha, 1988; Zaman and Leong, 1988; Chandra, 1994).

The parasitic infection disturbs the physiological as well as metabolic activities of the host inducing changes in blood parameters leading to the diseasepike anemia and eosinophiHa (Satpute and Agrawal, 1974; Roberts, 2001; Madhavi, 2003). The neeo'for establishment of standard normal haematological values with a view to aiding in diagnosis the state of health and disease in fishes has been emphasized by a number of workers owing to the growing interest in pisciculture (Hesser, 1960; Blaxhall, 1972; Blaxhall and Daisley, 1973; Roberts, 2001). Though distribution, life-history and parasitic infestation *of Procamallanus* have been recorded by several workers (Agarwal, 1958, 1966, Ali 1960; Yamaguti, 1961; Pande *et al.*, 1963; Sood, 1967, 1988; Rai, 1969; Sinha, 1988a,b, Zaman and Leong, 1988; Chandra, 1994; Martens and Moens, 1995; Bijukumar, 1996; Rajyalakshmi, 1996; Bharatha Lakshmi and Sudha, 1999; Bharatha Lakshmi, 2000a,b,c; Bharatha Lakshmi and Kumari, 2000; Gonzalez-Solis *et al.*, 2002; Moravec *et* al., 2003, 2004a,b,c), alterations in the various blood parameters of the fish due to the toxic substance(s) secreted by parasites in the gastrointestinal tract has not yet been clearly established (Sinha, 1992, 2000; Roberts, 2001). Therefore, an attempt has been made to record the alterations in haematological as well as biochemical parameters of *Clarias batrachus* due to experimental *Procamallanus* infection in order to determine the possible correlation with degree of susceptibility offish to the infection.

Materials and Methods

The adult catfish, *Clarias batrachus* (both sexes; average body weight 88.6±5.86 g) used in the present investigation, were procured from local freshwater ponds as well as markets of Meerut and adjoining districts of western Uttar Pradesh. They were acclimatized under the laboratory conditions for a week before initiating the experiment. The females *Procamallanus,* collected from longitudinally cut intestine of the catfish, were kept in watch glass filled with saline solution for natural egg laying at 24–27°C. The eggs were kept in Lock-Lewis solution for healthy embryonation. The solution was changed periodically and 0.1 per cent formalin added to the culture medium to protect the eggs from fungal contamination. 40 healthy catfish were randomly selected and divided into two equal groups. Catfishes of group 1 were not given any treatment and served as control whereas in catfish of group 2 experimental infection was induced by forcefully pushing 500 embryonated eggs of the nematode into the stomach of each catfish by means of a long-nozzled dropper (De and Maity, 2000; Ruheta *et al.*, 2006a).

Fresh blood samples from catfish of both the groups were collected by a sharp cut made near the caudal vein on day 15, 30, 45 and 60. Samples collected for the haematological analysis were kept in glass vials taking all necessary care to prevent haemolysis and clotting by using anticoagulant. Estimation of total erythrocyte count (TEC), haemoglobin content (Hb), total leucocyte count (TLC), packed cell volume (PCV) was done by haemoglobinometer, haematocrit tube and haemocytometer. Values of mean corpuscular volume (MCV), mean corpuscular haemoglobin (MCH) and mean corpuscular haemoglobin concentration (MCHC) were calculated as these parameters depend on the corresponding values of Hb, TEC and PCV. For biochemical estimations, fresh blood was centrifuged at 3000 rpm and sera were separated. Serum glucose, total protein, cholesterol, urea, SGPT and SGOT were estimated using standard kits- Variations in the haematological as well as biochemical values were evaluated for statistical significance by using Students 't' test.

Results

Haematological Parameters

Variations in TEC, Hb, TLC, PVC, MCV, MCH and MCHC of the control and infected catfish have been summarized in Table 14.1.

Progressive decrease in total erythrocyte count (TEC) was observed in *C. batrachus* due to *Procamallcmus* infection till day 45 followed by normal value on day 60. These values were 2.47±0.32, 2.33±0.24, 2.22±0.24 and 3.58±0.29 million/mm^3 on day 15, 30, 45 and 60 post-infection while the corresponding values for control catfish were 3.58±0.29, 3.48±0.44, 3.41±0.20 and 3.4±026 minion/mm^3, respectively. Similarly, a significant decrease in haemoglobin (Hb) content was also recorded in the infected catfish as these values were found to be 11.10±0.21, 11.05±0.15, 9.42±0.29 and 11.40±0.36 mg/dl on day 15, 30, 45 and 60 post-infection. The corresponding values for control fish were 12.76±0.46, 12.52±0.53, 11.97±0.52 and 12.0±0.58 mg/dl, respectively.

Table 14.1: Alterations in Haematological Parameters of *Clarias batrachus* Induced by *Procamallanus* Infection

Parameters	15 Days		30 Days		45 Days		60 Days	
	Control	Infected	Control	Infected	Control	Infected	Control	Infected
TEC (million/mm^3)	3.58±0.29	2.47±0.32[c]	3.48±0.44	2.33±0.24[b]	3.41±0.20	2.22±0.24[d]	3.49±0.26	3.58±0.29
Hb (mg/dl)	12.76±0.46	11.10±0.21[b]	12.52±0.53	11.05±0.15[a]	11.97±0.52	9.42 ±0.29[c]	12.00±0.58	11.40±0.36[c]
TLC (10^3/mm^3)	32.50±0.57	38.00±0.29[d]	31.10±0.46	37.50±0.58[d]	33.00±0.29	36.40±0.15[d]	30.90±0.58	35.00±0.32[d]
PCV (per cent)	40.14±0.69	36.86±0.80[b]	39.94±0.54	36.00±0.29[d]	32.67±0.33	28.00±0.58[d]	28.00±0.29	30.30±0.85[c]
MCV (μ^3)	114.68±0.51	143.42±1.43	122.89±1.26	136.88±0.99[d]	112.26±1.00	108.10±1.33[b]	80.22±0.89	90.17±1.00[d]
MCH (pg)	36.45±0.59	43.19±1.63[d]	36.98±0.34	40.01±0.33[d]	37.69±0.67	32.50±1.04[d]	34.38±0.79	33.92±0.79
MCHC (per cent)	32.98±0.59	28.21±0.58[a]	30.99±1.15	30.69±1.10	33.58±0.94	30.07±1.51[a]	43.85±1.37	37.62±0.87[c]

Values are mean±standard error (SE) of five specimens. Significant response: a: $P < 0.05$, b: $P < 0.02$, c: $P < 0.01$, d: $P < 0.001$.

Significant increase in total leucocyte count (TLC) was observed in *C. batrachus* due to *Procamallanus* infection as the values were 38.00±0.29, 37.50±0.58, 36.40±0.15 and 35.00±0.32 10^3/mm on day 15, 30, 45 and 60 post-infection. The corresponding values for control catfish were 32.50±0.57, 31.1±0.46, 33.00±0.29 and 30.90±0.58 10^3/ mm^3, respectively. The packed cell volume (PCV) of the control *C. batrachus* was 40.14±0.69, 39.94±0.54, 32.67±0.33 and 28.0±0.29 per cent respectively on day 15, 30, 45 and 60. A significant decrease in PCV value was observed in the infected catfish on day 15 (36.86±0.80 per cent), 30 (36.00±0.29 per cent) and 45 (28.00±0.58 per cent) followed by an increase (30.30±0.85 per cent) on day 60 post-infection.

There was an increase in mean corpuscular volume (MCV) in the infected *C. batrachus* on day 15 (143.42±1.43 μ^3) and 30 (136.88±0.99 μ^3) followed by a progressive decline on day 45 (108.10±1.33 μ^3) and an increase (against the respective control) by day 60 (90.17±1.00 μ^3). The corresponding values of the control catfish were 114.68±0.51, 122.89±1.26, 112.26±1.00 and 80.22±0.89 μ^3, respectively. A significant increase in mean corpuscular haemoglobin (MCH) was observed in the infected *C. batrachus* till day 30 as the values were 43.19±1.63 pg on day 15 and 40.01±0.33 pg on day 30 that decreased on day 45 (32.50±1.04 pg) while the values was non-significant (33.92±0.79 pg) by day 60. The corresponding values for control catfish were 36.45±0.59, 36.98±0.34, 37.69±0.67 and 34.38±0.79 pg, respectively. The mean corpuscular haemoglobin concentration (MCHC) of control *C. batrachus* was 32.98±0.59, 30.99±1.15, 33.58±0.94 and 43.85±1.37 per cent on day 15, 30, 45 and 60. A significant decrease in the MCHC value was observed in the experimental catfish on day 15 (28.21±0.58 per cent), 45 (30.07±1.51 per cent) and 60 (37.62±0.87 per cent).

Biochemical Observations

Variations in serum glucose, total protein, cholesterol, urea, SGPT and SCOT levels of the control and infected catfish have been summarized in Table 14.2.

Serum glucose levels of control *C. batrachus* were 55.25±3.18, 51.10±2.04, 54.19±2.22 and 60.28±1.51 mg/dl, respectively on day 15, 30, 45 and 60. A significant decline in serum glucose levels of infected catfishes were recorded as 36.01±0.50, 30.10±1.07,

Table 14.2: Alterations in Serum Biochemical Parameters (mg/dl) of *Clarias batrachus* Due to *Procamallanus* Infection

Parameters	15 Days		30 Days		45 Days		60 Days	
	Control	*Infected*	*Control*	*Infected*	*Control*	*Infected*	*Control*	*Infected*
Serum glucose (mg/dl)	55.25±3.18	36.01±0.50[c]	51.10±2.04	30.10±1.07[c]	54.19±2.22	40.02±3.61[b]	60.28±1.51	52.18±0.43[c]
Serum protein (mg/dl)	3.59±0.18	2.80±0.22[a]	4.73±0.16	2.92±0.08[c]	5.00±0.58	3.40±0.42[b]	4.16±0.57	4.10±0.59
Serum cholesterol (mg/dl)	370.70± 3.03	325.00± 4.51[c]	377.70± 2.06	342.25± 0.73[c]	352.00± 1.15	292.05± 5.14[c]	391.01± 13.21	364.28± 2.07[c]
Serum urea (mg/dl)	4.40±0.28	6.15±0.32[b]	5.37±0.74	6.15±0.52	6.02±0.59	8.19±0.60[b]	6.30±0.21	7.40±0.22[b]
SGPT (µg/l)	1.28±0.26	2.69±0.37[b]	1.30±0.05	2.46±0.18[b]	1.59±0.07	1.75±0.61	1.96±0.12	4.08±0.48[b]
SGOT (µg/l)	1.80±0.23	2.45±0.10[c]	1.28±0.27	2.93±0.31[c]	1.79±0.47	1.34±0.15	1.49±0.21	2.17±0.15[b]

Values are mean±S.E. of 5 specimens. Significant response; a: $P < 0.05$, b: $P < 0.01$, c: $P < 0.001$.

40.02±3.61 and 52.18±0.43 mg/dl, respectively during the corresponding post-infection period.

Serum total protein levels of control C *batrachus* were 3.59±0.18, 4.73±0.16, 5.0±0.58 and 4.16±0.57 mg/dl, respectively on day 15, 30, 45 and 60. A significant decline in serum protein level of infected catfishes was observed on day 15 (2.80±0.22 mg/dl), 30 (2.92±0.08 mg/dl) and 45 (3.40±0.42 mg/dl) while the level approaches normal value (4.10±0.59 mg/dl) on day 60.

Serum cholesterol levels of control *C. batrachus* were 370.70±3.03, 377.70±2.06, 352.00±1.15 and 391.01±13.21 mg/dl, respectively on day 15, 30, 45 and 60. A significant decline in serum cholesterol level in the infected fish was observed as these values were 325.00±4.51, 342.25±0.73, 292.05±5.14 and 364.28±2.07 mg/dl, respectively during the corresponding post-infection period.

Serum urea levels of control *C. batrachus* were 4.40±0.28, 5.37±0.74, 6.02±0.59 and 6.30±0.21 mg/dl respectively on day 15, 30, 45 and 60. A significant increase in serum urea level was found in infected catfish as the values were 6.15±0.32, 6.15±0.52, 8.19±0.60 and 7.40±0.22 mg/dl, respectively during the corresponding period.

The SGPT level of control *C. batrachus* were found to be 1.28±0.26, 1.30±0.05, 1.59±0.07 and 1.96±0.12 μg/l, respectively on day 15, 30, 45 and 60 whereas an increase in the level was observed in the infected fish as the corresponding values were 2.69±0.37, 2.46±0.18, 1.75±0.61 and 4.08±0.48 μg/l (Table 14.2). The SGOT level of control *C. batrachus* were 1.80±0.23, 1.28±0.27, 1.79±0.47 and 1.49±0.21 μg/l, respectively on day 15, 30, 45 and 60. The levels recorded an increase in the infected catfish (except on day 45) as the respective values were 2.45±0.10, 2.93±0.31, 1.34±0.15 and 2.17±0.15 μg/l on the corresponding day.

Discussion

The present observations provide an evidence for macrocytic anemia in *Clarias batrachus* as there was a marked reduction in TEC and Hb content in the catfish due to *Procamallanus* infection. Totterman (1944, 1947) also observed decrease in erythrocyte count and haemoglobin value to less than 20 per cent due to *Diphyllobothrium latum* and noticed bothriocephalous anemia due to fish tapeworm infection in man. The observed anemic state in the catfish may probably be due to: (*i*) vitamin B_{12} deficiency which may result in arresting maturation of RBC and formation of large but few RBC, (*ii*) deficiency of vitamin 13n caused by the nematodes, (*iii*) deficiency of folic acid as a result of its utilization by the parasites and the consequent slowing down of maturation process of RBC, (*iv*) failure to utilize the RBC maturation factor (vitamin B_{12} or an intrinsic factor) released by the stomach (Sinha, 1992, 2000) and (*v*) sidetracking of iron to the affected tissues instead of being sent to the bone marrow as there is a great need for iron in the tissues infected with parasites thereby impairing haemopoiesis. Sinha (1992) also observed pernicious anemia in *C. batrachus* due to *Procamallanus* infection. Sinha and Sinha (1988) confirmed the macrocytic hypochronic anemia in *H. fossilis* infected with *Procamallanus spiculogubernaculus.*

Hypohaemoglobinaemia followed by a fall in total erythrocyte count (TEC) and haemoglobin (Hb) content has been observed in the catfish during the post-infection period. The parasites present in the lumen of gastrointestinal tract of the host absorb vitamin $B_{12}$from the intestinal contents. There is competition between the worms and the host for vitamin B_{12} ingested with the food and the other substances necessary for blood formation. It is therefore probable that location of the parasites at level of the gastrointestinal tract playa significant role in pathogenesis of the anemia (Sinha, 2000). Significant fall in the PCV, TEC and Hb values in infected catfish seem to develop anaemia (Blaxhall and Daisley, 1973; Sinha 1992).

In *C. batrachus,* marked increase in TLC was also observed due to experimental *Procamallanus* infection throughout the experiment. The increase in TLC might be due to neutralization of harmful foreign bodies and/or adaptive nature to cope up for removing the debris of damaged tissue. It may probably be due to the toxicants (endotoxins/exptoxins) secreted by the parasites evoking change in the homeostatic mechanisms of *C. batrachus* termed as leucocytosis.

The decrease in PCV noticed in the catfish during the post-infection period suggests that the observed anemic state in catfish is an iron deficiency macrocytic anemia within the normal healthy range. The decrease level of PCV in the present study may probably be due to the presence of parasites in lumen of intestine that produces endotoxins which are absorbed by intestinal vein and go through blood of the host. The increase in PCV value on day 60 post-infection may be due to combined effects of RBC swelling and erythropoietically over-compensation mechanism of the fish to neutralize the reduction in oxygen carrying capacity of blood.

Mean corpuscular volume (MCV) in *C. batrachus* depicted increased value due to experimental *Procamallanus* infection at all the intervals. The increase in MCV related with the condition called macrocytosis is associated with macrocytic anemia. The production of a large number of large lymphocytes might have also contributed to the elevation in mean corpuscular volume.

The mean corpuscular haemoglobin (MCH) in *C. batrachus* infected with *Procamallanus* exhibited an increase till day 30 followed by a significant decline on day 45 leading to almost normal value by day 60. The higher MCR values in infected *Clarias batrachus* may probably be due to the toxic effect of substances secreted by the parasites on haematological parameters. The reduction in MCR in the infected catfish on day 45 reveals that damage to corpuscular membrane brings about some leakage of haemoglobin causing an increase in plasma haemoglobin and reduction in MCR. The MCV and MCH values also increase after *T. congolense* and *T. vivax* infection in cattles and mice (Groofenhuis *et al.*, 1990; Mbagwu and Opara, 1998, 2001). Variations in the mean corpuscular haemoglobin concentration (MCHC) at different intervals of infection appear to be due to stress caused by the parasites. The MCHC values below normal indicate that the red blood cells are not fully saturated with haemoglobin.

Serum biochemical parameters of this experiment indicate a significant decline in serum glucose, protein and cholesterol and increase in urea levels of *C. batrachus* experimentally infected with *Procamallanus*. Hypoglycemia observed in the infected catfish at all the intervals was apparent during period of first wave of parasitaemia (15 days post-infection). The observed hypoglycemia in the catfish owing to the experimental nematode infection appears to be due to excessive utilization of blood sugar by the parasites for their metabolism. Gupta and Gupta (1986) also recorded decrease in serum glucose level in fishes during infection wvith trypanosomes. Srivastava *et al.* (1988) reported decrease in serum glucose level of the cattle immunized with tick tissue extract of *Baophilus microplus*. Hypoglycaemia was also noticed in *H. fossilis* with *Trypanosoma karelensis* and *Myxosoma fossilis* infections (Gupta *et al.*, 2000).

Parasitic infestation causes disturbances in protein digestion in host and changes in serum protein level is characteristic feature depending on the stages or severity of an infection (von Brand, 1973). Sadun and Williams (1966) reported decrease in serum protein level of albino rats during experimental *Schistosomiasis*. During *Strongylus vulgaris* infestaion in horse, Vibe-Peterson and Nielson (1979) also reported a decline in the serum protein level. However, Khatoon and Ansari (1979) observed increased serum total protein level in buffalo calves with *Setaria cervi* infection. Nielson (1999) found hypoproteinemia in *Anguilla anguilla* infected with swim-bladder nematode, *Anguillicola crassus*. The decreased plasma protein level may either be due to loss by renal excretion, the impaired protein synthesis due to malnutrition or deficiency of vitamins involving the digestive organ and liver (Oser, 1976).

Gupta and Gupta (1998) observed hypocholesterolemia in catfish infected with haemoprotozoan. Taiwo *et al.* (2003) reported hypocholesterolemia in sheep infected with *Trypanosoma bruci* and *T. congolense*. A decrease in serum cholesterol level has also recorded by Rao (1991) and Chauhan (2002) in WLH chicks with *A. galli* infection. Sharma (1997) reported decreased level of cholesterol infected with sheep-whip worms. Khatoon and Ansari (1979) observed decrease cholesterol level in albino rats with an infection of *Setaria cervi*. Rani (1986) also reported decreased cholesterol level in chicks with experimental ascaridiasis. Kumar (1983) observed increase cholesterol level in the blood cells of ancylostomiasis patients infected with *Schistosoma mansoni*.

The parasites excrete/secrete certain toxic substances, the ingestion of which may cause nephrotoxicity leading to elevated serum urea level (Oser, 1976). The main contributors to nephrotoxicity are the vasoactive amines and histamines produced due to antigen-antibody interaction during infection. The present observation revealed a significant elevation in serum urea level of *C. batrachus* with experimental *Procamallanus* infection. The elevation in serum urea level also induce increased nitrogen metabolism due to presence, of *Procamallanus* in the gastrointestinal tract of host. Kumar (1983) and Gupta (1991) recorded increased level of serum urea in albino rats infected with *Bunostomum trigonocephallim*. Chauhan (2002) also found an increase in serum urea level in WLH chicks with *A. galli* infection.

Serum glutamic pyruvate transaminase (SGPT) and serum glutamic oxaloacetic transaminase (SGOT) are enzymes that are normally present in liver, heart cells, kidney, muscle and pancreas and released into blood circulation when there organs are damaged. In present study, significant increase in SGPT and SGOT levels were observed in *C. batrachus* due to experimental *Procamallanus* infection. Elevation of both the enzymes has also been recorded in animals infected with helminth parasites. Gray (1963) reported increased SGPT levels in the cattle and sheep infected with *Trypanosoma vivax*. Adah *et al.* (1992) also observed increase in level of transaminase in goats experimentally infected with *Trypanosoma congolense*. Taiwo *et al.* (2003) reported increase in the SGPT level in sheep infected with *T. congolense* and *T. brllcei*. In fish, Joshi (1979) recorded biochemical changes in the liver and blood of *Rita rita* infected with *Opisthorchis pedicellata*. Zsigmond *et al.* (2002) observed increase in the activity of transaminases (SGOT and SGPT) in infected fish suggesting the pathological processes induced by penetrating cercariae. The observed elevation in SGPT and SGOT levels of *C. batrachus* due to *Procamallanus* infection may probably be due to the tissue breakdown (necrosis) and inflammation in the host, particularly of liver, heart, muscle, intestine and kidney as varying degrees of degenerative changes were noticed in liver, kidney and intestine of the catfish at different time intervals due to the nematode infection (Ruhela *et al.*, 2006b,c, 2008). Another possibility of the increase in the serum enzyme levels of the catfish caused by *Procamallanus* infection might be due to the various toxic substances (exotoxins/endotoxins) secreted/excreted by the parasites harbouring intestine, the leakage of which into plasma affect the haematological as well as physiological activities of the host (Roberts, 2001).

Acknowledgements

We are grateful to the Head, Department of Zoology, Meerut College, Meerut for providing the necessary laboratory facilities to carry out the present investigation.

References

Adah, M.J., Otesile, E.B. and Joshua, R.A., 1992. Changes in level of transaminases in goats experimentally infected with *trypanosoma congolense*. *Revue Elev. Med. Vet. Pays Trop.*, 45: 284–286.

Agarwal, S.C., 1958. On a new species of *Procamallanus* Baylis, 1923 (Nematoda). *Curr. Sci.*, 27: 348–349.

Agarwal, V., 1966. On a new nematode, *Procamallanus muelleri* n.sp. from the stomach of a freshwater fish, *Heteropneustesfossilis*. In: *Proc. Helminthol. Soc., Washington.*, 33: 204–208.

Ali, S.M., 1960. On two new species of *Procamallanus* Baylis, 1923, from India, with a key to the species. *J. Helminthol.*, 43: 129–138.

Anon, 2006. *Handbook of Fisheries and Aquaculture.* Directorate of Information and Publications in Agriculture, Indian Council of Agricultural Research (ICAR), New Delhi.

Bashirullah, A.K.M. and Hafizuddin, A.K.M., 1974. Two new nematode species of *Procamallanus* Baylis 1923 from freshwater fishes of Dacca, Bangladesh. *Norw. J. Zool.*, 22: 53–55.

Bharatha Lakshmi, B., 2000a. *Procamallanus icakinadensis* n.sp. (Nematoda : Camallanidae) from the intestine of a marine fish, *Nebea soldado* from Kakinada, A.P., India. *Uttar Pradesh J. Zool.*, 20: 137–142.

Bharatha Lakshmi, B., 2000b. *Procamallanus* spp. (Nematoda : Camallanidae) from a marine fish, *Polynemus sextarius* (Bloch and Scheneider). *Uttar Pradesh J. Zool.*, 20: 147–150.

Bharatha Lakshmi, B., 2000c. On a new nematode *Procamallanus lutjanusi* n.sp. from the intestine of marine fish at Kakinada, India. *Uttar Pradesh J. Zool.*, 20: 159–164.

Bharatha Lakshmi, B. and Sudha, M., 1999. Redescription of *Procamallanus malhurai* Pande, Bhatia and Rai, 1963 (Camallanidae : Nematoda). *Bol. Chile Parasitol.*, 54: 115–117.

Bharatha Lakshmi, B. and Kumari, R., 2000. First record of the male of *Procamallanus spirits* (Baylis, 1923), Khan and Begum, 1971 from the intestine of marine fish *Johnieops macrorhynus* (Mohan) from Kakinada Bay. *Uttar Pradesh J. Zool.*, 20: 105–109.

Bijukumar, A., 1996. Nematode parasites associated with the flatfishes (Order : Pleuronectiformes) off the Kerala Coast. *J. Mar. Biol. Assoc., India*, 38: 34–39.

Blaxhall, P.C., 1972. The haematological assessment of the health of freshwater fish. *J. Fish Biol.*, 4: 593–605.

Blaxhall, P C. and Daisley, K.W., 1973. Routine haematological methods for use K.W. 1973 with fish blood. *J. Fish Biol.*, 5: 771–781.

Chandra, K.J., 1994. Infections, concurrent infections and fecundity of *Procamallanus heteropneustes* Ali, parasitic to the fish, *Heteropneustes fossilis. Environ. and Ecol.*, 12: 679–684.

Chandra, K.J. and Modak, P-C., 1995. Activity, aging and penetration of first stage larvae of *Procamallanus heteropneustes. Asian Fish. Sci.*, 8: 95–101.

Chauhan, D.K., 2002. Immunopathological studies in experimental ascoridiasis with heavy metal toxicity in W.L.H. Chick. *Ph.D. Thesis*, C.C.S. University, Meerut.

De, N.C., 1995. On the development and life-cycle of *Spirocamallanus mysti* (Nematoda : Camallanidae). *Folia Parasitol.*, 42: 135–142.

De, N.C. and Maity, R.N., 1999. Larval development of *Onchocamallanus hogaraii* (Nematoda : Camallanidae) in copepods. *Folia. Parasitol.*, 46: 53–58.

De, N.C. and Maity, R.N., 2000. Development of *Procamallanus saccobranchi* (Nematoda : Camallanidae), a parasite of a freshwater fish in India. *Folia. Parasitol.*, 47: 216–226.

De, N.C., Sinha, R.K. and Mazumdar, G., 1986. Larval development of *Procamallanus spiculogubemaculus* Agarwal, 1958 (Nematoda : Camallanidae) in copepods. *Folia. Parasitol.*, 33: 51–60.

Furtado, J.I. and Low, T.K., 1973. Incidence of some helminth parasites in the Malaysian catfish, *Glorias batrachus* (Linnaeus). (Congress in USSR 1971, p. 3). *Verb. Int. Theor. Angew. Limnol.*, 18: 1674–1685.

Gonzalez-Solis, D., Moravec, F. and Vidal-Martinez, M., 2002. *Procamallanus* (*Spirocamallanus*) *chetualensis* n.sp. (Nematoda : Camallanidae) from the Mayan Sea catfish, *Ariopsis assimilis*, off Caribbean Coast of Mexico. *J. Parasitol.*, 88: 765–768.

Gray, A.R., 1963. Serum transaminase levels in cattle and sheep infected with *Trypanosoma vivax. Exp. Parasitol.*, 14: 374–381.

Groofenhuis, J.G., Dwinger, R.H., Dolon, R.B., Mallo, S.K. and Murray, M., 1990. Susceptibility of African buffalo and Boran Cattle to *Trypanosoma congolense* transmitted by *Glossina morsitans centralis. Vet. Parasitol.*, 35: 219–231.

Gupta, N. and Gupta, D.K., 1986. Trypanosome infectivity and changes in the glucose level of two freshwater fishes. *Indian J. Parasitol.*, 10: 213–215.

Gupta, N. and Gupta, D.K., 1998. Hypocholesterolemia as a consequential event of haemo-protozoan infection in a catfish. *Flora and Fauna*, 3: 59–60.

Gupta, N., Gupta, D.K. and Saraswat, H., 2000. Hypoglycemia in *Heteropneustes fossilis* parasitized by two species of parasites (*Trypanosoma karelensis* n.sp. and *Myxosoma fossilis* n. sp.). In: *National Symposium on Fish Health Management and Sustainable Aquaculture*. College of Fishery Science, G.B. Pant University of Agriculture and Technology, Pantnagar, p. 67.

Gupta, V., 1991. Microecological studies effect of heavy metal (cadmium acetate) on certain helminth parasites and on infected host. *Ph.D. Thesis*, Meerut University, Meerut.

Hesser, E.F., 1960. Methods for routine fish haematology. *Progr. Fish-Cult.*, 122: 164–171.

Joshi, B.D., 1979. Biochemical changes in the liver and blood of a freshwater fish, *Rita rita*, infected with a trematode parasite, *Opisthorchis pedicellata*. *Folia Parasitol.*, 26: 143–144.

Katoch, K., 2002. *Procamallanus bilaspurensis*, a new record from *Heteropneustes fossilis* (Bloch). In: *Coldwater Fish Genetic Resources and their Conservation*, (Eds.) P. Das, S.R. Verma, J.R. Dhanze and D.S. Malik. Nature Conservators, Muzaffarnagar, pp. 261–264.

Khatoon, H. and Ansari, J.A., 1979. Biochemical studies on blood alterations in experimental *Solaria cervi* infection. *Helminthologia*, 17: 167–206.

Kumar, S., 1983. Cytochemical, biochemical and histochemical studies on *Bunostomum trigonocephalum* (Rud. 1808) Railliet, 1902. *Ph.D. Thesis*, Vol II, Meerut University, Meerut.

Madhavi, R., 2003. Metazoan parasites in fishes. In: *Aquaculture Medicine*, (Eds.) I.S. Bright Singh, S.S. Pai, R. Philip and A. Mohandas. Cochin University of Science and Technology, Cochin, pp. 64–88.

Martens, E. and Moens, J., 1995. The metazoan ecto- and endo-parasites of the rabbit fish, *Siganus sutor* (Cuvier and Valenciennes, 1835), off the Kenya Coast. *Afr. J. Ecol.*, 33: 405–416.

Mbagwu, H.O. and Opara, K.N., 1998. Therapeutic effects of Sarmorin® in experimental murine trypanosomiasis. *Nig. J. Parasitol.*, 19: 51–58.

Mbagwu, H.O. and Opara, K.N., 2001. Therapeutic and prophylactic effects of Antrycide Pro-Salt® in murine trypanosomiasis. *J. Parasitic Dis.*, 25: 34–39

Moravec, F., 1975. The development of *Procamallanus laeviconchus* (Wedl, 1862) (Nematoda : Camallanidae). *Adv. Soc. Zool., Bohemoslov.*, 39: 23–38.

Moravec, F. and Vargas-Vazquez, J., 1996. The development of *Procamallanus (Spirocamallanus) neocaballeroi* (Nematoda : Camallanidae), a parasite of *Astyanax fasciatus* (Pisces) in Mexico. *Folia Parasitol.*, 43: 61–70.

Moravec, F., Cohen, A. and Fernandes, B.M.M., 1993. Nematode parasites of fishes of the Panama River, Brazil. Part 3: Camallanoidea and Dracunculoidea. *Folia Parasitol.*, 40: 211–229.

Moravec, F., Nie, P. and Wang, G., 2003. Some nematodes of fishes from Central China, with the re-description of *Procamallanus (Spirocamallanus) fulvidraconis* (Camallanidae). *Folia. Parasitol.*, 50: 220–230.

Moravec, F., Chara, J. and Sinha, A.P., 2004a. Two nematodes, *Dentinema trichomycteri* n.g. n. sp. (Cosmocercidae) and *Procamallanus chimusensis* Freitas and Ibanez, 1968 (Camallanidae), from catfishes *Trichbmycterus* spp. (Pices) in Colombia. *Syst. Parasitol.*, 59: 189–197.

Moravec, F., E.R. Cruz–Lacierda and K. Nagaswa, 2004b. Two *Procamallanus* sp. (Nemataoda : Camallanidae) from fishes in the Philippines. *Ada. Parasitol.*, 49: 309–318.

Moravec, F., Nie, P. and Wang, G., 2004c. Reduction of *Camallanus hypothalmichthys* Dogel and Akhmerov, 1959 (Nematoda :

Camallanidae) and its first record from fishes in China. *J. Parasitol.,* 90: 1463–1467.

Nielson, M.E., 1999. An enhanced humoral immune response against the swimbladder nematode *Anguillicola japonica* compared with the European eel, *Anguilla anguilla. J. Helminthol.,* 73: 227–232.

Oser, B.L., 1976. *Hawk's Physiological Chemistry.* Tata McGraw Pub. Co., New Delhi.

Pande, B.P., Bhatia, B.B. and Rai, P., 1963. On the camallanid genus *Procamallanus* Baylis 1923 in two of the freshwater fishes. *Indian J. Helminthol.,* 16: 105–118.

Pandian, T.J., 2001. *Sustainable Indian Fisheries.* National Academy of Agricultural Sciences, New Delhi.

Rai, P., 1969. On some of the hitherto known and unknown nematode parasites in some of the freshwater siluroid fishes. *Indian J. Helthinthol.,* 21: 94–108.

Rajyalakshmi, I., 1996. *Procamallanus seenghali* n. p. (Nemtoda : Camallandae) from the intestine of freshwater catfish, *Mystus senghala* (Sykes) of Visakhapatnam. *Riv. Parassitol.,* 57: 161–168.

Rani, K., 1986. Biochemical and immunological responses induced by *Ascaridia galli* in experimental infection of WLH, chicks, *M.Phil. Thesis,* Meerut University, Meerut.

Rao, M.V.S., 1991. Immunological and immunopathological studies in experimental ascaridiasis. *Ph.D. Thesis,* Meerut University, Meerut.

Roberts, R.J., 2001. *Fish Pathology,* 2nd Edn. W.B. Saunders, Philadelphia and London.

Ruhela, S., Pandey, A.K. and Khare, A.K., 2006a. Effect of experimental *Procamallanus* infection on certain blood parameters of the freshwater catfish, *Clarias batrachus* (Linnaeus). *J. Ecophysiol. Occup. Hlth.,* 6: 73–76.

Ruhela, S., Pandey, A.K. and Khare, A.K., 2006b. Histopathological changes in intestine of catfish, *Glorias batrachus,* induced by experimental *Procamallanus* infection. *J. Ecophysiol. Occup. Hlth.,* 6: 111–117.

Ruhela, S., Pandey, A.K. and Khare, A.K., 2006c. Histopathological changes in liver of catfish, *Clarias batrachus,* induced by experimental *Procamallanus* infection. *J. Appl. Biosci.,* 33: 36–41.

Ruhela, S., Pandey, A.K. and Khare, A.K., 2008. Histopathological manifestations in kidney of *Clarias batrachus* (Linnaeus), induced by experimental *Procamallcmus* infection. *J. Environ. Biol.* (in press).

Sadun, E.H. and Williams, J.S., 1966. Biochemical aspects of *Schistosoma rncmsoni* in mice in relation to worm burdens and duration of infection. *Exp. Parasitol.,* 18: 266–273.

Satpute, L.R. and Agrawal, S.M., 1974. Parasitic effects on its haematology and histopathology. *Indian J. Exp. Biol.,* 12: 584–586.

Sharma, S., 1997. Immunopathological and biochemical changes in experimental infection induced by sheep-whip worms. *Ph.D. Thesis.* Meerut University, Meerut.

Sinha A.K., 1988a. On the life-cycle of *Procamallanus spiculogubernaculus* (Camallanidae) (Agrawal, 1958) a nematode parasite of fishes. *Riv. Parasitol.,* 49: 111–116.

Sinha, K.P., 1988b. *Procamallanus* (Camallanidae : Nematoda) infection in the fish, *Clarias batrachus. Environ. and Ecol.,* 6: 1035–1037.

Sinha, K.P., 1992. Macrocytic anaemia in the catfish, *Clarias batrachus. Bio Journal,* 4: 229–230.

Sinha, K.P., 2000. Haematological manifestation in *Clarias batrachus* carrying helminth infections. *J. Parasit. Dis.,* 24: 167–170.

Sinha, A.K. and Sinha, C., 1988. Macrocyte hypochromic anaemia in *Heteropneustes fossilis* infected by blood sucker, *Procamallanus spiculogubernaculus. Indian J. Parasitol.,* 12: 93–94.

Sood, M.L., 1967. On some species of the genus *Procamallanus* Baylis, 1929 from freshwater fishes of India. *Proc. Natl. Acad. Sci., India,* 37B: 291–303.

Sood, M.L., 1988. Some nematode parasites from freshwater fishes of India. *Indian J. Helminthol.,* 29: 83–110.

Srivastava, P.S., Das, S.S., Murtuza, Md. and Sinha, S.R.P., 1988. Biochemical changes in the sera of cattle immunized with tick tissue extract of *Baophilus microplus. Indian J. Anim. Sci.*, 58: 548–551.

Taiwo, V.O., Olaniyi, M.O. and Ogunsanmi, A.O., 2003. Comparative plasma biochemical changes and susceptibility of erythrocytes to *in vitro* peroxidation during experimental *Trypanosoma congolense* and *T. brucei* infection in sheep. *Israel Vet. Med. Assoc.*, 58: 245–255.

Totterman, G., 1944. On the occurrence of pernicious tapeworm anaemia in *Diphyllobothrium* carriers. *Acta. Med. Scand.*, 118: 410–416.

Totterman, G., 1947. Is the broad tapeworm the causal agent of hypochromin anaemia? *Ann. Med. Intern. Fenniae* (*Helsinki*)., 36: 185–190.

Vibe-Peterson, G. and Nielson, K., 1979. *Vermious enteritis* and *Thromboembolic celic* in the horse: A description of 36 cases. *Nordisk Vet. Med.*, 31: 385–391.

von Brand, T., 1973. *Biochemistry of Parasites,* 2nd Edn. Academic Press, NewYork.

Yamaguti, S., 1961. *Systema Helminthum: The Nematodes of Vertebrates.* Interscience Publishers, New York and London, Vol. 3(1 and 3): 1–1261.

Zaman, Z. and Leong, T.S., 1988. Occurrence of *Procamallanus malaccensis* Femando and Furtado 1963 in *Clarias batrachus* and *C. macrocephalus* from Kedah and Perak, Malaysia. *Asian Fish. Set.*, 2: 9–16.

Zsigmond, J.E., Valtonen, T., Galina, J.E. and Jokinen, I., 2002. Effect of pulp and paper mill effluent (BKME) on physiological parameters of roach (*Rutilus rutilus*) infected by the digenean, *Rhipidocotyle fennica. Folia Parasitol.*, 49: 103–108.

Chapter 15

Amphibian and Reptilian Diversity in Western Himalaya: Current Status and Future Need

Deep N. Sahi
Department of Zoology, University of Jammu, Jammu – 180 006

Western Himalaya, with a considerable heterogeneity in its topography, altitude and climate, harbours a rich biodiversity. The area stretching from Kashmir to Himachal is ingressed by Western-Himalayan chain which is drawn into three longitudinal ranges namely Outer, Middle and Great Himalayas. Both the states share common geological history and have more or less similar topography, climate and vegetational pattern and thus have great resemblance in their faunal elements so much so that a deer Hangul (*Cervus elaphus hanglu*), a Kashmir endemic has been reported in some parts of Himachal Pradesh.

The natural vegetation in Western Himalaya has a climatic altitudinal zonation. The altitudinal variation within this region has led to great ecological diversity from subtropical plains and foot hills in Outer Himalayan (Shivalik) zone to temperate in Middle Himalaya and cold arid mountainous desert in Great Himalaya.

The herpetological elements (Amphibia and Reptilia) of the western Himalayan region has not drawn the attention of scientific community as has been shown to other faunal elements such as mammals, birds, fishes and butterflies. Casual reference to the herptiles of this region has appeared from time to time in fauna of British India (Boulenger, 1890; Smith, 1931, 1935, 1943). After independence, contribution to herpetology of Jammu and Kashmir region has been made by Das, Malhotra and Duda (1964), Sharma and Sharma (1976), Dubois (1978), Duda, Dutta and Sahi (1977), Sahi (1978) and Sahi and Duda (1983). Himachal Pradesh has remained quite neglected as far as reptilian fauna is concerned but significant contribution has been done in the field of batrachology of the state (Tilak Hussain, 1976; Tilak and Mehta, 1977, 1983; Mehta, 1983). Since Kashmir Himalaya is in no way different from Himachal Himalaya, it is expected that reptilian elements in both these regions will be similar.

The present work deals with the distribution, current status and habitat preferences of 17 sps. of amphibians and 63 sps. of reptiles in all the three longitudinal zones in the West Himalaya *i.e.* Outer, Middle and Great Himalaya.

Outer (Shivalik) Himalayan Zone

The area upto 1000 mt elevation falls in this zone. In Himachal Pradesh, it stretches from Poanta Sahib to Pathankot bordering Punjab and Haryana, in Sirmur, Solan, Bilaspur, Hamirpur, Kangra and Una Districts. In Jammu and Kashmir state, it is spread over from foothills of Kathua and Jammu districts, Reasi, Ramnagar in Udhampur and Nowshera, Sunderbani and Kalakot in Rajouri district.

Middle (Lesser) Himalayan Zone

This zone contain the area between 1000 mt to 2000 mt elevation in the districts of Shimla, Sirmur Mandi, Kulu, Kinaur and Chamba in Himachal Pradesh. This Himalayan belt is commonly referred as Dhouladhar range in H.P. and Pir-Panjal in Kashmir. The area like Mohore, Arnas in Udhampur District, Mender, Surankot and Haweli in Poonch district and whole of Doda district of J&K falls under this category.

Great (Inner) Himalayan Zone

This zone contains area upto 2000 mt to snowline, high mountainous area in Shimla, Kullu, Manali, Kinaur and Kangra in

Himachal Pradesh, whole of Kashmir region and Kargil and Leh districts of Ladakh region falls under this zone.

Why Conservation is Needed? (Uses and Exploitation)

Amphibians

Most of the species of frogs and toads live in or near agricultural fields, they are one of the major predators of insect pests in the agricultural fields. It is a best method of Biological control of insect pests.

Indian Bull Frog, *Hoplobatrachus tigerinus* and some other species are edible and their meat especially hindlegs is regarded as 'Luxury protein'. Due to their economic value, there has been tremendous commercial exploitation of these frogs from nature.

Chelonians

The freshwater turtles perform a valuable service as scavengers in the rivers and lakes. Some of the species are vegetarian while others are more strictly carnivorous. Fishermen on the banks of the Chambal and Narmada rivers report that the soft shell turtles are normal scavengers on human corpses. In this way, turtles play a useful functioning of cleaning the rivers and lakes and thus significantly controlling pollution in the freshwater systems (Verma, 1992).

Besides this, turtles provide an important source of protein and many people eat turtle meat and eggs. They are also kept as pets. Some turtles are caught in local fishing nets, where they slowly suffocate and die.

Lacertilians

In nature, they act as predators of harmful insects.

Monitor Lizards are hunted for many purposes *e.g.* the skins are used for leather and a few "Nat" tribals eat the flesh and drink the blood (often for medicinal purposes such as relief of rheumatism). The skin is also widely used in making drum heads for percussion instruments called "Dholak" and "Dug-Dugi". The large abdominal fat found in the monitor is used as a salve for epidermal bacterial infection (Auffenberg *et al.*, 1989).

Ophidia

Snakes are important for agriculture as they feed on rats, mice and birds which act as predators of agricultural crops.

Snakes are commercially very important, particularly for their skins which are used for making bags, belts, shoes, boxes, and other decorative items.

Furthermore, about 80 per cent of the snakes are non-poisonous, still their population is declining because they are killed due to man's natural fear and the lack of knowledge about the value of snakes. Example is of Indian Python which is the largest of the non-poisonous snakes and so easily detectable and killed. At present, this snake is listed in schedule-I of wildlife (Protection) Act 1972.

In addition to all this, the good number of herptiles are also used by colleges and universities as 'display specimens' for scientific research and for dissection purpose.

What Steps were Taken Till Today?

(In Indian and at Global Level)

Amphibians

H. tigerinus was placed in Appendix II of cites. The export of *H. tigerinus* (dead, alive, parts of derivatives) will require export permit under CITES. Furthermore, the Indian Wildlife (Protection) Act 1972 requires capture permits for all the frogs of the family Ranidae.

Lacertilians

Varanus flavescens is placed in Schedule-I and *V. benghalensis* in Scheduled-II and thus are protected by Indian wildlife (Protection) Act 1972.

Ophidia

The Indian Wildlife (Protection) Act 1972 has included the Indian Python in Scheduled-I; *Xenochrophis piscator, Ptyas mucosus, Naja naja* and *Naja oxiane, Viper russelli* in Schedule-II and all other Indian snakes in Schedule-III (capture permit required).

Indian Python is also listed in the IUCN Red Data Book.

Chelonians

T. gangeticus, K. tecta, T. hurum and *G. hamiltoni, L. punctata* are listed in Schedule-I of Wildlife (Protection) Act 1972.

CITES has included *T. gangeticus, L. punctata* and *K. tecta* in Schedule-I. Schedule-IV of CITES enlists the family Trionychidae and so capture permits are required for all turtles of this family.

What will be the Future Approaches to Conservation?

Although India's herptiles have the support of wildlife laws, it continues to be depleted at unprecedented rates. The values of herpeto fauna are least appreciated and in economic terms, poorly valued natural resource.

Suggested Conservation Measures are:

1. *Herpetological Parks*: These should be established where people could be educated in all aspects of herptiles. These parks should be established in areas easily accessible to the public, students and tourists.
2. *Educational Awareness*: Design and develop herptile conservation and educational programs in the local schools and prepare educational material on a participatory basis. Popularise the importance of herpetofauna through the display of wildlife hoardings, hand-bills, projecting wildlife oriented movies in open-air theaters.
3. As there are only five species of poisonous snakes in Western Himalayas, so an easy identification key based on striking morphological features should be presented to the general public and doctors.
4. Setting up turtle hatcheries, frog production and snake protection research centres on commercial scale needs immediate consideration. This would serve the dual purpose of continued export earning as well as reduced pressure on the wild stocks.
5. During the breeding season, gravid females and mature males can be picked up and sent to hatcheries and eggs or nestlings can be reared there.
6. More and more antivenom production centres must be established like the one in Kasoli (H.P.) and tribal people and snake charmers should be involved in capturing and maintaining them to reduce pressure on wild stock for their earnings.

7. Long term research and monitoring should be carried out to define the actual status of herptiles in West Himalayas.
8. In nature, their habitats must also be prevented from degradation. For this, identification of activities which have or are likely to have a deletrious effect on habitat is necessary.
9. Trade in wildlife if any, should be made at the government level and within the framework of wildlife conservation.
10. Stringent measure should be taken against poachers.
11. Integrate local, regional, national, and international efforts through advertisement and advocacy.
12. *In situ*-management techniques, where the core area is given protection, maintaining a breeding population and allowing a limited harvest at the peripheral region is another possible approach.
13. A pragmatic management strategy should be designed to facilitate their reintroduction in areas where they formerly occur.

In the light of these facts, it is suggested that protected areas should be set up in each zone of West Himalayan belt *i.e.* Outer, Middle and Great Himalaya.

Amphibians of West Himalaya

Zoological Name	*Distributional Range*	*Habitat*	*Status*
Order–Anura			
Family–Bufonidae			
1. *Buto beddomii* Gunther	G.H.	AZ, SAq	VR
2. *Buto himalayanus* Gunther	OH-MH	TF, SAq	R
3. *Buto latastii* Boulenger	GH	AZ, SAq	VR
4. *Buto melanostictus* Schneider	OH-MH	StF, SAq	C
5. *Buto stomaticus* Lutken	OH	StF, SAq	VC
6. *Buto viridis* Laurenti	MH-GH	TF-AZ, SAq	C
7. *Uperodon systoma* (Schneider)	OH	StF, SAq	VR

Contd...

Contd...

Zoological Name	*Distributional Range*	*Habitat*	*Status*
Family–Ranidae			
8. *Euphlyctis cyanophylyctis* (Schneider)	OH-MH	StF-TF, Aq	VC
9. *Hoplobatrachus tigerinus* Daudin	OH-M H	Aq	C
10. *Limnocetes limnocharis* (Gravenhorst)	OH	StF, Aq	VR
11. *Paa liebigii* (Gunther)	MH	Az, Aq	VR
12. *Paa minica* (Dubois)	MH	Az, Aq	VR
13. *Paa sternosignata* (Murraya)	MH	TF, Aq	VR
14. *Paa vicina* (Stoliczka)	MH	TF, Aq	VR
15. *Tomopterna breviceps* (Schneider)	OH	StF, Aq	R
Family–Pelobatidae			
16. *Scutiger occidentalis* (Dumeril and Bibron)	GH	Az, SAq	VR
Family–Microhylidae			
17. *Microhyla ornata* Dubois	OH-MH	SAq	C

Reptiles of West Himalaya

Zoological Name	*Distributional Range*	*Habitat*	*Status*
Class–Reptilia			
Order–Chelonia (7 sps., 5 genera, 2 families)			
Family–Emydidae			
Geoclomys hamiltoni	OH	Aq	VR
Kachuga smithi	OH	Aq	UC
Kachuga tectum tectum	OH	Aq	UC
Family–Trionychidae			
Chitra indica	OH	Aq	VR
Lissemys punctata	OH	Aq	UC
Trionyx gangeticus	OH	Aq	UC
Trionyx hurum	OH	Aq	VR

Contd...

Contd...

Zoological Name	*Distributional Range*	*Habitat*	*Status*
Order–Lacertilia (24 sps., 14 genera, 5 families)			
Family–Gekkonidae			
Cyrtodactylus himalayanus	MH	Rd.	VR
C. lawderanus	GH	Nv.	C
C. montium salrorum	GH	Nv.	VVR
C. scaber	OH	Rk	VR
C. stoliczkai	GH	Nv.	C
Eublepharis macularius	OH	Rk.	VR
Hemidactys brooki	OH	Sd, Rd, Rk	VC
H. flaviviridis	OH	Sd, Rd, Rk	VC
Family–Agamidae			
Agama agronensis	OH	Rk	C
A. himalayana	GH	Nv	UC
A. tuberculata	OH-MH-GH	Rk	C
Calotes versicolor	OH	Sd, Rd.	C
Sitana ponticerina	OH	Sd, Gr.	UC
Phrynocephalus reticulates	GH	Nv.	VR
P. theobaldi	GH	Nv.	VR
Family–Scincidae			
Eumeces taeniolatus	OH-MH	Sd, Rk.	UC
Mabuya dissimilis	OH	Gr.	UC
Scincella himalayanum	M H-G H	Tj.Nv.	C
S. ladacense	GH	Nv.	UC
Family–Lacertidae			
Aplepharus pannonicus	OH	Sd.	VV
Acanthodactylus cantoris cantoris	OH	Sd.	WR
Ophisops jerdoni	OH	Sd, Rk.	UC

Contd...

Contd...

Zoological Name	*Distributional Range*	*Habitat*	*Status*
Order–Ophidia (32 sps., 21 Genera and 5 families)			
Family–Typhlopidae			
Ramphotyphlops bramina	OH	Rd	C
Typhlops porreatus	OH	Sd.	R
Family–Boidae			
Eryx conicus	OH-MH	Sd.	VR
Eryx johni	OH	Sd.	VR
Python molurus	OH	StF	VR
Family–Colubridae			
Amphiesma platyceps	OH-MH	T.F.	VR
A. stolata	OH-MH	StF, TF	C
Boiga multifasciata	OH-MH	Sd, StF, Rd.	R
B. trigonata	ÖH-MH	Sd, StF, Rd	VR
Coluber rhodorachis	OH-MH-GM	TF, Rd.	VR
C. ventromaculatus	OH-MH	TF. RK	VR
Dendrelophis tristis	OH	Sd.	VR
Elaphe helena	OH	Sd.	UC
E. hodgsoni	MH-GH	TF	UC
Lycodon aulicus	OH	Sd, Rd.	C
L. striatus	OH-MH	Rd.St.F, TF	VR
Oligodon arnersis	OH	Sd, Rd.	R
Psamophis leithi	OH	Sd.	VR
P. schokari	MH	T.F	VR
Ptyas mucosus	OH-HM-GH,	Sd, StF, Rd, TF, AF	VC
Sibynophis saggitarius	OH	Sd,	VR
Sphalerosophis arenaris	OH-MH	Sd, Rd,	R
S. diadema diadema	OH-MH	Sd, Rd, StF,	OH-MH
Sd. D. atriceps	OH-MH	Sd, Rd.	VR
Xenochrophis piscator	OH-MH	Aq. StF, Grass field	VC

Contd...

Contd...

Zoological Name	Distributional Range	Habitat	Status
Family–Elapidae			
Bangarus caeruleus	OH-MH	StF., Rd.	C
Naja naja oxiana	OH	Rd, StF.	VR
N. naja naja	OH	Rd, StF.	R
Family–Viperidae			
Agkistrodon himalayanus	MH	TF	VR
Echis carinatus	OH	Sd, Rk, Rd.	VC
Vipera lebetina	MH	TF	VR
Vipera russelli	OM-MH	Sd, Rk, Rd	VC

References

Auffenberg, W., Rehman, H., Iffet, F. and Perveen, Z., 1989. Notes on the biology of *Varanus griseus koniecznyi* Mertens saura: Varanida. *J. Bomb. Nat. Hist. Soc.*, 87: 26–36.

Boulenger, G.A., 1890. *The Fauna of British India: Reptilia and Batrachia.* Taylor and Francis, London, xviii+541 pp.

Das, S.M., Malhotra, Y.R. and Duda, P.L., 1964. The palearctic elements in the fauna of Kashmir. *Kashmir Science,* 1–2: 100–111.

Dubois, A., 1978. Une espece nouvelle de Scutiger Theobald 1868 de 1'Himalaya Occidental (Anura–Pelobatidae). *Senckeubergiana Biol.,* 59(3–4): 163–171.

Duda, P.L., Dutta, S.P.S. and Sahi, D., 1977. Some tortoises from Jammu. *Geobios,* 4: 263–264.

Mehta, H.S., 1983. Some observations on the development stages of *Rana (Poa) minica* (Amphibia) at high altitude. *Ent. and Wildlife Eco. Zool. Surv.,* India, p. 291–298.

Sahi, D.N. and Duda, P.L., 1983. Checklist and keys to the herpetiles of Jammu and Kashmir State. *Chic. Herp. Soc.,* 20(3–4): 86–97.

Sahi, D.N., 1978. A contribution to the herpetiles of Jammu and Kashmir state. *Ph.D. Thesis,* University of Jammu, Jammu, p. 360.

Sharma, B.D. and Sharma, T., 1976. Some new trends of snakes (Reptilia : Serpentes) from Jammu and Kashmir state. *Curr. Sci.*, 44(17): 646–647.

Smith, M.A., 1931. *The Fauna of British India including Ceylon and Burma: Reptilia and Amphibia, Vol I: Loricata and Testudines.* Taylor and Francis, London, xxviii+185 pp.

Smith, M.A., 1935. *The Fauna of British India including Ceylon and Burma: Reptilia and Amphibia, Vol II: Sauria.* Taylor and Francis, London, xii+440 pp.

Smith, M.A., 1943. *The Fauna of British India including Ceylon and Burma: Reptilia and Amphibia, Vol III: Ophidia.* Taylor and Francis, London, xii+583 pp.

Tilak, R. and Hussain, A., 1976. Extension of the range of distribution of a Microhydid frog *Uperodon systoma* (Schneider). *J. Bomb. Nat. Hist. Soc.*, 73: 407–408.

Tilak, R. and Mehta, H.S., 1977. Report on a collection of amphibians from district Kangra, Himachal Pradesh. *Newl. Zool Surv.*, Kolkata India, 3(4): 196–198.

Verma, A.K., 1992. On the male reproductive cycle of some emydid turtles of Jammu. *Ph.D. Thesis,* University of Jammu, Jammu, p. 169.

Chapter 16

Seasonal Changes in Plasma Calcium and Inorganic Phosphate Levels in Relation to Parathyroid and Ultimobranchial Glands of the Grey Quail, *Coturnix coturnix coturnix*

R.R. Dhande[1], *S.A. Suryawanshi*[2] *and A.K. Pandey*[3]

[1]*Department of Zoology, Amravati University, Amravati – 444 602*

[2]*Swami Ramanand Teerth Marathwada University, Nanded – 431 606*

[3]*Central Institute of Freshwater Aquaculture, Kausalyaganga, Bhubaneswar – 751 002*

ABSTRACT

Plasma calcium (Ca) concentration (annual mean) in *Coturnix coturnix coturnix* males was 10.27±0.14 mg/100 ml while it was slightly higher (11.85±0.15 mg/100 ml) among females. Plasma inorganic phosphate (P*i*) levels (annual means) in males and females were 5.62±0.12 mg/100 ml and 6.52±0.20 mg/100 ml, respectively. The pattern of seasonal variations in plasma Ca and P*i* levels were different in male and female grey quail. While the males did not exhibit significant fluctuation in plasma Ca and P*i* levels either in winter or in summer (breeding season), the female recorded significant elevation in the levels of both these electrolytes during breeding season. The peak value of plasma Ca (17.66±0.38 mg/100 ml) and P*i* (8.64±0.22

mg/100 ml) were observed during the month of June. Parathyroid and ultimobranchial glands of the grey quail exhibited hyperactivity (hypertrophy and hyperplasia) during breeding season, however, the activity was more conspicuous among females than in males. The maximum increase in cell and nuclear diameter in both the glands were observed during May to July in females. The follicles were also filled with AF- and PAS-positive materials during these months. The parathyroid and ultimobranchial glands depicted signs of hypoactivity during winter months as evident by decrease in cell and nuclear diameter as well as vacuolation in the parenchymal cells. The cystic follicles and lymphocytic invasion were observed in the ultimobranchial gland of quail during the peak winter (December–January).

Introduction

Endocrine regulation of plasma calcium (Ca) in non-laying birds is achieved by the interactions of parathyroid hormone (PTH), calcitonin (CT) and active metabolite of vitamin D_3 (1,25-dihydroxycholecalciferol) (Dacke, 1979; Feinblatt, 1982; Taylor, 1984; Pang and Schreibman, 1989; Suryawanshi *et al.*, 1997; Dhande *et al.*, 1995a, b, 1997a, b, c, 1998). Parathyroid glands made their first phylogenetic appearance only in tetrapods (Clark *et al.*, 1986; Pandey, 1991, 1992), probably to protect against the hypocalcemia and maintain skeletal integrity in terrestrial animals (Wendalaar Bonga and Pang, 1991; Pandey, 1992). Parathyroid hormone (PTH) is a predominant hypercalcemic and hypophosphatemic factor in birds and mammals (Pang and Schreibman, 1989; Pandey, 1991, 1992; Aurbach *et al.*, 1992; Dacke *et al.*, 1996; Dhande *et al.*, 1997a; Dacke, 2000). Calcitonin (CT), the hypocalcemic and hypophosphatemic hormone, is secreted by C cells which are embryologically derived from the neural crest and migrate during development into thyroid gland of mammals and ultimobranchial gland (UBG) of submammalian vertebrates (Nunez and Gershon, 1978; Robertson, 1986). Though UBG of birds are rich source of CT (Copp *et al.*, 1967; Moseley *et al.*, 1968; Wittermann *et al.*, 1969; Nieto *et al.*, 1973; Cutler *et al.*, 1977; Honma *et al.*, 1986), its role in normal Ca metabolism of birds has not yet been clearly defined (Barlet, 1982; Feinblatt, 1982; Pang and Schreibman, 1989; Dacke *et al.*, 1996; Dacke, 2000). Mammalian or avian CT failed to induce hypocalcemia in intact chicken. Even ultimobranchialectomized (UBGX)+ parathyroidectomized (PTGX) chicken did not exhibit hypocalcemic response to UBG extract administration (Gonnerman *et al.*, 1972). Loyd *et al.* (1970) also recorded that UBG extract had no effect on plasma Ca of laying hens. Interestingly, CT injected into laying hens when eggshell was not being calcified caused hypocalcemia, however, during eggshell formation it was ineffective (Luck *et al.*, 1980). Contrary to these findings, Calamy and Barlet (1970) recorded slight but statistically significant hypocalcemia in intact hen and cockerels following porcine CT infusion. Furthermore, salmon CT administration resulted in depression of plasma Ca level in the normal as well as PTGX grey quails (Dhande *et al.*, 1997b, c). Enhanced secretion of CT has also been observed in geese, hens, turkeys and Japanese quails following hypercalcemic blood perfusion (Bates *et al.*, 1969; Ziegler *et al.*, 1970; Boelkins and Kenny, 1973).

Birds do have very high rate of Ca turnover during egg-laying which is associated with the concomitant increase in plasma Ca and Pi levels (Baldini and Zarrow, 1952; Simkiss, 1961, 1967; Sturkie, 1976; Mori and George 1978; Scanes *et al.*, 1982; Dacke, 2000). A domestic hen may utilize Ca equivalent to its 10 per cent body weight per day during this period (Bentley, 1982; Dacke, 2000). The large storage of Ca in yolk or eggshell is subsequently utilized for the ossification of developing embryo (Feinblatt, 1982). Among birds, egg production is associated with the concomitant increase in plasma Ca and Pi levels (Mori and George, 1978; Bentley, 1982; Pang and Schreibaman, 1989). Estrogen has specific effects upon Ca metabolism in birds related to the ovulatory cycle and of physiological importance in yolk accumulation in the ovarian follicles (Baldini and Zarrow, 1952; Urist, 1976; Wallace, 1985). Administration of estradiol in pigeon, cocks, capons and Japanese quail induced marked increase in non-ultrafiltrable (protein-bound) fraction of plasma Ca (Baldini and Zarrow, 1952; Urist *et al.*, 1960; Simkiss, 1961, 1967; Dacke, 1979; Dhande *et al.*, 1997d, 1999). Speers *et al.* (1970) reported that UBGX hens consumed significantly less feed and produced smaller eggs with a trend toward reduced cell thickness. Though parathyroid hormone (PTH) appears to play major role in synchronizing egg formation by altering the concomitant changes in skeletal metabolism (Mueller *et al.*, 1973), no major seasonal changes in the avian parathyroid structure have been observed except lipid inclusions which are more common in winter birds (Stoeckel and Porte 1970, Roth and Schiller, 1976; Clark *et al.* 1986). Though there exist reports on enhanced cytological activity in parathyroid (PTG) and UBG of the laying birds but annual cyclic variations in the plasma electrolytes has not yet been correlated with the activity of these glands (Urist, 1967; Roth and Schiller 1976; Clark *et al.* 1986). An attempt has, therefore, been made to record the annual variations in total plasma Ca and Pi levels in relation to histomorphology of the PTG and UBG of *Coturnix coturnix coturnix.*

Materials and Methods

Adult quails, *Coturnix coturnix coturnix* Linnaeus weighing 60–75 gm (both sexes) were procured from the local markets of Nagpur (Maharashtra). They were kept under the laboratory conditions for a week and fed on millet (*Sorghum vulgare*) grains. Five birds (both sexes) were killed in the fourth week of every month throughout the year for recording the seasonal changes. Blood samples from both the sexes were collected in heparinized centrifuge tubes from the cut made on the jugular vein. Plasma Ca and inorganic phosphate (Pi) levels were estimated by the methods described by Wootton (1974).

Parathyroid (PTGs) and ultomobranchial glands (UBGs) of the birds from both the sexes were surgically removed and fixed immediately in freshly prepared Bouin's solution. After 24 hours, the tissues were washed thoroughly in running tap water, dehydrated in ascending series of alcohol, cleared in xylene and embedded in paraffin wax at 60°C. Serial sections were cut at 6 µm and stained in hematoxylin-eosin (H&E) and periodic acid-Schifs reagent (PAS) as well as specific stains for C cells like Devenport's silver impregnation (Kameda, 1968) and lead-hematoxylin (Pb-H) (Solicia *et al.*, 1969). Cell and nuclear diameters were measured with the help of ocular micrometer (PZO, Poland) along its long and short axes and the mean values were calculated. 250 cells and nuclear measurements (25 from each birds) of both the glands were recorded for both the sexes at monthly intervals.

Results and Discussion

Plasma Ca levels in birds other than egg-laying hens are much the same as those in mammals (Simkiss, 1961, 1967). The plasma Ca concentration in normal male *Coturnix coturnix coturnix* was 10.27±0.14 mg/100 ml while in female it was 11.85±0.15 mg/100 ml (annual mean). The average value of plasma Ca concentrations in males and females during non-breeding season were 9.27±0.12 mg/100 ml and 10.44±0.15 mg/100 ml, respectively (Table 16.1). These values are comparable to those of non-reproducing female or male pigeon (*Columbia livia,* 9.0 mg/100 ml), ring dove (*Columbia palumbus,* (9.00 mg/100 ml), house sparrow (*Passer domestiells,* 10.00 mg/100 ml), Canada goose (*Branta canadensis,* 10–11 mg/100 ml), male or immature domestic fowl (*Gallus domestieus,* 12.00 mg/100 ml), Japanese quail (*Coturnix japonica,* 10.4 mg/100 ml) and rain quail (*Coturnix coromondalica,* 9.5 mg/100 ml). However, the value seems to be lower as compared to Bobwhite quail (*Colinus virginiamus,* 13.0 mg/100 ml) and pullet (17.4 mg/100 ml) (Riddle and Reinhart, 1926; Baldini and Zarrow, 1952; Simkiss, 1961; Mori and George, 1978). As the pattern of cyclic variations in plasma Ca was different in male and female grey quail hence values for both the sexes were treated separately. During breeding season (April–July), plasma Ca level of female grey quail reached to maximum (17.66±0.38 mg/100 ml) in June which is less as compared to laying hen (25–29 mg/100 ml), Bobwhite quail (29 mg/100 ml), Canada goose (22 mg/100 ml) and Japanese quail (26 mg/100 ml). In the present study, only the female grey quails showed marked increase in plasma Ca and *Pi* levels during breeding season. Increase in the serum levels of Ca and Pi has been reported during breeding season or during egg-laying period in several birds (Simkiss, 1961, 1967; Mori and George, 1978; Dacke, 1979). As avian oviduct is not able to store Ca, most of the metabolite for formation of egg may be obtained from the circulating Ca levels.

Generally, birds possess two pairs of parathyroid glands but in some species the glands on each side are fused to form a single mass. However, considerable variations occur in its precise location. In *Coturnix coturnix coturnix,* there were two pairs of parathyroid glands-anterior (parathyroid III) and posterior (parathyroid IV), closely situated near caudal region of thyroid glands just above the junction of common carotid with subclavian artery. The arterial blood was supplied by branches of common carotid artery and venous blood collected by the jugular vein (Figures 16.1 and 16.2). Our observations are in agreement with those reported for chicken, pigeon, duck and starlings (Benoit *et al.,* 1944a, b; Smith, 1945; Benoit, 1950; Schrier and Hamilton, 1952; Ramanoff, 1960; Hodges, 1970; Clark and Wideman, 1977). In chicken, a pair of each side consists of a larger and smaller lobe. Sometimes, smaller lobe may be in contact with thyroid and larger lobe behind the smaller (Nonidez and Goodale, 1927). In pigeon and starlings, the two parathyroid on each side are located posterior to thyroid and separated trom each other (Smith, 1945; Clark and Wideman, 1977) while in laying hen, both the parathyroid of each side are usually fused to form a glandular mass (Nevalainen, 1968; Taylor, 1971; Dacke, 1979).

Table 16.1: Annual Variations in the Plasma Calcium and Inorganic Phosphate Levels (mg/100 ml) in *Coturnix coturnix coturnix*

Months	*Female*		*Male*	
	Plasma Calcium	*Plasma Inorganic Phosphate*	*Plasma Calcium*	*Plasma Inorganic Phosphate*
January	10.78±0.09	5.82±0.07	9.86±0.06	5.18±0.08
February	11.54±0.07	6.84±0.08	10.60±0.17	6.60±0.06
March	11.78±0.13	6.80±0.06	11.22±0.08	6.06±0.09
April	12.52±0.10	7.56±0.41	10.70±0.17	6.50±0.09
May	13.88±0.12	8.24±0.16	11.26±0.13	6.80±0.08
June	17.66±0.38	8.64±0.22	11.20±0.18	5.78±0.22
July	13.64±0.22	6.54±0.15	11.50±0.29	5.10±0.21
August	11.46±0.20	5.94±0.12	10.62±0.16	4.92±0.09
September	9.62±0.13	5.66±0.15	8.50±0.11	4.90±0.22
October	10.14±0.12	5.84±0.13	9.42±0.12	5.20±0.15
November	9.80±0.12	5.14±0.45	9.64±0.15	5.54±0.13
December	9.46±0.15	5.24±0.10	8.82±0.17	4.88±0.06
Annual mean	11.85±0.15	6.52±0.20	10.27±0.14	5.62±0.12

Values are shown in mg/100 ml mean±SE for 5 samples.

Histologically, the parathyroid gland of bird appears to be similar to those of mammals comprising chief cells but lacking oxyphil cells (Benoit *et al.*, 1944a, b; Benoit, 1950; Doiron, 1973; Roth and Schiller, 1976; Clark *et al.*, 1986). Each parathyroid gland of the grey quad was a small, oval or spherical yellowish structure encapsulated by connective tissue sheath. The chief cells were arranged in elongated and branching cords separated by thin connective tissue stroma, capillaries and sinusoids. Sometimes cords were arranged in whorls forming follicles with a central cavity filled with AF- and PAS-positive material (Figures 16.3 and 16.4). The cords of chief cells are either elongated as in duck (Benoit, 1950) or irregular and anastomose as in chicken and laying hen (Nonidez and Goodale, 1927; Nevalainen, 1969). The oxyphil cells were totally absent in the parathyroid glands of *Coturnix coturnix coturnix*. The parenchymal chief cells of parathyroid were ovoid, elongated with little cytoplasm and poor cellular boundaries. Each cell contained a distinct centrally situated large nucleus occupying major portion of the cell (Figure 16.5). The average cell and nuclear diameter ranged from 8.20 to 10.85 µm and 3.77 m to 6.64 µm, respectively (Table 16.2).

Table 16.2: Annual Changes in the Cell and Nuclear Diameter (µm) of the Parathyroid Gland of *Coturnix coturnix coturnix*

Months	*Female*		*Male*	
	Cell Diameter	*Nuclear Diameter*	*Cell Diameter*	*Nuclear Diameter*
January	9.32±0.34	4.69±0.30	8.70±0.22	4.28±0.26
February	9.28±0.28	4.99±0.42	9.12±0.34	4.88±0.30
March	9.72±0.40	5.94±0.26	8.16±0.14	3.77±0.16
April	9.94±0.24	6.24±0.28	8.20±0.18	4.80±0.18
May	10.55±0.36	6.64±0.36	8.90±0.37	5.10±0.22
June	10.50±0.26	6.64±0.22	9.47±0.24	5.85±0.22
July	10.85±0.18	6.26±0.16	9.70±0.20	5.68±0.32
August	9.84±0.40	5.35±0.20	8.67±0.43	5.40±0.38
September	18.75±0.32	5.12±0.24	8.62±0.42	5.10±0.11
October	8.67±0.25	4.62±0.40	8.34±0.26	4.42±0.20
November	8.72±0.11	4.46±0.26	8.56±0.21	4.35±0.14
December	8.56±0.11	4.30±0.18	8.42±0.20	4.30±0.24
Annual mean	9.52±0.27	5.35±0.27	8.73±0.26	4.82±0.24

Histological study of parathyroid glands of *Coturix coturnix coturnix* throughout the year indicated marked cyclic changes, particularly during the breeding season (April-July). Seasonal variations in parathyroid structure have been recorded extensively in poikilothermic animals (Isono *et al.*, 1959; Isono, 1960; Sidkey, 1965; Dubewar and Suryawanshi, 1978; Pandey, 1990a, Swarup

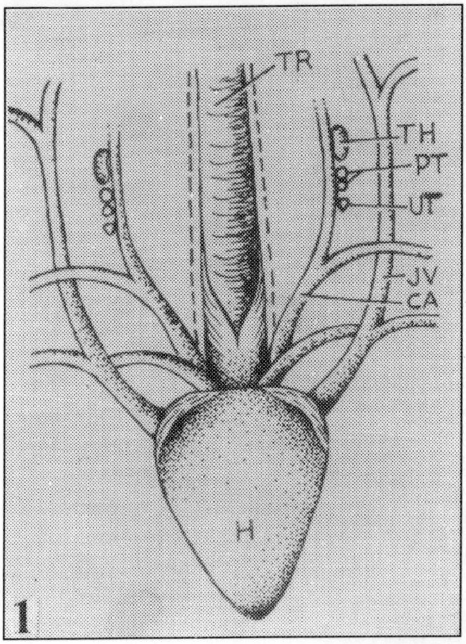

Figure 16.1: Diagrammatic Representation of the Neck Region of *Coturnix coturnix coturnix* Showing the Location of Thyroid (TH), Parathyroid (PT) and Ultimobranchial (UT) Glands (TR: Trachea; CA: Carotid Artery; JV: Jugular Vein; H: Heart).

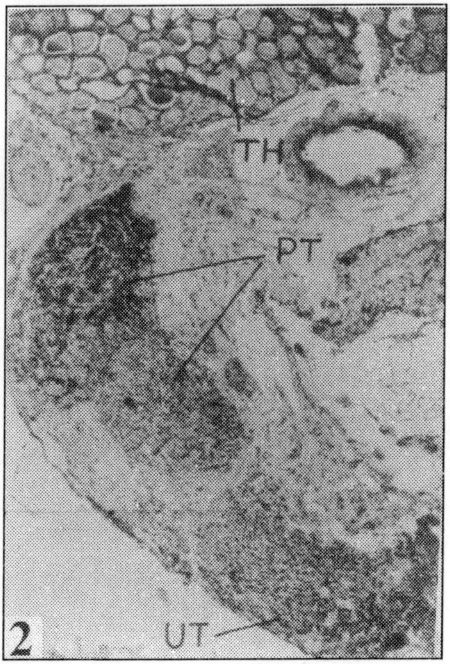

Figure 16.2: Section of the Neck Region of Grey Quail Showing Relationship of Parathyroid Gland with Thyroid and Ultimobranchial Gland. H&E. x 75.

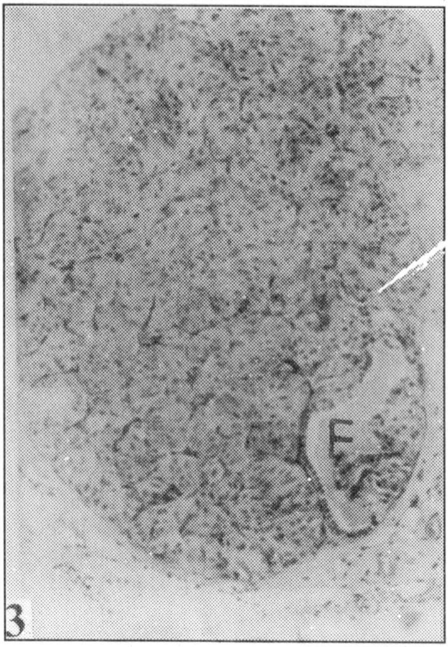

Figure 16.3: Parathyroid Gland of Male Grey Quail in April Showing Follicular Arrangement of Chief Cells. Note the follicle (F) with homogenous colloid-like material in the lumen. H&E. x 200.

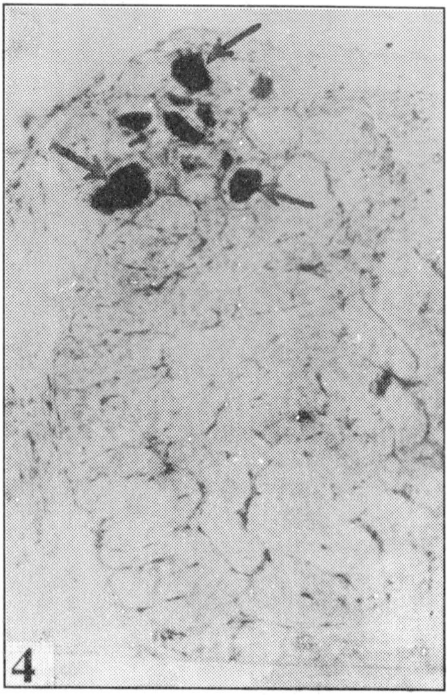

Figure 16.4: Parathyroid Gland of Female Grey Quail in June Showing Follicular Arrangement of Chief Cells and AF-Positive Material in the Lumina (Arrow). H&E. x 200.

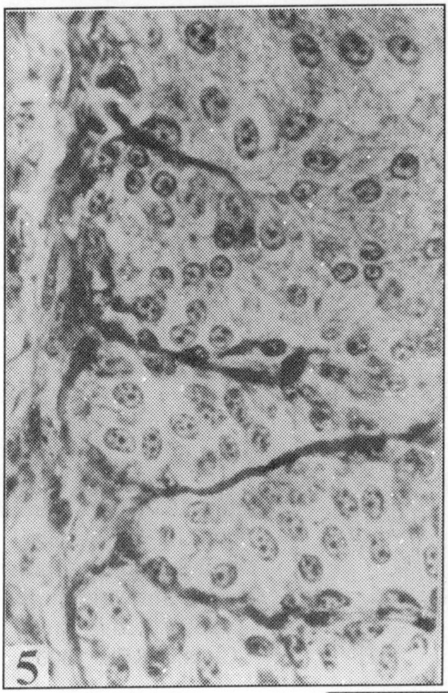

Figure 16.5: Parathyroid Gland Female Grey Quail in May Exhibiting the Connective Tissue Sheath and Arrangement of Chief Cells in Cords. Mark the hyperplasia in the gland. H&E.x 800.

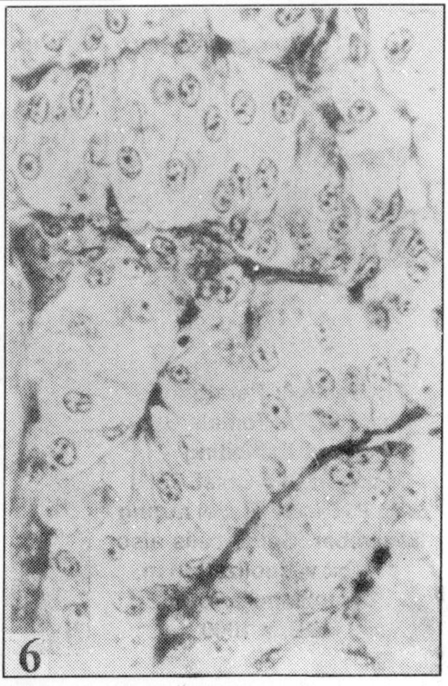

Figure 16.6: Parathyroid Gland of Female Grey Quail in July Exhibiting Cord-Like Arrangement of Hypertrophied Chief Cells. H&E. x 1000.

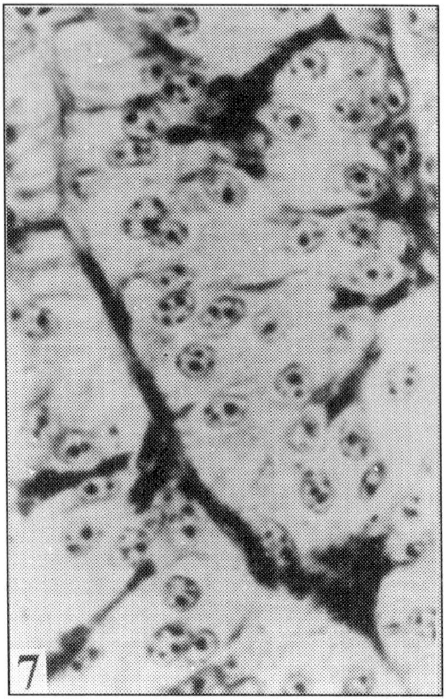

Figure 16.7: Parathyroid Gland of Female Grey Quail during Breeding Season Showing Single Type of Active Chief Cells. H&E.x 1000.

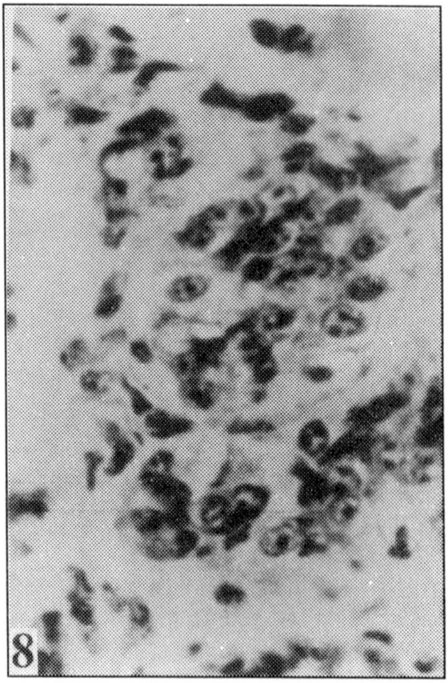

Figure 16.8: Parathyroid Gland of the Female Grey Quail Exhibiting Degenerating Chief Cells with Pycnotic Nuclei during December. Some cells also exhibit vacuolization in cytoplasm (arrow). H&E. x 1000

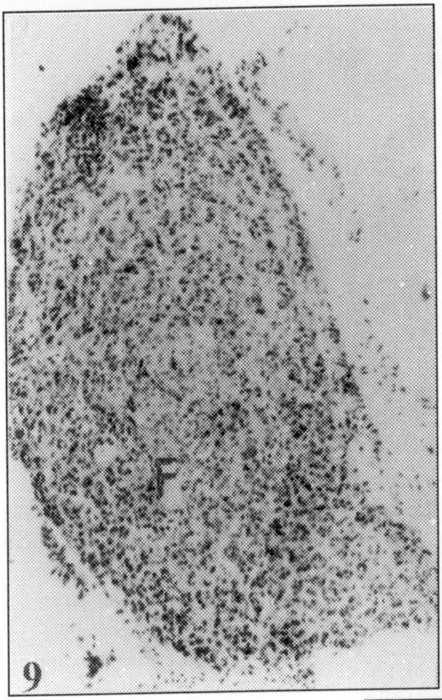

Figure 16.9: Ultimobranchial Gland (UBG) of the Female Grey Quail during April Showing Epithelial Cells Arranged in Cords and Follicles (F). Mark the discontinuous framework of connective tissue stroma. H&E. x 200.

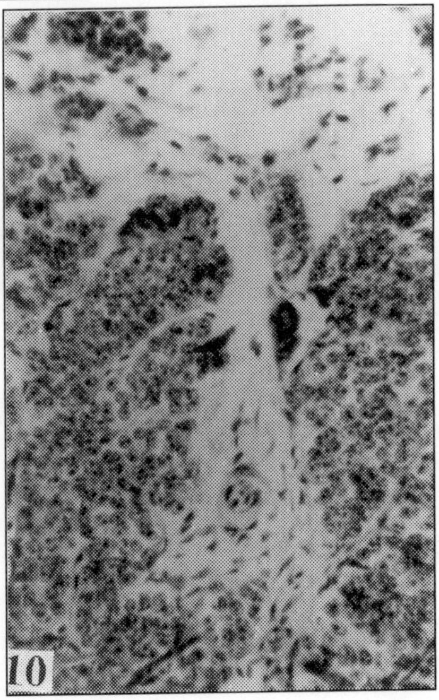

Figure 16.10: UBG of the Female Grey Quail in May Depicting Hyperplastic Glandular Epithelial Cells. H&E. x 400.

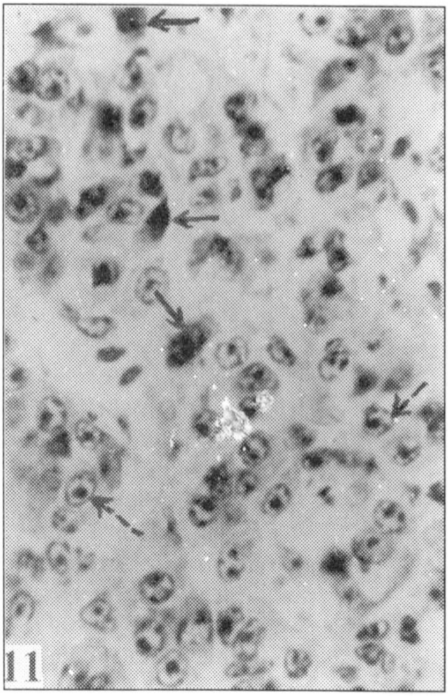

Figure 16.11: UBG of the Female Grey Quail in June Exhibiting Dark (Arrow) and Light (Broken Arrow) Parenchymal Cells. H&E. x 400.

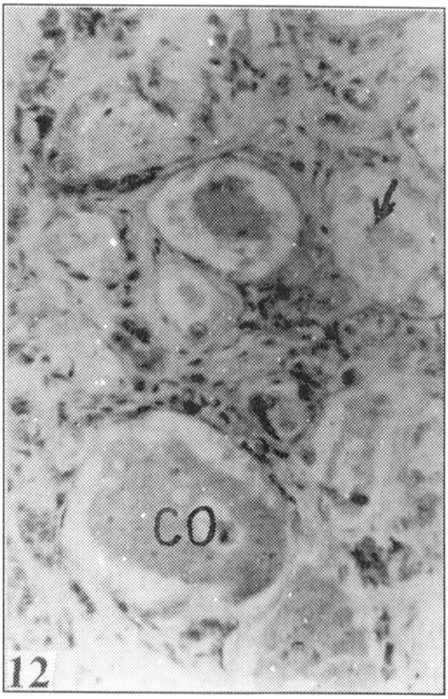

Figure 16.12: UBG of the Female Grey Quail in July Showing Large Number of Follicles Containing Varying Amount of Colloid Material (CO). Mark the presence of desquamated cells in the follicular lumia (arrow). H&E.x 400.

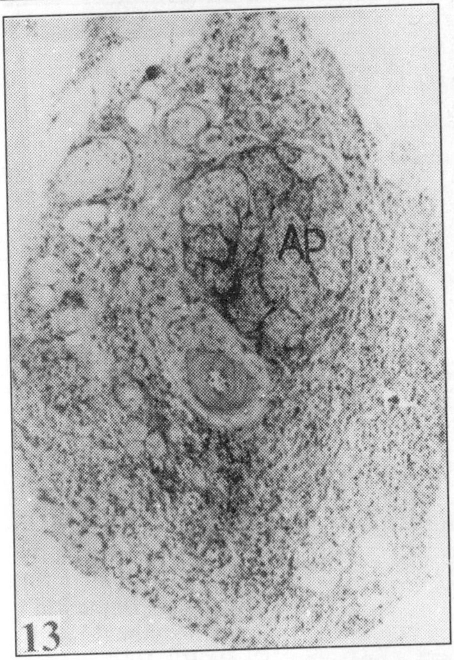

Figure 16.13: UBG of the Female Grey Quail during August Displaying Accessory Parathyroid Tissue (AP). H&E.x 150.

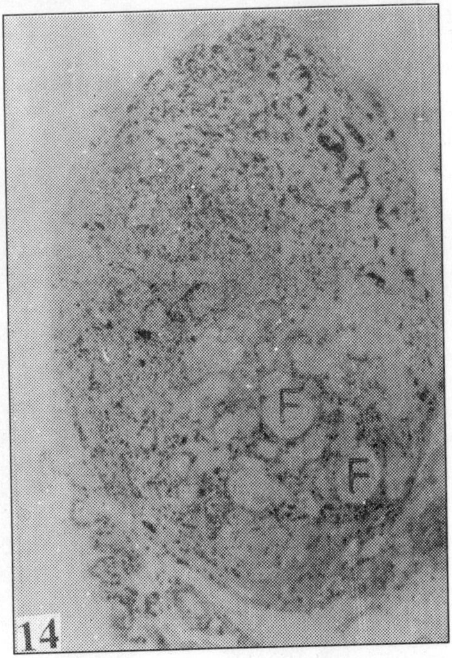

Figure 16.14: UBG of the Male Grey Quail in October Exhibiting Large Number of Follicles (F) with Homogenous Colloid Material. H&E.x 140.

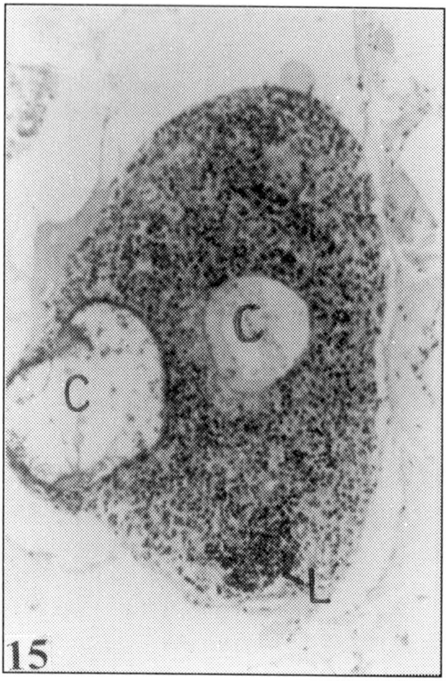

Figure 16.15: UBG of the Female Grey Quail during December Depicting Lymphocytes (L) and Cysts (C). H&E. x 200.

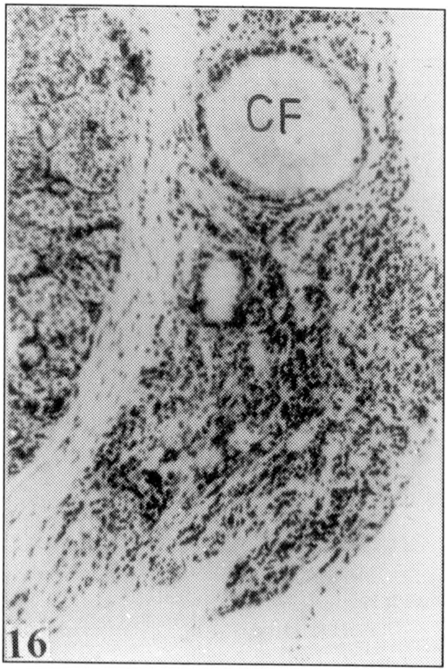

Figure 16.16: UBG of the Female Grey Quail in January (Peak Winter) Showing of Cystic Follicles (CF). H&E. x 200.

and Pandey, 1990), however, such cyclic changes, though of less magnitude, are also observed in homeothermic animals (Roth and Schiller, 1976; Clark *et al.*, 1986). Varieak (1956) reported that birds parathyroid were vacuolated in winter and similar changes also occurred in parathyroid of the ground squirrel and hedgehog (Kayser, 1961). The cytoplasm of chief cells of the grey quail usually lacked secretory granules but during hyperactivity, it exhibited the presence of granules that stained pink with eosin, a feature consistent with a low level of PTH secretion in birds (Kenny, 1986). The histological observation also revealed uneven staining of the parenchymal chief cells as some of them stained dark giving false appearance of two cell types. Electron microscopic studies of the avian parathyroid have conclusively shown the presence of single type of cells in two different secretory phases–the dark and light cells (Nevalainen, 1969; Fuji and Isono, 1972; Isono, 1972, 1974, 1975; Narbaitz, 1972; Shizuko, 1974; Fuji, 1975; Chan, 1977; Isono *et al.*, 1979). Parathyroid glands of the grey quail exhibited heightened activities such as hypertrophy (increase in cell and nuclear diameter) and hyperplasia during breeding season which was more marked in females than in males (Figures 16.5–16.7) similar to that of egg-laying hen (Marine, 1913; Taylor, 1971). van der Vield *et al.* (1984) have also recorded enhanced PTH levels in hen during eggshell formation. While in winter (November–January), parathyroid gland of the grey quail showed sign of hypoactivity (decrease in cell and nuclear diameter) and degeneration (Figure 16.8; Table 16.2). The cord-like arrangement of chief cells in parathyroid gland was disturbed and cytoplasm became vacuolated, possessed pycnotic nuclei and lacked cytoplasmic granules (Figure 16.8). This is in agreement with the findings of Varieak (1956) and Kaysér (1961). Though presence of group of ultimobranchial cells in parathyroid gland has been reported by Le Douarin and Le Leivre (1970), no such cells was encountered within parathyroid gland of the grey quail.

Ultimobranchial gland of *Coturnix coturnix coturnix* comprised epithelial cells arranged in cords and follicles (Figures 16.9 and 16.10) similar to those reported in other avian species (Chan *et al.*, 1969; Stoeckel and Porte, 1969, 1970; Hodges, 1970; Be'langer, 1971; Isler, 1973; Swarup and Das, 1974; Chan, 1978; Robertson, 1986). However, recent immunofluorecent studies have localized CT immunoreactivity in the cells arranged in cords whereas follicular epithelial cells and colloid displayed either mild CT immunoreactivity or even lacked the response (Yamada *et al.*, 1983; Treilhou-Lahille *et al.*, 1984). The main glandular cells in UBG of the grey quail were of two light and dark types (Figure 16.11) probably representing two secretory phases of the same cell (Be'langer, 1971; Swarup and Das, 1974). The light cells are said to be secretory or active in nature whereas the dark cells having dense cytoplasm represent inactive phase (Be'langer, 1971; Swarup and Das, 1974). Ultrastructural studies of UBG in chickens (Chan *et al.*, 1969; Malmquist *et al.*, 1968; Stoeckel and Porte, 1969), fowl, pigeon and turtle doves (Stoeckel and Porte, 1970) confirm that the avian UBG possessed two types of glanduar parenchyma cells and duct-like follicles. However, Moseley *et al.* (1968) did not find 'true follicles' in the ultmobranchial gland of the chicken and pigeon.

During breeding season (April–July), the UBG of female grey quail showed cell and nuclear hypertrophy as well as hyperplasia (Figure 16.10–16.12; Table 16.3) similar to the laying hens (Urist, 1967), the hyperactivity being more pronounced in females than in males. The changes observed in the UBG during breeding season correspond to the alterations in the plasma Ca and Pi levels (Tables 16.1 and 16.3). The UBG of *Coturnix coturnix coturnix* exhibited declined activity during September-October with decrease in the cell and nuclear size and marked increase of the follicles with homogenous colloid-like material (Figure 16.14). During peak winter (November–January), the UBG possessed numerous large cystic follicles with desquamated epithelial cells in their lumina (Figures 16.15 and 16.16). There were lymph tissue as well as pseudoeosinophic invasions in the UBG of the grey quail during the peak winter (Figure 16.15). Heightened activity during breeding season and hypoactivity as well as degenerative changes in UBG have also been recorded in the yellow monitor, *Varanus flavescens* (Pandey, 1990b). The results may be explained as during breeding season, endogenous level of estrogen is increased resulting in elevation of blood Ca (Simkiss, 1967; Sturkie, 1976; Scanes *et al.*, 1982; Dacke *et al.*, 1979). It appears that to counteract the hypercalcemia caused by high levels of circulating estrogen, UBG of the female birds might be activated for production of CT, the hypocalcemic or anti-hypercalcemic hormone. Similar activation in the gland of male grey quails was also observed after estradiol benzoate induced hypercalcemia (Dhande *et al.*, 1997d, 1999).

Table 16.3: Annual Changes in the Cell and Nuclear Diameter (μm) of the ultimobranchial gland of *Coturnix coturnix coturnix*

Months	*Female*		*Male*	
	Cell Diameter	*Nuclear Diameter*	*Cell Diameter*	*Nuclear Diameter*
January	5.36±0.54	3.86±0.26	5.63±0.30	3.37±0.12
February	5.48±0.32	3.89±0.28	5.36±0.22	13.16±0.16
March	6.28±0.42	4.11±0.33	5.72±0.40	3.55±0.24
April	6.46±0.70	4.48±0.60	6.36±0.34	4.04±0.18
May	0.64±0.34	4.56±0.32	5.83±0.42	4.44±0.16
June	6.66±0.40	4.50±0.42	5.67±0.28	4.26±0.10
July	6.85±0.37	4.32±0.10	5.68±0.26	3.96±0.40
August	5.97±0.26	3.36±0.30	5.30±0.22	3.48±0.30
September	5.81±0.44	3.66±0.45	5.52±0.38	3.68±0.28
October	5.38±0.32	3.82±0.56	5.10±0.12	3.30±0.15
November	5.44±0.30	3.12±0.16	4.43±0.24	3.38±0.21
December	5.30±0.28	3.30±0.26	5.00±0.26	3.21±0.16
Annual mean	5.96±0.39	3.91±0.33	5.46±0.28	3.63±0.20

Values are mean±SE for 5 samples.

Inclusion of accessory parathyroid tissue is observed in UBG of the grey quail (Figure 16.13). The presence of such accessory parathyroid tissue has also been reported in the UBG of several avian species (Terni, 1924; Dudley, 1942; Chan *et al.*, 1969; Hurst and Newcomer, 1969; Swarup and Das, 1974). Besides UBG, the accessory parathyroid tissue has also been observed in the caudal lobe of thymus of the birds (Nonidez and Goodale, 1927; Dudley, 1942), such tissue was not observed in the thymic lobe of *Coturnix coturnix coturnix*. Occasionally, the lymph tissue has been encountered in UBG of the grey quail similar to other species of birds (Be'langer, 1971; Swarup and Das, 1974) but the functional aspect of this is not known. Pseudoeosinophils have been found occasionally in UBG of the grey quail (Figure 16.15). Though such cells are also reported in chicken (Be'langer, 1971) and *Stumus contra* (Swarup and Das, 1974), no specific significance related to endocrine function of this structure has been assigned. However, Dudley (1942) remarked that UBG might act as temporary hemopoetic centre for production of these pseudoeosinophils.

There exist reports about the presence of hypocalcemic activity in the thyroid and parathyroid glands of birds (Moseley *et al.*, 1968; Hoyt *et al.*, 1972; Kapoor and Chhabra, 1981) but we could not locate C cells in these glands of *Coturnix coturnix coturnix* even after specific stains. Furthermore, thyroid extract was ineffective in altering plasma Ca level in the grey quail (Suryawanshi *et al.*, 1995).

The source Ca for eggshell formation in birds is either diet or medullary bone (Simkiss, 1967; Dacke, 1979; Dacke *et al.*, 1993). In the present study, only females showed marked hypercalcemia during breeding season and this may be due to formation of the medullary bone (Taylor, 1971; Dacke, 1979; Dacke *et al.*, 1993). It appears that the female sex steroid hormone synergistically acts with the parathyroid hormone and increases resorption of Ca from the bones (Dacke *et al.*, 1993; Dacke, 2000). The grey quail showed hyperactivity in the ovaries during breeding season (Saxena and Saxena, 1980) suggesting increase in endogenous estrogen level due to increased number of rapidly developing ova. Similarly, exogenous administration of estrogen in egg-laying period increased the plasma Ca and P*i* levels (Riddle and McDonald, 1945; Urist *et al.*, 1960; Simkiss, 1961, 1967; Dacke, 1979). The parathyroid gland of the female grey quail also showed signs of hyperactivity during breeding season. It may be concluded that both estrogen and parathyroid increased mobilization of Ca and Pi from the long bones resulting in elevation in level of these electrolytes which may be transported through blood to the uterus for egg-shell formation. It appears that to counteract the hypercalcemia caused by high levels of circulating estrogen, UBG of the female birds might be activated for production of CT. There exist reports on the elevation of CT levels during breeding period of fishes, probably to protect the skeleton from excessive Ca mobilization (Bjornsson *et al.*, 1986, 1989).

Acknowledgement

The senior author (RRD) is grateful to the University Grants Commission, New Delhi for financial support.

References

Aurbach, G.D., Marx, S.J. and Spiegel, A.M., 1992. Parathyroid hormone, calcitonin, and the calciferols. In: *Williams Handbook of Endocrinology,* (Eds.) J.D. Wilson and D.W. Foster. W.B. Saunders, Philadelphia, p. 397–1476.

Baldini, J.T. and Zarrow, M.X., 1952. Estrogen and serum calcium levels in Bobwhite quail. *Poultry Sci.,* 31: 800–804.

Bartet, J.-P., 1982. Comparative physiology of calcitonin. In: *Endocrinology of Calcium Metabolism,* (Ed.) J.A. Parson. Raven Press, New York, pp. 235–270.

Bates, R.F.L., Bruce, J. and Care, A.D., 1969. Measurement of calcitonin secretion in goose. *J. Endocr.,* 45: 14–15.

Be'langer, L.F., 1971. The ultimobranchialgland of birds and effects of nutritional variarions. *J. Exp. Zool.,* 178: 125–138.

Benoit, J., 1950. Les glandes endocrines. In: *Traite de Zoologie,* (Ed.) P.P. Grasse. Oiseaux, Masson, Paris, pp. 290–334.

Benoit, J., Clavert, T. and Cabannes, R., 1944a. Etude histo-physiologique de la Parathyroid du canard domestique. I. Conditionnment partiel de son activite par la Prehypophysie. *Compo Rend Soc. Biol.,* 138: 1071.

Benoit, J., Clavert, T. and Cabannes, R., 1944b. Etude histo-physiologique de la Parathyroid du canard domestique. II. Modifications histologiques determinees. Par le traitement o'la folliculine. *Compo Rend Soc. Biol.,* 138: 1074.

Bentley, P.J., 1982. *Comparative Vertebrate Endocrinology.* Cambridge University Press, Cambridge.

Bjornsson, B.Th., Haux, C., Forlin, L. and Deftos, L.J., 1986. The involvement of calcitonin in the reproductive physiology of the rainbow trout. *J. Endocr.,* 108: 17–23.

Bjornsson, B.Th., Haux, C., Bern, H.A. and Deftos, L.J., 1989. Estradiol-17β increases plasma calcitonin in salmonid fish. *Endocrinology,* 125: 1754–1760.

Boelkins, J.N. and Kenny, A.D., 1973. Plasma calcitonin levels in Japanese quail. *Endocrinology,* 92: 1754–1760.

Calamy, H. and Barlet, J.P., 1970. Etude des effects de la calcitonin exogene sur la calcemic de la poule. *C.R. Acad Sci. (Paris),* 271D: 2152–2156.

Chan, A.S., 1977. Ultrastructure of parathyroid glands in the chicks. *Acta Anal.,* 97: 205–212.

Chan, A.S., 1978. Ultrastructure of the ultimobranchial follicles of the laying chicken. *Cell Tissue Res.*, 195: 309–316.

Chan, A.S., Cipera, J.D. and Be'langer, L.F., 1969. The ultimobranchial gland of the chick and its response to high calcium diet. *Rev. Can. Biol.*, 2: 19–31.

Clark, N.B. and Wideman, R.F., 1977. Renal excretion of phosphate and calcium in parathyroidectomized starlings. *Am. J. Physiol.*, 233: F138–FI44.

Clark, N.B., Kaul, K. and Roth, S.J., 1986. The parathyroid glands. In: *Vertebrate Endocrinology: Fundamentals and Biomedical Implications, Vol. 1: Morphological Considerations,* (Eds.) P.K.T. Pang and M.P. Schreibman. pp. 207–234. Academic Press, New York and San Diego.

Copp, D.H., Cockroft, D.W. and Kueh, Y., 1967. Calcitonin from the ultimobranchial glands of dogfish and chicken. *Science,* 158: 924–925.

Cutler, G.B., Habener, J.F. and Potts, J.T., 1977. Biosynthesis and secretion of calcitonin by ultimobranchial glands. *Endocrinology,* 100: 537–548.

Dacke, C.G., 1979. *Calcium Regulation in Submammalian Vertebrates.* Academic Press, London.

Dacke, E.G., 2000. The parathyroids, calcitonin, and vitamin D. In: *Avian Physiology,* 5th Edn. (Ed.) P.D. Sturkie. Academic Press, London and New York, pp. 473–488.

Dacke, E.G., Arkle, S., Cook, D.J., Warmstone, I.M., Jones, S. and Askal, Z.A., 1993. Medullary bones and avian calcium regulation. *J. Exp. Biol.,* 184: 63–88.

Dacke, E.G., Danks, J., Caple, I. and Flik, G., 1996. *The Comparative Endocrinology of Calcium Regulation.* The Journal of Endocrinology Press, Bristol.

Dhande, R.R., Suryawanshi, S.A. and Pandey, A.K., 1995a. Parathyroid and ultimobranchial glands of the grey quail, *Coturnix coturnix coturnix,* in response to experimental hypercalcemia. *Proc. Natl. Acad. Sci. (India),* 66B: 81–88.

Dhande, R.R., Suryawanshi, S.A. and Pandey, A.K., 1995b. Parathyroid and ultimobranchial glands of the grey quail,

Coturnix coturnix coturnix, in response to experimental hypocalcemia. *J. Adv. Zool.,* 16: 96–103.

Dhande, R.R., Suryawanshi, S.A. and Pandey, A.K., 1997a. Effects of parathyroid extract administration on plasma calcium and inorganic phosphate levels in intact and parathyroidectomized grey quail, *Coturnix coturnix coturnix* Linnaeus. *J. Adv. Zool.,* 18: 46–50.

Dhande, R.R., Suryawanshi, S.A. and Pandey, A.K., 1997b. Effects of salmon calcitonin administration on plasma calcium and inorganic phosphate levels of the intact and parathyroidectomized grey quail, *Coturnix coturnix coturnix* Linnaeus. *Indian J. Anim. Sci.,* 67: 906–909.

Dhande, R.R., Suryawanshi, S.A. and Pandey, A.K., 1997c. Response of plasma calcium and inorganic phosphate levels of the grey quail, *Coturnix coturnix coturnix,* to salmon calcitonin adminstration. *J Adv.Zool.,* 19: 37–40.

Dhande, R.R., Suryawanshi, S.A. and Pandey, A.K., 1997d. Response of plasma calcium and inorganic phosphate levels of the intact and parathyroidectomized grey quail, *Coturnix coturnix coturnix,* to estradiol benzoate administration. *Proc. Zool. Soc., Calcutta,* 50: 120–128.

Dhande, R.R., Suryawanshi, S.A. and Pandey, A.K., 1998. Response of plasma calcium and inorganic phosphate levels, parathyroid and ultimobranchial glands of the grey quail, *Coturnix coturnix coturnix,* to sodium fluoride administration. *Proc. Zool. Soc., Calcutta,* 51: 101–110.

Dhande, R.R., Suryawanshi, S.A. and Pandey, A.K., 1999. Effects of estradiol benzoate adminstration on plasma calcium and inorganic phosphate levels and parathyroid gland of male grey quail, *Coturnix coturnix coturnix* Linnaeus. *J. Adv. Zool.,* 20: 80–84.

Doiron, W.F., 1973. Studies on the structure and function of the parathyroid glands of the European starlings, *Sturnus vulgaris. Amer. Zool.,* 13: 1286–1287.

Dubewar, D.M. and S.A. Suryawanshi, 1978. Seasonal variations in the parathyroid gland of the lizard, *Uromastix hardwickii* (Gray). *Z. mikrosk.-Anat. Forsch.,* 92: 298–304.

Dubewar, D.M., 1978. Function of parathyroid gland in oestrogen-produced hypercalcemia in the lizard, *Uromastix hardwickii. Indian J. Exp. Biol.,* 16: 1324–1325.

Dudley, J., 1942. The development of the ultimobranchial body of the fowl, *Gallus domesticus. Am. J. Anal.,* 71: 65–97.

Feinblatt, J.D., 1982. The comparative physiology of calcium regulation in submammalian vertebrates. *Adv. Comp. Physiol. Biochem.,* 8: 73–110.

Fuji, H., 1975. Electron microscopic studies of the parathyroid glands of the domestic fowl, *Gallus domesticus,* during various developmental stages. *Acta Sch. Med Uni.,* 23: 545–574.

Fuji, H. and Isono, H., 1972. Ultrastructural observations on the parathyroid glands of the hen (*Gallus domesticus*). *Arch. Histol. Jpn.,* 34: 155–165.

Gonnerman, W.A., Breitenbach, R.P., Erfling, W.F. and Anast, C.S., 1972. An analysis of ultimobranchial gland function in the chicken. *Endocrinology,* 91: 1423–1429.

Honma, T., Watanabe, M., Hirose, S., Kanai, A., Kanagawa, K. and Matsuo, H., 1986. Isolation and determination of amino acid sequence of chicken calcitonin I from chicken ultimobranchial glands. *J. Biochem., Tokyo,* 100: 459–467.

Hodges, R.D., 1970. The structure of the fowl's ultimobranchial gland. *Ann. Biol. Anim. Biochem. Biophys.,* 10: 255–279.

Hoyt, R.F., Tashjian, A.H. and Hamilton, D.W., 1972. Distribution of thyroid, parathyroid and ultimobranchial hypocalcemic factor in birds: Thyroid and ultimobranchial calcitonins in pigeons and pullet. *Endocrinology,* 91: 770–783.

Hurst, I.G. and Newcomer, W.S., 1969. Functional accessory parathyroid tissue in ultimobranchial bodies of chickens. *Proc. Soc. Exp. Biol. Med.,* 132: 555–557.

Isler, H., 1973. Fine structure of the ultimobranchial body of the chick. *Anat. Rec.,* 177: 441–460.

Isono, H., 1960. Histological study of the parathyroid gland in the toad (*Bufo vulgaris japonicus*). *Acta Sch. Med Gifu,* 8: 277–293.

Isono, H., 1972. Ultrastuctural observation of the parathyroid gland of the hen. *Arch. Histol., Jpn.,* 34: 155–165.

Isono, H., 1974. Electron microscopic study of parathyroid gland of the pet birds. *Acta Sch. Med. Uni. Gifu*, 22: 300–325.

Isono, H., 1975. Electron microscopic studies of parathyroid glands of the domestic fowl. *Acta Sch. Med. Uni. Gifu*, 23: 545–574.

Isono, H., Isono, S. and Komura, H., 1959. On the seasonal cyclic changes of the glycogen contents of the parathyroid gland of the toad (*Bufo vulgaris japonicus*). *Acta Sch. Med Gifu*, 7: 1696–1706.

Isono, H., Shizuko, S. and Kazuko, H., 1979. Ultrastuctural of the parathyroid gland of the quail, *Coturnix coturnix japonicus. Okajimas Folia Anat. Jpn.*, 55: 241–254.

Kameda, Y., 1968. Parafollicular cells of the thyroid as studied with Devenport's silver impregnation. *Arch. Histol. Jpn.*, 30: 83–91.

Kapoor, A.S. and Chhabra, S.K., 1981. Relative physiological activity of calcitonin in the thyroid and ultimobranchial glands of pigeon, *Columbia livia* Gmelin. *Gen. Comp. Endocrinol.*, 44: 307–313.

Kayser, C.H., 1961. *The Physiology of Natural Hibernation*. Pergaman Press, Oxford.

Kenny, A.D., 1986. Parathyroid and ultimobranchial glands. In: *Avian Physiology*, 4th edn. (Ed.) P.O. Sturkie. Springer-Verlag, New York, pp. 466–478.

Le Douarin, N. and Lievre, G., 1970. Demonstration de l'origine neurale cellules a calcitonine du corps ultomobranchial chez l'embryon de Poulet. *C.R. Acd. Sci., Paris*, 270D: 2857–2860.

Loyd, J.W., Peterson, R.A. and Collins, W.E., 1970. Effects of an avian ultimobranchial extract in domestic fowl. *Poultry Sci.*, 49: 1117–1120.

Luck, M., Sommerville, B. and Scannes, C., 1980. The effect of eggshell calcification on the response of plasma calcium activity to parathyroid hormone and calcitonin in the domestic fowl (*Gallus domesticus*). *Comp. Biochem. Physiol.*, 65A: 151–154.

Malmquist, E., Ericson, L.E., Almquist, S. and Ekholm, R., 1968. Granulated cells, uptake of amine precursors, and calcium-lowering activity in the ultimobranchial body of the domestic fowl. *J. Ultrstruc. Res.*, 23: 457–461.

Marine, D., 1913. Parathyroid hypertrophy and hyperplasia in fowls. *Proc. Soc. Exp. Biol. Med.*, 11: 118.

Mori, J.G. and George, J.C., 1978. Seasonal changes in serum levels of certain metabolites, uric acid and calcium in migratory Canada goose (*Branata canadensis interior*). *Comp. Biochem. Physiol.*, 59B: 263–269.

Moseley, I.M., Matthews, E.W., Breed, R.H., Galante, L., Tse, A. and MacIntyre, I., 1968. The ultimobranchial origin of calcitonin. *Lancet*, 1: 108–110.

Mueller, W.J., Hall, K.L., Maurer, C.A. and Joshua, J.G., 1973. Plasma calcium and inorganic phosphate response of laying hens to parathyroid hormone. *Endocrinology*, 92: 853–856.

Narbaitz, R., 1972. Submicroscopical aspects of chick embryo parathyroid glands. *Gen. Comp. Endocrinol.*, 19: 253–258.

Nevalainen, T., 1969. Fine structure of the parathyroid gland of the laying hen (*Gallus domesticus*). *Gen. Comp. Endocrinol.*, 12: 561–567.

Nieto, A., Moya, F. and Candella, J.L.R., 1973. Isolation and properties of two calcitonins from chicken ultimobranchial glands. *Biochim. Biophys. Acta*, 322: 383–391.

Nonidez, I.F. and Goodale, H.D., 1927. Histological studies on endocrines of chickens derived of ultraviolet light. Parathyroid. *Am. J. Anat.*, 38: 310–347.

Nunez, E.A. and Gershon, M.D., 1978. Cytophysiology of parafollicular cells. *Intern. Rev. Cytol.*, 52: 1–80.

Pandey, A.K., 1990a. Seasonal changes in the ultimobranchial body, parathyroid gland, serum calcium and inorganic phosphate levels of *Bufo melanostictus*. *Biol. Struct. Morphogen., Paris*, 3: 101–106.

Pandey, A.K., 1990b. Seasonal changes in the ultimobranchial body, serum calcium and inorganic phosphate levels of *Varanus flavescens*. *Biol. Struct. Morphogen., Paris*, 3: 66–70.

Pandey, A.K., 1991. Endocrinology of calcium regulation in reptiles: A comparative aspect in lower vertebrates. *Biol. Struct. Morphogen., Paris*, 3: 159–176.

Pandey, A.K., 1992. Endocrinology of calcium metabolism in amphibians with emphasis on the evolution of hypercalcemic regulation in tetrapods. *Biol. Struct. Morphogen, Paris,* 4: 102–126.

Pandey, A.K. and Swarup, K., 1989. Seasonal changes in the ultimobranchial body and parathyroid gland of *Bufo andersoni* in relation to serum calcium and inorganic phosphate levels. *Bolm. Fisiol. Anim., Sao Paulo,* 13: 55–65.

Pang, P.K.T. and Schreibman, M.P., 1989. *Vertebrate Endocrinology: Fundamentals and Biomedical Implications, Vol. 3: Regulation of Calcium and Phosphate.* Academic Press, San Diego.

Riddle, O. and McDonald, M.R., 1945. The partition of plasma calcium and inorganic phosphorus in estrogen treated normal and parathyroidectomized pigeons. *Endocrinology,* 36: 41–50.

Riddle, O. and Reinhart, W.H., 1926. Studies on the physiology of reproduction in birds. *Am. J. Physiol.,* 76: 660–687.

Robertson, D.R., 1986. The ultimobranchial body. In: *Vertebrate Endocrinology: Fundamentals and Biomedical Implications. Vol. 1: Morphological Considerations,* (Eds.) P.K.T. Pang and M.P. Schreibman. Academic Press, New York and San Diego, pp. 235–259.

Romanoff, A.L., 1960. *The Avian Embryo: Structural and Functional Development.* MacMillan, New York.

Roth, S.I. and Schiller, A.L., 1976. Comparative anatomy of parathyroid glands. In: *Endocrinology, Vol. 7,* (Eds.) R.O. Greep, E.B. Estwood and G.D. Aurbach. American Physiological Society, Washington, pp. 281–311.

Saxena, A. and Saxena, A.K., 1980. Gonadal cycle and secondary sex characters of the grey quail, *Coturnix coturnix coturnix.* In: *The Second All India Symposium on Life Sciences.* Institute of Science, Nagpur, p. 34.

Scanes, E.G., Ottinger, M.A., Kenny, A.D., Bulthazart, J., Crowshaw, J. and Jones, I. Chester, 1982. *Aspects of Avian Endocrinology: Practical and Theoretical Implications.* Texas Technical University Press, Lubbock.

Schrier, J.E. and Hamilton, H.L., 1952. An experimental study of the origin of the parathyroids and thymus glands in the chick. *J. Exp. Zool.*, 119: 165–178.

Shizuko, S., 1974. Electron microscopic study of parathyroid glands of the pet birds. Fine structure of the parathroid glands of the love birds and Australian love birds. *Acta Sch. Med. Uni. Gifu*, 22: 300–325.

Sidky, Y., 1965. Histological studies on the parathyroid gland of lizards. *Z. Zellforsch.*, 65: 760–769.

Simkiss, K., 1961. Calcium metabolism and avian reproduction. *Biol. Rev. Cambridge Phil. Soc.*, 36: 321–396.

Simkiss, K., 1967. *Calcium in Reproductive Physiology.* Chapman and Hall, London.

Smith, G.C., 1945. Techniques for parathyroidectomy in pigeons. *Anat. Rec.*, 92: 81–87.

Solcia, E.G., Capella, C. and Vassallo, C., 1968. Lead-hematoxylin as a stain for endocrine cells. *Histochemie*, 20: 116–126.

Speers, G.M., Percy, D.Y.E. and Brown, D.M., 1970. Effect of ultimobranchialectomy in the laying hen. *Endocrinology*, 87: 1292–1297.

Stoeckel, M.E. and Porte, A., 1969. Etude ultrastructurale des corps ultimobranchiauz du poulet. I. Aspect normal et developpement embryonnire. *Z. Zellforsch.*, 94: 495–512.

Stoeckel, M.E. and Porte, A., 1970. A comparative electron microscopic study on the fowl, the pigeon and the turtle dove of the C cells localized in the ultimobranchal body and the thyroid. In: *Calcitonin 1969: Proceedings of the Symposium on Calcitonin and C-Cells*, (Ed.) S. Taylor. William Heinemann Medical Books, London, pp. 327–338.

Sturkie, P.D., 1976. *Avian Physiology*, 3rd edn. Springer-Verlag, New York.

Suryawanshi, S.A., Dhande, R.A. and Pandey, A.K., 1995. Plasma calcium and inorganic phosphate regulation in the grey quail, *Coturnix coturnix coturnix*: Role of thyroid and adrenal glands. *J. Adv. Zool.*, 16: 52–56.

Suryawanshi, S.A., Dhande, R.R. and Pandey, A.K., 1997. Effects of parathyroidectomy on plasma calcium and inorganic phosphate levels of the grey quail, *Coturnix coturnix coturnix* Linnaeus. *Nat. Acad Sci. Letters*, 20: 14–18.

Swarup, K. and Das, V.K., 1974. The ultimobranchial body of pied myna, *Stumus contra* (Linn.). *Arch. Anat. Microsc.*, 63: 203–215.

Swamp, K. and Pandey, A.K., 1990. Seasonal changes in the parathyroid gland of the yellow monitor, *Varanus flavescens* (Grey), in relation to serum calcium and inorganic phosphate levels. *Proc. Nat. Acad Sci., India*, 60B: 229–236.

Taylor, C.W., 1984. Calcium regulation in vertebrates: An overview. *Comp. Biochem. Physiol.*, 82A: 249–255.

Taylor, T.G., 1971. The parathyroid glands. In: *Physiology and Biochemistry of Domestic Fowl, Vol.* 1, (Eds.) D.D. Bell and B.M. Freeman. Academic Press, London and New York, pp. 473–480.

Terni, T., 1924. Ricerche sulla eosinofilopoiseie degli ucceli. *Arch. Ital. Anat. Embryol.*, 21: 533–561.

Treilhou-Lahille, F., Lasmoles, F., Taboulet, J., Barlet, J.-P., Milhaud, G. and Mukhtar, M.S., 1984. Ultimobranchial gland of the domestic fowl: Two types of secretory cells involved in calcitonin metabolism. *Cell Tissue Res.*, 235: 439–448.

Urist, M.R., 1967. Avian parathyroid physiology, including special comment on calcitonin. *Am. Zool.*, 7: 883–895.

Urist, M.R., 1976. Biogenesis of bone: Calcium and phosphorus in the skeleton and blood in vertebrate evolution. In: *Handbook of Physiology, Vol. 7: Endocrinology*, (Eds.) R.O. Greep, E.B. Estwood and G.D. Aurbach. American Physiological Society, Baltimore, pp. 183–213.

Urist, M.R., Deutsch, N.M., Pomperantz, G. and McLean, F.C., 1960. Interrelations between actions of parathyroid hormone and oestrogen on bone and blood in avian species. *Am. J Physiol.*, 199: 851–855.

van der Velde, J.P., Loveridge, N. and Vermeiden, J.P., 1984. Parathyroid hormone responses to calcium stress during eggshell calcification. *Endocrinology*, 115: 1901–1904.

Varieak, T.H., 1956. Uber die vakudarisation und reticularisation de parathyroidea bei einigen voglin. *Anat Anz.*, 103: 122–131.

Wallace, R.A., 1985. Vitellogenesis and oocyte growth in non-mammalian vertebrates. In: *Developmental Biology,* (Ed.) L.W. Browder. Plenum Press, New York, pp. 127–177.

Wendelaar Bonga, S.E. and Pang, P.K.T., 1991. Control of calcium regulating hormones in vertebrates: parathyroid hormone, calcitonin, prolactin and stanniocalcin. *Intern. Rev. Cytol.*, 128: 139–213.

Wittermann, E.R., Cherian, G. and Riddle, L.G., 1969. Calcitonin content of ultomobranchal body tissue in chicks, pullets, and laying hens. *Can. J. Physiol. Pharmacol.*, 47: 175–180.

Wootton, I.D.P., 1974. *Microanalysis in Medical Biochemistry*, 5th edn. Churchill Livingston, Edinburg and London.

Yamada, K., Takagi, I., Kundo, Y., Karasawa, N. and Nagatsu, I., 1983. Immunoflurescence studies on catecholamine-synthesizing enzymes and calcitonin in ultimobranchial bodies of grass parakeets and quails. *Biomed. Res.*, 4: 1–8.

Ziegler, R., Delling, G. and Pfeiffer, E.F., 1970. The secretion of calcitonin by the perfused ultimobranchial gland of the hen. In: *Calcitonin 1969: Proceedings of the Symposium on Calcitonin and C-Cells,* (Ed.) S. Taylor. William Heinemann Medical Books, London, pp. 301–310.

Chapter 17

Effect of Sublethal Heroin Administration on Serum Acid Phosphatase and Alkaline Phosphatase Activities as well as Histopathology of Liver and Kidney of *Rattus norvegicus*

S.R. Barai[1], *S.A. Suryawanshi*[2] *and A.K. Pandey*[3]

[1]*Department of Zoology, Udai Pratap College, Varanasi–221 002*

[2]*Swami Ramanand Teerth Marathwada University, Nanded–431 606*

[3]*Central Institute of Freshwater Aquaculture, Kausalyaganga, Bhubaneswar–751 002*

ABSTRACT

Serum acid phosphatase and alkaline phosphatase activities of *Rattus norvegicus* ranged between 1.02±0.08–1.08±0.06 and 10.53±0.48–10.94±0.63 KA unit/100 ml. Daily intramuscular administration of heroin (16.4 mg/kg body weight; 0.75 LD_{50} dose for 96 hours) in rats induced a progressive increase in serum acid phosphatase and alkaline phosphatase activities throughout the experimental duration. The maximum value for acid phosphatase and alkaline phosphatase activities were 2.82±0.04 and 33.71±1.04 KA unit/100 ml, respectively on day 30 of the treatment.

Liver of *Rattus norvegicus* is comprised of polygonal hepatocytes having a large nucleus with acidophilic cytoplasm. These cells were arranged in lobules which were divisible into central, middle and peripheral zones. Branches of hepatic artery and hepatic portal vein ramified throughout the liver, the larger vessels tending to be concentrated at the lobule margins in the portal tracts. Blood from the portal tracts percolated towards a central vein of each lobule via sinusoids. Sublethal heroin administration for 30 days elicited degenerative changes in the cytoplasm and nuclei of the hepatocytes. At places, signs of focal necrosis were also observed in liver of the experimental rats. No remarkable changes could be seen in the Kupffer cells. Kidney of the control rats consisted of nephrons which were divisible into renal capsule (Bowman's capsule and glomerulus) and renal tubules (proximal convoluted tubules, loop of Henle, distal convoluted tubules and collecting tubules). Sublethal heroin administration for 30 days inflicted degenerative changes in the kidney. The lobules of glomeruli showed diffused glomerulosclerosis, enlargement of Bowman's capsule and thickening of the capillary walls. Fibrosis of the hematopoetic tissue and a marked increase in mesangial matrix were observed in the kidney of heroin treated rats.

Keywords: *Heroin, Acid phosphatase, Alkaline phosphatase, Liver, Kidney, Rattus norvegicus.*

Introduction

Heroin (Brown Sugar) abuse is a burning problem of the society (Kringsholm *et al.*, 1994; Cami and Farrie, 2003). The drug is taken into systemic circulation generally by smoking, chasing, intravenous injections and sometimes orally (Martin, 1984; Sawynok, 1986; Neri-Semeri and Modesti, 1991; Cami and Farrie, 2003). Heroin (diacetylmorphine) is metabolized into 6-acetylmorphine and subsequently to morphine in the human body (Sawynok, 1986, Goldberger *et al.*, 1994; Sporer, 1999). Chronic abuse of heroin has diverse effects on various body systems due to widespread distribution of specific receptors in many tissues and organs (Martin, 1984; Sawynok, 1986). This drug produces nausea, vomiting. dizziness, mental clouding, dysphoria, constipation and increased pressure in billiary tracts. There exist reports on the effects of heroin abuses on liver, kidney, lungs, brain and a few endocrine glands of human subjects and other laboratory mammals (Brambilla *et al.*, 1980; Charuvastra *et at.*, 1980; Weller *et al.*, 1984; Novick *et al.*, 1986, 1987; Gomez-Lechon *et al.*, 1987; Kringsholm and Christoffersen, 1989; Barai, 1996, Barai *et al.*, 2004, 1996). Sublethal heroin administration resulted in elevation of serum cholesterol, serum triglyceride and serum total rapid and inflicted injuries to liver of mammals (Hussain and Kumar, 1989; El-Daly, 1994; Sharma *et al.*, 2001). Acid phosphatase and alkaline phosphatase play important role in carbohydrate metabolism (Miller and Crane, 1961) and oxidative phosphorylation (Goodman and Rothstein, 1957). As stimulation or inhibition in the activity of phosphatases disturbs the normal metabolism among vertebrates (Moriarty, 1975; Hayes, 1982), an attempt was made to record the effects of sublethal heroin administration on serum acid phosphatase (ACP) and alkaline phosphatase (ALP) activities of *Rattus norvegicus*. Histopathological changes in liver and kidney of the rat under above treatment were also studied.

Materials and Methods

Healthy male rats (*Rattns norvegicus*) weighing 200–250 g were procured from the Bombay Municipal Corporation, Mumbai. They were acclimatized under the ambient laboratory conditions (temperature 28±2°C; photoperiod 14L : 10D) for 10 days, fed on rat feed (Lipton, Bangalore) and clean water was provided *ad libitum*. 50 rats were randomly selected and divided into two equal groups-experimental and control The experimental group rats were given intramuscular injection of heroin 16.4 mg/kg body weight (0.75 LD_{50} dose for 96 hours) (the drug was initially dissolved in alcohol and the desired dose was prepared in physiological saline) while the control rats received equal volume (0.2 ml/kg body weight) of the physiological saline. Animals from both the groups were dissected on day 1, 7, 15 and 30 of the treatment. Blood samples were collected in sterilized glass syringe and serum was separated by centrifugation at 3,500 rpm. Serum acid phosphatase and alkaline phosphatase activities were assayed by the Kits manufactured by M/S Mediprob Laboratory Pvt. Ltd, Hyderabad (India) and Span Diagnostic Limited, Surat (India) based on Kind and King (1954) and King and Jagatheesan (1959) methods, respectively. The results were evaluated for statistical significance using Students 't' test.

Liver and kidney of five rats from both the groups were surgically removed on day 30 of the treatment and fixed immediately in the freshly prepared Bouin's solution. After 24 hours, the tissue was washed thoroughly in running tap water, dehydrated in ascending series of alcohol, cleared in xylene and embedded in paraffin wax at 60°C. Serial sections were cut at 6 pm and stained in hematoxylin and eosin (H and E).

Results and Discussion

Phosphatases play important role in carbohydrate metabolism and oxidative phosphorylation (Goodman and Rothstein, 1957; Miller and Crane, 1961). Acid phosphatase (ACP) is a key enzyme involved in autolytic degradation of tissues (Eto and Okhawa, 1970). Serum acid phosphatase and alkaline phosphatase activities of the control *Rattus norvegicus* ranged between 1.02±0.08–1.08±0.06 and 10.53±0.48–10.94± 0.63 KA unit/100 ml which were in the normal range of mammals. In the present study, we found that the daily intramuscular administration of sublethal dose of heroin in rat induced a progressive increase in serum acid phosphatase and alkaline phosphatase activities throughout the experimental duration. The maximum value for acid phosphatase and alkaline phosphatase activities were 2.82±0.04 and 33.71±1.04 KA unit/100 ml, respectively on day 30 of the treatment. Pedrazzoni *et al.* (1993) studied the effects of chronic heroin abuse on bone and mineral metabolism and found that bone alkaline phosphatase and osteocalcin values were not significantly different from the control. Novick *et al.* (1981, 1986) reported lower levels of serum alkaline phosphatase among the abusers of alcohol and parenteral drugs including heroin.

Hepatocytes of *Rattus norvegicus* were polyginal in shape, having a large round nucleus with acidophilic cytoplasm. These cells were arranged in lobules which were divisible into central, middle and peripheral zones. The roughly hexagonal structure of the lobule maximized contact of hepatocytes with blood flowing through the liver. Within the lobules were seen the laminae or plates of hepatic cells separated by sinusoids. The plates of hepatocytes were usually one walled thick and anastomosed to form a structure like a honeycomb (Figure 17.1). Branches of the hepatic artery and hepatic portal vein ramified throughout the liver, the large vessels tending to be concentrated at the lobule margins in the portal tracts. Blood

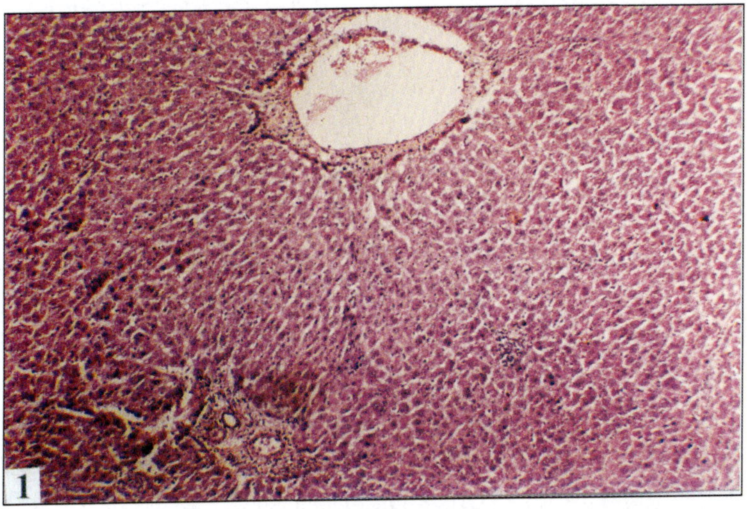

Figure 17.1: Liver of the Control *Rattus norvegicus* Showing the Central Vein and Arrangement of Parenchymal Cells (H and E. x 100).

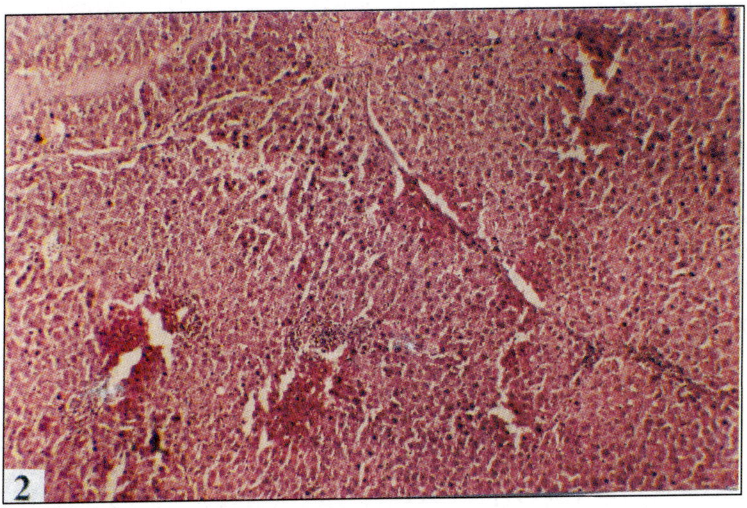

Figure 17.2: Liver of the Rat Treated with Sublethal Dose of Heroin for 30 Days Exhibiting Patchy Areas of Necrosis (H and E. x 100).

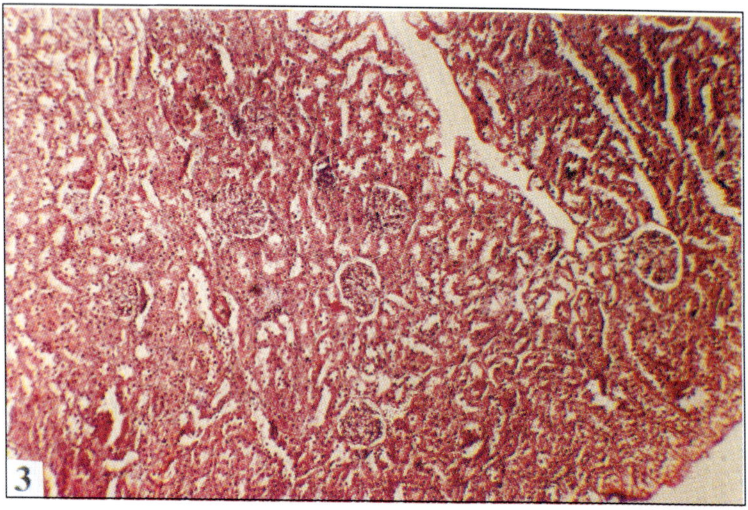

Figure 17.3: Kidney of Control *Rattus norvegicus* Showing Normal Glomerulus and Bowman's Capsule (H and E. x 100).

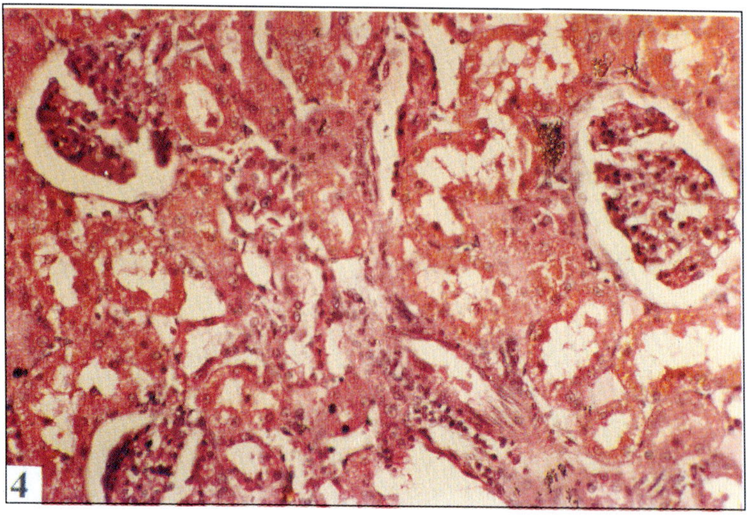

Figure 17.4: Kidney of the Rat Treated with Sublethal Dose of Heroin for 30 Days Exhibiting Damaged Glomerulus and Bowman's Capsule (H and E. x 240).

from the portal tracts percolated towards a central vein of each lobules via sinusoids. The sinusoides were lined by Kuffer cells. The portal tracts or triads contained three main structures-the portal vein, the hepatic artery and the bile duct in very minimal fibrous stroma. Sublethal heroin administration in the rats for 30 days elicited damage to hepatocytes in the central, middle as well as peripheral zones. The liver showed patchy areas of liquefaction of hepatocytes with loss of parenchyma (focal necrosis), condensation and fragmentation (Figure 17.2). No remarkable changes could be seen in the Kupffer cells of the experimental rats.

Table 17.1: Effect of 0.75 LD_{50} Heroin Administration on Serum Acid Phosphatase and Alkaline Phosphatase Activities (KA unit/100 ml) of *Rattus norvegicus*

Group	*Duration*			
	24 hours	*7 days*	*15 days*	*30 days*
	Acid Phosphatase			
Control	1.02±0.08	1.04±0.03	1.08±0.06	1.05±0.09
Experimental	1.37±0.06**	1.78±0.05**	2.13±0.05**	2.82±0.04**
	(+34)	(+71)	(+97)	(+169)
	Alkaline Phosphatase			
Control	10.72±0-39	10.53±0.48	10.94±0.63	10.65±0.91
Experimental	12.98±0.52*	17.26±0.74**	24.38±0.79**	33.71±1.04**
	(+21)	(+64)	(+123)	(+217)

Each value represents means±S.D. of 5 determinations.

Significant response: *: P < 0.05; **: P < 0.001. Value in parenthesis indicates per cent increase (+) over the control.

Chronic liver diseases in abusers of alcohol and parenteral drugs have been reported by Weller *et al.* (1984) and Novick *et al.* (1985, 1986). They found that chronic active hepatitis and chronic persistent hepatitis were more frequent in abusers of parenteral drugs alone whereas cirrhosis was most frequent in abusers of both alcohol and parenteral drugs. Recently, Shanna *et al.* (2001) recorded elevation in serum cholesterol, serum triglyceride and serum total lipids as well as injuries to liver of the rabbits subjected to sublethal heroin administration. Our observations in *Rattus norvegicus* support the findings of above workers in human subjects and rabbit.

The kidney of *Rattus norvegicus* consisted of nephrons which had two major components-renal capsule and renal tubules. The renal capsule, responsible for filtration of the plasma, was the combination of two structures-Bowman's capsule and glomerulus. Bowman's capsule comprised a single layer of flattened cells resting on the basement membrane and opened into the proximal convoluted tubule. The glomerulus was a tightly coiled network of anastomising capillaries which invaginated the Bowman's capsule. Within the capsule, the glomerulus was invested by a layer of epithelial cells which constituted the visceral layer of Bowman's capsule. The space between the visceral layer and Bowman's capsule is known as Bowman's space. The renal tubule was divisible into (*i*) proximal convoluted tubules, (*ii*) loop of Henle, (*iii*) distal convoluted tubules and (*iv*) the collecting duct. The proximal convoluted tubules constituted bulk of the renal cortex. They were relatively small, had an uneven lumen and were composed of large cuboidal cells. The distal convoluted tubules were fewer in number with large lumen, the cell lining was smaller and more distinctly cuboidal (Figure 17.3). The lobules of glomeruli of sublethal heroin treated rats exhibited strikingly diffused glomerulosclerosis. There were enlargement of Bowman's capsule and thickening of the capillary wall. Periphery of the kidney had glomerular lesions in the form of brightly staining, necrotic brown material which was lacking from the central part of the kidney. Fibrosis of the hematopoetic tissue and a marked increase in the mesangial matrix were observed in the heroin treated rats. All the renal tubules exhibited vacuolar degeneration with cloudy swelling (Figure 17.4). There exist reports that heroin addicts are more at the risk of chronic nephropathy Amyloidotic nephropathy is the main cause of renal failure among the drug addicts (New *et al.*, 1991-cited by Barai, 1996). Patients with long history of heroin addiction manifested nephrotic syndrome and renal biopsy showed various types of glomerulonephritis without glomerular sclerosis (Campistol *et al.*, 1988; Volochine *et al.*, 1988). Johnson *et al.* (1987-cited by Barai. 1996) have also observed degenerative changes in the kidney of rats receiving subcutaneous pellets containing 75 mg morphine followed 3 days later by implantation of 2 additional pellets.

Acknowledgements

One of us (SRB) is thankful to the Council of Scientific and Industrial Research, New Delhi for providing Fellowship. We are grateful to Hon'ble (Mrs). K.K. Baam, former Judge, Bombay High Court, Mumbai for permission to carry out the work on heroin and to Mr. Rahul Rai Sur, the then Deputy Commissioner of Police, Mumbai for his help in obtaining the drug.

References

Barai, S.R., 1996. Effects of heroin on electrolyte metabolism in the rat, *Rattus norvegicus*. *Ph.D. Thesis*, University of Bombay, Bombay.

Barai, S.R., Suryawanshi, S.A. and Pandey, A.K., 2004. .Effect of sublethal heroin administration on thyroid gland of *Rattus norvegicus*. *J. Appl. Zool. Res.*, 15: 195–197.

Brambilla, F., Nobile, P., Zanoboni, A., Muciaccia, W. and Meroni, P.L., 1988. Effects of chronic heroin addiction on pituitary-thyroid function in man. *J. Endocrinol. Invest.*, 3: 251–255 (1980).

Cami, J. and Farrie, M., 2003. Drug addiction. *N. Eng. J. Med.*, 349: 975–986.

Campistol, J.M., Montoliu, J., Soler-Amigo, J., Darnell, A. and Revert, L., 1988. Renal amyloidosis with nephrotic syndrome in a Spanish subcutaneous heroin abuser. *Nephrol. Dial. Transplant*, 3: 471–473.

Charuvastra, V.C., Ouren, J. and Powel, B., 1980. Heroin addiction, renal failure, and methadone maintenance: A follow-up report. *Drug Alcohol Depend.*, 6: 137–139.

El-Daly, E.S., 1994. Effect of morphine and stadol on lipid content in Hver of rat. *Life Sci.*, 55: 1419–1426.

Eto, M. and Okhawa, H., 1970. *Biochemical Toxicology of Insecticides.* Academic Press, New York.

Goldberger, B.A., Cone, E.J., Grant, T.M., Caplan, Y.H., Levine, B.S. and Smiaiek, J.E., 1994. Disposition of heroin and its metabolites in heroin-related deaths. *J. Anal. Toxicol.*, 18: 22–28.

Goodman, J. and Rothstein, A., 1957. The active transport of phosphate into the yeast. *J. Gen. Physiol.*, 40: 915–925.

Gomez-Lechon, M.J., Ponsoda, X., Jover, R., Pabra, R., Trullenque, R. and Castell, J.V., 1987. Hepatotoxicity of the opioids morphine, heroin, mepridine and methadone to cultured human hepatocytes. *Mol. Toxicol.*, 1: 453–463.

Hayes, W.J. Jr., 1982. *Pesticide Studies in Man.* Williams and Wilkins, Baltimore.

Hussain, K.A. and Kumar, A., 1988. Physiological, hematological, biochemical and clinical effects of epidural morphine in dogs. *Indian Vet. J.*, 65: 491–495.

Kind, P.R.N. and King, E.J., 1954. Estimation of plasma phosphatase by determination of hydrolysed phenol with amino-antipyrine. *J. Clin. Pathol.*, 7: 322–326.

King, E.J. and Jagatheesan, K.A., 1959. Estimation of serum acid phosphatase. *J. Clin. Pathol.*, 12: 85–89.

Kringsholm, B. and Christoffersen, P., 1989. Morphological findings in fatal drug addiction: An investigation of injection marks, endocrine organs and kidneys. *Forensic Sci. Int.*, 40: 15–24.

Kringsholm, B., Kaa, E., Steentoft, A., Worm, K. and Simonsen, K.W., 1994. Deaths among drug addicts in Denmark in 1877–1991. *Forensic Sci. Int.*, 67: 185–195.

Martin, W.R., 1984. Pharmacology of opioids. *Pharmacol. Rev.*, 35: 283–323.

Miller, D. and Crane, R.D., 1961. The digestive function of the epithelium of small intestine. I. An intracelluar locus of disaccharide and sugar phosphate ester hydrolysis. *Biochim. Biophys. Acta*, 57: 281–293.

Moriarty, F., 1975. *Organochlorine Pesticides: Persistant Organic Pollutants.* Academic Press, New York.

Neri-Semeri, G.G. and Modesti, P.A., 1991. Medical complications connected with the use of drugs. *Ann. Ital. Med. Int.*, 6: 313–324.

Novick, D.M., Enlow, R.W., Gelb, A.M., Stenoer, R.J., Fotino, M., Winter, J.W., Yancovitz, S.R., Schoenberg, M.D. and Kreek, M.J., 1985. Hepatic cirrhosis in young adults: Association with adolescent onset of alcohol and parenteral heroin abuse. *Gut*, 26: 8–13.

Novick, D.M., Stenoer, R.J., Gelb, A.M., Most, J., Yancovitz, S.R. and Kreek, M.J., 1986. Chronic liver diseases in abusers of alcohol and parenteral drugs; a report of 204 consecutive biopsy proven cases. *Alcohol Clin. Exp. Res.*, 10: 500–505.

Pedrazzoni, M., Vescovi, P.P., Maninetti, L., Michelini, M., Zaniboni, G., Pioli, G., Costi, D., Alfanso, F.S. and Passeri, M., 1993. Effect of chronin heroin abuse on bone and mineral metabolism. *Acta Endocrinol. (Kbh)*, 129: 42–45.

Sawynok, J., 1986. The therapeutic use of heroin: a review of the pharmacological literature. *Can. J. Physiol. Pharmacol.*, 64: 1–6.

Sharma, S., Mohite, V., Rangoonwala, S.P. and Suryawanshi, S.A., 2001. Heroin induced changes in the lipid profiles in rabbit, *Lepus cuniculus. Biochem. Cell. Arch.*, 1: 109–113.

Sporer, K.A., 1999. Acute heroin overdose. *Ann. Intern Med.*, 130: 584–590.

Volochine, M.L., Rondeau, E., Viron, B., Mougenot, B., Beaufils, H., Pourriat, J.L. and Chauveau, P., 1988. Renal disease associated with heroin abuse. *Nephrologie*, 9: 217–221.

Weller, I.V., Cohen, D., Sierralta, A., Mitcheson, M., Ross, M.G., Mantano, L., Scheuer, P. and Thomas, H.C., 1984. Clinical, biochemical, serological, histological and ultrastructural features of liver disease in drug abusers. *Gut*, 25: 417–423.

Section II

Animal Reproduction

Chapter 18

Seasonal Growth and Reproduction in Two Brachyuran Species Inhabiting Diverse Ecosystems

*V.P. Syama[1], N.T. Supriya[1], K. Sudha[1] and G. Anilkumar[2]**

[1]Postgraduate Department of Zoology and Research Centre, Sree Narayana College, Kannur – 670 007, Kerala

[2]School of Biotechnology, Chemical and Biomedical Engineering VIT University, Vellore – 632 014, Tamil Nadu

ABSTRACT

Sesarma quadratum and *U. triangularis*, inhabiting the mangroves and the estuarine intertidal zone respectively along the Malabar coast (N. Kerala, India), exhibit distinct patterns in programming their somatic growth and reproduction. In both the species, the annual physiological cycle shows definite seasons devoted for reproduction and growth. In *S. quadratum*, August–January is devoted mainly for reproduction. The breeding activity, however, declines as a sizable proportion of the females enters moult cycle from February onwards; the species releases 12–16 broods a year. In *U. triangularis*, mid August–January is devoted essentially for reproduction, and

---

Corresponding author: E-mail: gopianilkumar@hotmail.com;
Tel: 0497-2733303, 9447406197

February–May is the season for both moult and reproduction. In June–July, ~55 per cent of the females involves neither in reproduction nor moult, while rest of the population engages in moulting; the species releases as many as 16–18 broods a year. The present article also reveals that the interrelation between growth and reproduction varies in both the species. Premoult and reproduction are antagonistic in *S. quadratum* inasmuch as majority of the reproductive females are in intermoult ($P < 0.005$). On the contrary, in *U. triangularis*, moult and vitellogenesis occur synergistically ($P > 0.1$). Our study further reveals that the breeding activity in *S. quadratum* is strongly correlated with lunar rhythms, while such a correlation does not seem to exist in *U. triangularis*.

Keywords: *Sesarma quadratum, Uca triangularis, Growth, Reproduction, Moulting, Spawning.*

Introduction

Growth and reproduction are the two major physiological events, but diversely programmed in various crustacean groups. Previous investigators have opined that such diversities, existing at inter- and intra-specific levels, could be essentially due to the complex interactions of extrinsic and intrinsic factors (Sastry, 1983; Adiyodi, 1988; Cobo and Fransozo, 2003; Litulo, 2005). It is now increasingly being realized that a concrete understanding of the patterns of breeding and growth in crustaceans is a requisite not only from academic stand-points, but also for facilitating successful management of the commercially important varieties. Surveying the seasonal frequencies of major physiological events in several crustacean groups have helped us understand that striking differences exist in the patterns of growth and reproduction, even in closely allied phylogenic groups (Erdman and Blake, 1988; Erdman *et al.*, 1991), and in some instances, by the same species but occupying different latitudes and ecological niches (Anilkumar and Adiyodi, 1983; Conan, 1985; Ituarte *et al.*, 2006). Such wide discrepancies do not allow us to draw a common design for brachyurans and hence, warrant more systematic studies among various taxa. This inadequacy of information has encouraged us to undertake a comparative study on the seasonal physiology of two highly fecund brachyuran species (the grapsid crab, *Sesarma quadratum* and the ocypodid fiddler crab, *Uca triangularis*) inhabiting different ecological niches. *S. quadratum* inhabits the mangroves, while *U. triangularis* is an inhabitant of the estuarine intertidal zone of North Malabar (Kerala, India). The present study is devoted to a thorough assessment of the seasonal fluctuations in the frequencies of breeding and moulting of these two brachyuran species. Although their territories are quite close in terms of geographical latitudes, *S. quadratum* and *U. triangularis*, as they inhabit two diverse ecosystems, show distinct patterns in programming breeding and growth. A comparative analysis of our results obtained from the two species, also brings out certain unique and interesting features, unreported so far, in the context of growth-reproduction interrelation in brachyuran crabs.

Materials and Methods

Collection and Maintenance of Animals

All the adult females of *Sesarma quadratum* (N=1598) and *U. triangularis* (N=1013) were collected from mangrove swamps and the estuarine region respectively of Malabar area (North Kerala, India). *S. quadratum* was collected by baiting, while *U. triangularis* was trapped after being driven out of the burrows. Collections were made on a weekly basis. The crabs were maintained in the laboratory in aquaria containing water from the respective natural habitat and fed *ad libitum* on boiled mussel meat and/or egg yolk. The aquaria were also provided with mangrove leaves and stones or shells with a view to maintain the animals in near-natural conditions.

Classification of Ovarian Stages

Ovaries were dissected out by cutting open the dorsal part of the carapace, rinsed in normal (0.9 per cent) saline, placed on a clean glass slide and observed under a light microscope (Leica, Germany). Oocyte diameter (OD) was measured using an ocular micrometer to the nearest 1 μm. Stages in ovarian growth have been classified using OD and ovarian hue as the criteria. In *S. quadratum*, the entire process of vitellogenesis encompasses six stages characterized as per the following details: In Stage 1, ovary appears transparent, the OD being less than 50 μm. Stage 2 ovary appears light yellow in colour due to the onset of yolk accumulation, OD being 50–100μm. Stage 3 ovary is yellowish brown in colour with OD 101-150 μm. During Stage 4, ovary acquires brownish tint and appears more massive and lobulated due to intense yolk accumulation, OD being 151-200 µm. In Stage 5, ovary is dark brown in colour, with large oocytes (OD 201–250 µm) pending spawning. The post-spawned ovary (Stage 6) appears as translucent yellowish flaccid band of tissue with residual eggs.

In *U. triangularis*, on the other hand, vitellogenesis is classified into three stages: In Stage 1, ovary appears light yellow and the oocyte measures <90 µm. In Stage 2, it appears brown in colour and measures 91–150 µm and in Stage 3, it acquires dark brown hue and measures 151–200 µm.

Characterization of Moult Stages

Microscopic observation of setagenic events was taken as the criterion for accurate identification of moult stages in both *S. quadratum* and *U. triangularis*. Epipodite of the third maxilliped was snipped off with a pair of forceps, mounted on a clean glass slide with seawater and observed under a light microscope. The premoult stage D_1 is characterized by the separation of epipodite epidermis from the cuticular layer. Setal grooves from the epidermal layer become discernible during D_2. The tips of the new set of setae begin to appear during D_3 and the new set of setae becomes quite prominent during D_4 when the exoskeleton becomes paper-thin and brittle, and the animal gets ready for exuviating the old exoskeleton (Stage E). The immediately post-ecdysial crab is inactive, the carapace at this stage being soft and pliable. As the crab proceeds towards later stages of postmoult and subsequently the intermoult, the cuticle gets hardened, and the crab becomes agile with normal feeding activity.

Results

Our weekly samplings of the populations of *S. quadratum* and *U. triangularis* have allowed us to have a clear picture on the programming of growth and reproduction in these species inhabiting different ecological niche.

Programming of Growth and Reproduction in *S. quadratum*

August–January is devoted primarily for reproductive activity in *S. quadratum*. Out of the total of 670 females examined during this period, 292 (~44 per cent) were brood carrying. [The eggs are borne by the females in her broadened abdomen for a period of 14-15 days until hatching out as zoea larvae]. Among the 378 (56 per cent) non-berried females, 305 (~46 per cent of the total collection) were having growing ovaries, implicating that 89 per cent of the female population was engaged in reproduction during the season (Figure 18.1). In order to explore the possible existence of continuous breeding in the species, we had been closely examining the ovaries of the brood-carrying females. Accordingly, we were able to observe that majority of the brood carrying females were having growing ovaries, thus strongly suggesting for the occurrence of continuous breeding in the population. During August–January season, an average female (*S. quadratum)* releases 8–10 broods. Interestingly, we could also observe that there exists a synchronous relation between the ovarian stage and the stage of embryogenesis of the respective brood (Table 18.1). During August–January season, a small proportion of the females (~4 per cent) (26 nos.) was engaging in premoult. Significantly, only ~27 per cent of the premoult females (7 nos.) were having vitellogenic ovaries.

Table 18.1: Shows Synchrony Between Stages of Ovarian Growth and Embryogenesis in *S. quadratum*

Days	*Embryonic Stage*	*Oocyte Diameter (μm)*
0	Immediate post-spawn	up to 50
1	Blastula	51–100
2–5	Gastrula	101–150
6–10	Eyespot stage	151–200
11–13	Appendages	201–225
>14	Zoea release	Spawn

From February onwards, the reproductive activity of *S. quadratum* registered a tendency for decrease; the percentage of brood-carrying females declined to ~23 per cent in February and ~21 per cent in March. The percentages of non-berried females with active ovaries were ~26 in February and ~19 in March; thus the reproductive activity declines to 49 per cent and 40 per cent in February and March respectively. The breeding activity became less frequent during the succeeding months so as to reach the minimum levels in April (15 per cent) and May (8 per cent), but showed a tendency for increase

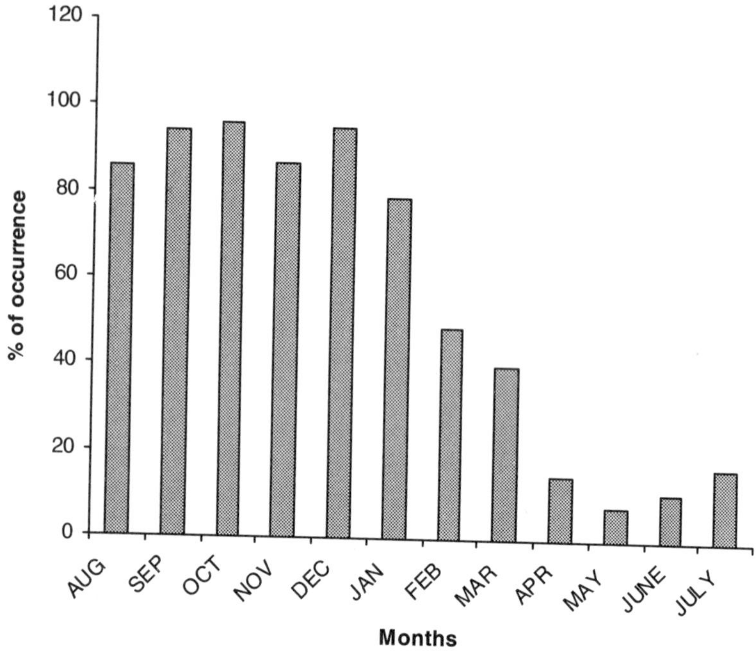

Figure 19.1: Shows Monthly Frequency of Reproductive Activity in *S. quadratum*

during June (11 per cent) and July (17 per cent) (Figure 18.1); the species releases 4–6 broods in February–July, thus releasing a total of 12–16 broods annually. From February onwards, the females, which were generally in intermoult, showed signs of premoult initiation. In February, 19 per cent of the females were seen to be involved in premoult–postmoult activities. From then on, the percentage of females involving in premoult–postmoult activities increased perceptibly to reach peak levels during May (36 per cent) and June (54 per cent), but to decline in subsequent months of July (39 per cent) and August (36 per cent) (Figure 18.2). As a rule, the premoult females of February–July season did not engage in vitellogenesis. However, ~2 per cent of the postmoult females showed the occurrence of growing ovaries during the season.

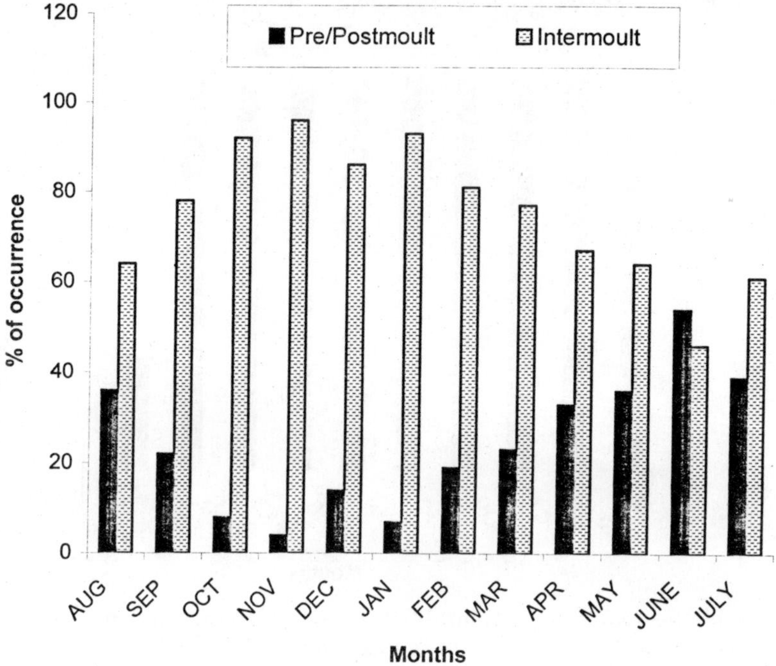

Figure 18.2: Shows Monthly Frequency of Moulting in *S. quadratum*

Programming of Growth and Reproduction in *Uca triangularis*

August–January season is devoted primarily for breeding activity in *U. triangularis* as well. A total of 573 female crabs were observed, out of which 455 individuals were brood carrying. Of the remaining 118 non-berried females, 106 crabs were having growing ovaries (Stages 2 and 3 of vitellogenesis); thus 561 females out of 573 (98 per cent) were seen to be active in reproduction during the season. The monthly occurrences of brood-carrying females and non-berried females with vitellogenic ovaries are shown in Figure 18.3. *U. triangularis* also showed signs of continuous breeding during August–January, judged by the occurrence of growing ovaries in berried females and the existence of a synchronous relation between the ovarian stage and the stage of embryogenesis in the respective

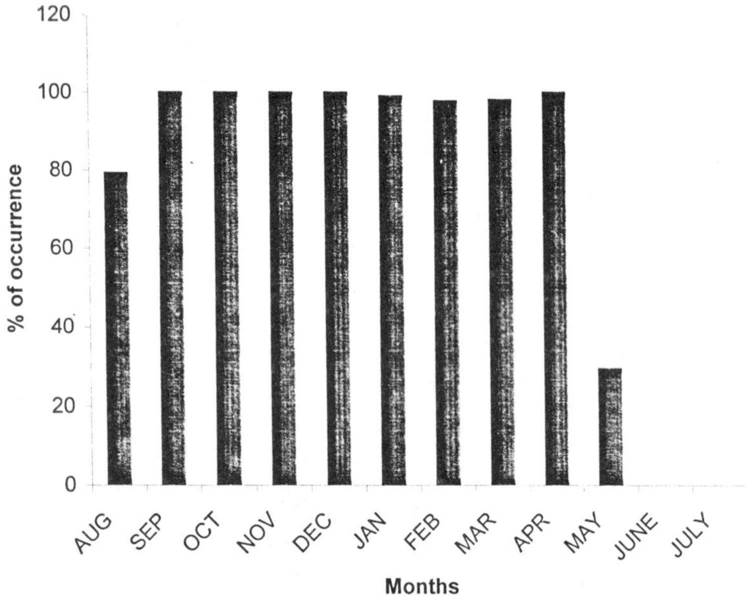

Figure 18.3: Shows Monthly Frequency of Reproductive Activity in *U. triangularis*

brood (Table 18.2); the spawned eggs would require ~10 days of embryogenesis before larval release. In the meantime, the ovary of the same crab would undergo vitellogenesis so as to ensue spawning in about 3-4 days after larval release.

While majority (98 per cent) of the females of *U. triangularis* was engaging in breeding activity and remaining in intermoult, a small proportion of the individuals (12 out of 573; ~2 per cent) were initiating premoult growth in August–January. Interestingly, out of the 12 premoult females, 9 were brood-carrying and two among the non-berried females were having growing ovaries (Stages 2); this means that 11 out of the 12 premoult females were reproductively active, indicating unusually high degree of synergistic relation between premoult and reproduction in the species.

The female population of *U. triangularis* thrusts on both breeding and moulting during February–May. The percentage of breeding activity, however, remains unchanged (from its frequency of August–January season) in February–April, but to decline from May onwards.

Out of the total of 339 animals collected during February–April, 271 were brood carrying and 63 were non-berried with active ovaries *i.e.* ~98.5 per cent of the female crabs were engaged in reproduction during February–April. A close examination of the population during August–April revealed that majority of the females were engaging in continuous release of the brood, one after the other, almost every 15 days, so as to release 16-18 broods during the period. The percentage of breeding females, however, declined considerably in May (29 per cent), and practically no breeding activity was observed during June-July. The females which were hitherto in intermoult until January, showed signs of premoult initiation in February. The percentage of premoult-postmoult individuals increased from 10 per cent in February to 18 per cent in March, and subsequently to reach peak levels (54 per cent) in July. In August, however, the percentage of moulting population decreased (17 per cent), and continued to decrease thereafter (Figure 18.4).

Table 18.2: Shows Synchrony Between Stages of Ovarian Growth and Embryogenesis in *U. triangularis*

Days	*Embryonic Stage*	*Oocyte Diameter (µm)*
0	Immediate post-spawn	Previtellogenic oocytes
2nd day	Blastula	< 90
3–6th day	Gastrula	91 -150
6-7th day	Eyespot stage	151- 165
8-9th day	Appendages	166–175
10th day	Zoea release	176–190
13/14th day		Spawn

Antagonism Between Premoult and Vitellogenesis

Our year-round survey on the seasonal fluctuation in a total of 1598 crabs showed that intervention of premoult could restrain reproduction in females of *S. quadratum*. While 95 per cent of the intermoult females were either brood-carrying or having growing ovaries during August–January season, only 27 per cent (7 out of 26) of the premoult females were engaging in reproduction during the season. Our χ^2 analysis reveals that this discrepancy in reproductive activity between the intermoult and the premoult females is statistically significant ($P < 0.005$), affirming that the

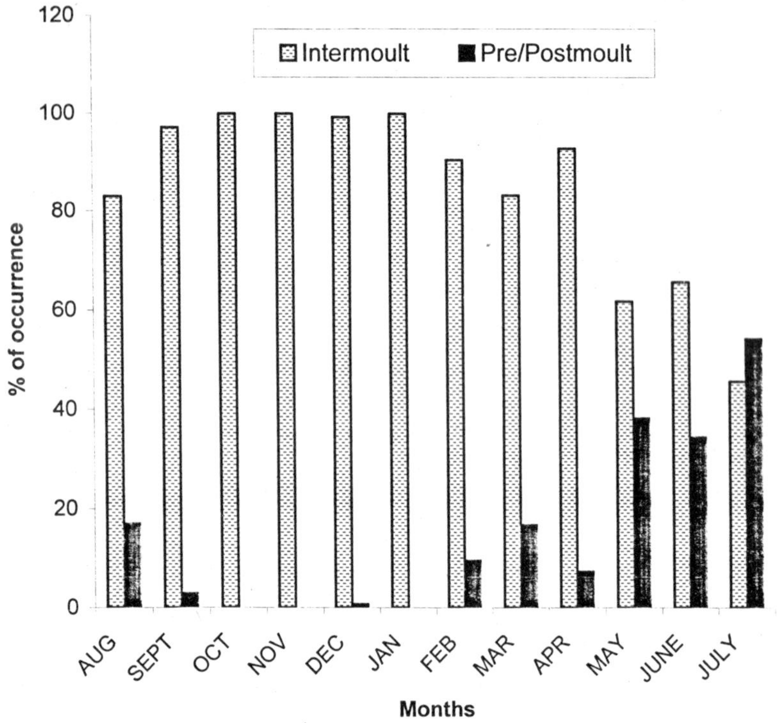

Figure 18.4: Shows Monthly Frequency of Moulting in *U. triangularis*

premoult is truly antagonistic with vitellogenesis in the species. Further, the degree of antagonism between premoult and vitellogenesis appears to be more in February–July, than that in August–January ($P < 0.05$).

Synchrony in Breeding Activity and Lunar Rhythms

We had very closely and continuously examined both the populations of *S. quadratum* and *U. triangularis* for the entire breeding season (August–January) for the possible existence of synchrony with respect to vitellogenesis and spawning. The data thus obtained were superimposed on the occurrence of New Moon/Full Moon. Resultantly, we could find that the spawning in *S. quadratum* is remarkably entrained with lunar rhythms. On the contrary, the breeding pattern in *U. triangularis* did not show any signs of correlation with lunar periodicity (Figure 18.5).

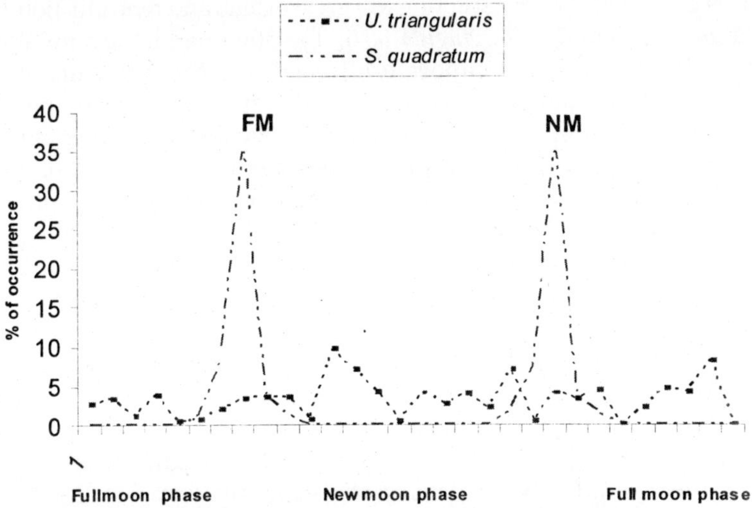

Figure 18.5: Spawning and Lunar Rhythms in *S. quadratum* and *U. triangularis*

Discussion

Our investigations on the breeding frequencies of *S. quadratum* and *U. triangularis* revealed that both the brachyuran crabs resort to continuous breeding and are highly fecund. *U. triangularis*, being the most prolific breeder of the two, releases up to 18 broods a year. Although there are previous reports on continuous breeding in crabs, especially among those inhabiting the tropical waters (Negreiros-Fransozo *et al.*, 2002), the fecundity rate as found in *U. triangularis* (present study), appears to be exceptionally high. Some of the multiparous crabs are known to release up to 12 broods annually (Hines, 1982; Perez, 1990; Skov, 2001; Kotb and Hartnoll, 2002), while one brood per season appears to be the rule in several species of *Cancer* (Hines, 1991; Shields, 1991).

The present study on the annual physiological cycles of *S. quadratum* and *U. triangularis* shows that there exists definite peak seasons for growth and reproduction in these brachyuran crabs. The programming of growth and reproduction occurs almost as semi-annual rhythms in *S. quadratum*; August–January is primarily devoted for reproductive activity in the species, while February–

July is prevailed over by moulting to a great extent, and reproduction to a lesser extent. In *U. triangularis*, on the other hand, the annual physiological cycle is temporally divided into three functional seasons, August–January being the reproductive season, February–April the moult–reproductive season, and May–July is devoted for moulting only. However, despite the onset of moulting season in February, the reproductive season continued unstinted up to April in *U. triangularis* (Figures 18.3 and 18.4), whereas the breeding frequency declined in *S. quadratum* from February onwards (Figure 18.1). This discrepancy in breeding pattern existing between *S. quadratum* and *U. triangularis* during February–April is very significant from physiological stand points in that premoult and reproduction are antagonistic events in *S. quadratum* [majority of the breeding females were in intermoult], while these two functions occur synergistically in *U. triangularis* [wherein almost all the premoult females were possessing growing ovaries]. In its antagonistic programming of premoult and vitellogenesis, *S. quadratum*, seems to resemble *Metopograpsus messor*, another member of the grapsid family (Sudha and Anilkumar, 1996), and a parathelphusid crab, *Paratelphusa hydrodromous* (Adiyodi, 1988). And significantly, the extent of synergistic relation between premoult and vitellogenesis as exhibited by *U. triangularis* during August–April (our present study), has never been reported in any of the brachyurans investigated so far.

The discrepancy in the absolute fecundity existing between *S. quadratum* and *U. triangularis*, and the very high fecundity shown by the latter (18 broods annually), encourage us to investigate into the pattern of reproductive investment (calculated by the brood weight as a percentage of female body weight, after Hartnoll, 2006) in these two brachyurans. The breeding investment in *S. quadratum* is found to be 10.7 per cent, a value that conforms very well with the mean investment predicted for free-living brachyurans by Hartnoll (2006), while that of *U. triangularis* is less (9.62 per cent), comparable with *Uca pugnax* (9.5 per cent) (Hines, 1982). However, while the question of annual investment is addressed, *U. triangularis* shows a higher percentage (153.9 per cent) than that of *S. quadratum* (117.7 per cent). This higher percentage of reproductive investment found in *U. triangularis* could be to offset the species' vulnerability to predation as they inhabit the intertidal area (Hartnoll, 2006 for review).

Programming of growth and reproduction in *S. quadratum* resembles that of *U. triangularis* in some respects in that August–January is devoted for reproduction in both the species, an observation that invites our further attention. A previous study from our laboratory has shown that August to December is the breeding season in the grapsid crab *M. messor* that inhabits the estuarine region of Malabar (Sudha and Anilkumar, 1996), so also with the peaneid shrimp *Penaeus monodon* (Krishnakumar, personal communication), implicating that August–December (postmonsoon season) is ideal for decapod breeding along this part of the Indian peninsula.

The role of ecdysteroids as growth promoter is very well established in crustaceans. Although there are convincing evidences to demonstrate the role of ecdysteroids in insect reproduction (Carney and Bender, 2000; Koslova and Thummel, 2000; Riddiford *et al.*, 2001), this aspect has not been successfully attempted in crustaceans (Subramoniam, 2000; Okumura, 2004; Diwan, 2005 for reviews). At this juncture, an understanding of the (antagonistic or synergistic) interactions between moult and reproduction in crustaceans would also allow us to address the question of regulation, if any, of reproduction through ecdysteroids. The antagonism between premoult and vitellogenesis, as exhibited by *S. quadratum* (present study) and *M. messor* (Sudha and Anilkumar, 1996), and a previous observation from our laboratory that the onset of premoult could restrain vitellogenesis in *M. messor* (Sudha and Anilkumar, 1989), could implicate that ecdysteroids might play an inhibitory role in brachyuran vitellogenesis. However, the synchronous occurrence of ovarian growth and premoult existing in *U. triangularis* (present study) seems to argue that a high ecdysteroid titre (of premoult stage) does not impede reproduction. In adhering to this contention, we are further encouraged by the results of our previous experiments on *M. messor*, where ovarian growth occurs at an elevated ecdysteroid titre induced by deeyestalking (Sudha and Anilkumar, 2007).

Lunar periodicity appears to be an extraneous factor that influences breeding activity in *S. quadratum*, evidenced from our field observations (Figure 18.5). Programming of spawning activity in tune with lunar rhythms has also been observed in other members of grapsid family such as *M. messor* (Sudha and Anilkumar, 1996) and *Chasmagnathus granulatus* (Ituarte *et al.*, 2004). It also invokes much interest in that the breeding pattern in *U. triangularis* (present study), a member of the ocypodid family, that occupies the same latitude, but a different ecosystem, is not entrained to lunar periodicity. The exact reason for such a discrepancy in spawning pattern is still enigmatic. However, this aspect seems to have an adaptive significance. The asynchronous spawning in *U. triangularis* would persuade the females to enter the sea at random (for the purpose of larval release), a factor that renders the population more vulnerable to predation; synchronous spawning, on the other hand, would allow the species to go to the sea in mass which in turn would limit the predation risk (Hartnoll, 2006; Liu and Jeng, 2007).

Acknowledgements

VPS thankfully acknowledges the University Grants Commission, New Delhi for the award of FIP Teacher Fellowship. NTS thanks the Kannur University for the award of a Junior Research Fellowship. Financial supports from the International Foundation for Science (Stockholm) (for the Project: RGA: A/3520-1 awarded to KS) and Ajit Balakrishnan Foundation (awarded to GA) are gratefully acknowledged.

References

Adiyodi, R.G., 1988. Reproduction and development. In: *Biology of Land Crabs*, (Eds.) W.W. Burggren and B.R. McMahon. Academic Press Inc., USA, p. 139–185.

Anilkumar, G. and Adiyodi, K.G., 1983. Reproductive strategies of two populations of the freshwater crab, *Parathelphusa hydrodromous* (Herbst). In: *IIIrd International Symposium on Invertebrate Reproduction*, Tubingen, W. Germany.

Carney, G. E. and Bender, M., 2000. The *Drosophila* Ecdysone receptor (EcR) gene is required maternally for normal oogenesis. *Genetics*, 154: 1203–1211.

Cobo, V.S. and Fransozo, A., 2003. External factors determining breeding season in the red mangrove crab *Goniopsis cruentata* (Latrelle) (Crustacea, Brachyura, Grapsidae) on the Sao Paulo State northern coast, Brazil. *Revista Brasileira De Zoologia*, 20: 248–261.

Conan, G.Y., 1985. Periodicity and phasing of moulting. In: *Factors in Adult Growth: Crustacean Issues,* (Ed.) A.M. Wenner, 3: 74–99.

Diwan, A.D., 2005. Current progress in shrimp endocrinology: A review. *Indian Journal of Experimental Biology,* 43: 209–223.

Erdman, R.B. and Blake, N.J., 1988. Reproductive ecology of female golden crabs, *Geryon fenneri* from southeastern Florida. *Journal of Crustacean Biology,* 8: 392–400.

Erdman, R.B., Blake, N.J., Lockhart, F.D., Linberg, W.J., Perry, H.M. and Waller, R.S., 1991. Comparative reproduction of deep sea crabs *Chaceon fenneri* and *C. quinquedens* (Brachyura, Geryonidae) from the northeast Gulf of Mexico. *Invertebrate Reproduction and Development,* 19: 175–184.

Hartnoll, R.G., 2006. Reproductive investment in Brachyura. *Hydrobiologia,* 557: 31–40.

Hines, A.H., 1982. Allometric constraints and variables of reproductive efforts in Brachyuran crabs. *Marine Biology,* 69: 309–320.

Hines, A.H., 1991. Fecundity and reproductive output in nine species of Cancer crabs (Crustacea, Brachyura, Cancridae). *Canadian Journal of Fisheries and Aquatic Sciences,* 48: 267–275.

Ituarte R.B., Spivak, E. and Luppi, T., 2004. Female reproductive cycle of the Southwestern Atlantic estuarine crab, *Chasmagnathus granulatus* Dana, 1851 (Brachyura: Grapsoidea: Varunidae). *Scientia Marina,* 68: 127–137.

Ituarte, R.B., Bas, C., Luppi, T. and Spivak, E., 2006. Interpopulational differences in the female reproductive cycle of the southwestern Atlantic estuarine crab *Chasmagnathus granulatus* Dana, 1851 (Brachyura: Grapsoidea: Varunidae). *Scientia Marina,* 70: 709–718.

Koslova, T., and Thummel, C.S., 2000. Steroid regulation of postembryonic development and reproduction in *Drosophila. Trends in Endocrinology and Metabolism,* 11: 276–280.

Kotb, M.M.A. and Hartnoll, R.G., 2002. Aspects of the growth and reproduction of the coral gall crab, *Hapalocarcinus marsupialis. Journal of Crustacean Biology,* 22: 558–566.

Litulo, C., 2005. External factors determining the reproductive periodicity in a tropical population of the hairy crab *Pilumnus vespertilio* (Crustacea, Brachyura, Pilumnidae). *The Raffles Bulletin of Zoology*, 53: 115–118.

Liu, H.C. and Jeng, M.S., 2007. Some reproductive aspects of *Gecarcoidea lalandii* (Brachyura: Gecarcinidae) in Taiwan. *Zoological Studies* (in press).

Negreiros-Fransozo, M.L., Fransozo, A. and Bertini, G., 2002. Reproductive cycle and recruitment of *Ocypode quadrata* (Decapoda, Ocypodidae) at a sandy beach in southeastern Brazil. *Journal of Crustacean Biology*, 22: 157–161.

Okumura, T., 2004. Perspectives on hormonal manipulation of shrimp reproduction. *Japan Agricultural Research Quarterly*, 38: 49–54.

Perez, O.S., 1990. Reproductive biology of the sand shore crab *Matuta lunaris* (Brachyura: Calappida). *Mar. Ecol. Prog. Ser.*, 59: 83–90.

Riddiford, L.M., Cherbas, P. and Truman, J.W., 2001. Ecdysone receptors and their biological actions. *Vitamins and Hormones*, 60: 1–73.

Sastry, A.N., 1983. Ecological aspects of reproduction. In: *The Biology of Crustacea, Vol. 8: Environmental Adaptations*, (Ed.) W.B. Vernberg. Academic Press, p. 197–270.

Shields, J.D., 1991. The reproductive ecology and fecundity of cancer crabs. In: *Crustacean Egg Production: Crustacean Issues*, (Eds.) A.M. Wenner and A.M. Kuris, p. 193–213.

Skov, M.W., 2001. Reproduction and feeding ecology of East African mangrove crabs, and their influence on forest energy flow. *Ph.D. Thesis*, Liverpool University.

Subramoniam, T., 2000. Crustacean ecdysteroids in reproduction and embryogenesis. *Comparative Biochemistry and Physiology*, 125(C): 135–156.

Sudha, K. and Anilkumar, G., 1989. Ovarian growth is restrained by the onset of premoult in the crab, *Metopograpsus messor*. In: *Proc. 58th Annual Meeting of the Society of Biological Chemists*, I.V.R.I., Izatnagar, India.

Sudha, K. and Anilkumar, G., 1996. Seasonal growth and reproduction in a highly fecund brachyuran crab, *Metopograpsus messor* (Forskal) (Grapsidae). *Hydrobiologia*, 319: 15–21.

Sudha, K. and Anilkumar, G., 2007. Elevated ecdysteroid titer and precocious molt and vitellogenesis induced by eyestalk ablation in the esturine crab, *Metopograpsus messor* (Brachyura: Decapoda). *Journal of Crustacean Biology*, 27: 304–308.

Chapter 19

Light Induced Ovarian Maturation in a Freshwater Crab, *Oziotelphusa senex senex*: A Biochemical Approach

M.G. Ragunathan and J. Jayanthi
*Department of Advanced Zoology and Biotechnology,
Guru Nanak College, Chennai – 42*

ABSTRACT

The biochemical investigation shows higher deposition of organic contents (Proteins, Carbohydrates and Lipids) from hepatopancreas to ovaries and spermatheca through haemolymph than control crabs. The same results further significantly increased when the crabs were concurrently treated with light plus Methyl Farnesoate. The results were confirmed statistically and were presented in the form of tables.

Keywords: Light, Gonadal index, Ovary, Spermatheca, Neurosecretion, Biochemical mobilization.

Introduction

Reproductive activity is a cyclical process in which biochemical composition of the animal occurs. In freshwater and marine molluscs, a fluctuation in the biochemical composition of reproductive organs in relation to reproductive activity has been reported by Giese (1969) and Webber (1970). Methyl Farnesoate (MF) plays a stimulatory role in crustaceans reproduction and acts as a gonadotropin and also as a morphogen (Laufer and Sagi, 1991; Laufer *et al.*, 1993). Sagi *et al.* (1993) have reported a strong relationship between MF levels, male morphology and reproductive system development of a male spider crab, *L.emarginata.* The present study is aimed at fulfilling the paucity of information existing on the combined effects of light and subsequent treatment of MF and their combined effects on the biochemical composition in haemolymph, hepatopancreas, ovary and spermatheca.

Materials and Methods

The *O. senex senex* used in the present study were collected from the Lake in Vandalur village near Chennai city. The animals were acclimatized for a week in the laboratory and were fed with beef mutton *ad libitum.* After acclimatization about 40 female intermoult stage I crabs were selected and were divided into 4 groups (A.B, C and D) of ten each. Group A is normal group and was sacrificed on the first day of the experiment. Group B crabs were untreated, Group C crabs were injected with crustacean saline and the Group D crabs were injected with MF and the 3 groups B, C and D were placed inside a wooden chamber fitted with a light of 1100 Lux intensity.

The injection of saline and MF were given at a dose of 25 µl/ crab and the concentration of MF was 5µg. The crabs were injected on 1st, 5th and 10th days of the experiment. On the 15th day all the 3 groups of crabs were weighed and sacrificed, tissues such as ovary, spermatheca were removed and weighed.

For the biochemical studies tissues such as haemolymph, hepatopancreas, ovary and spermatheca were taken and total Protein content was quantified following Lowry *et al.* (1951) method, Carbohydrates by Roe (1955) and Lipids by Barnes and Blackstock (1973). The results were statistically analyzed using RSS 7.0 student's package.

Results and Discussion

The normal quantity of protein and their changes in all the four groups of crabs are shown in the Table 19.1, likewise, the Table 19.2

Table 19.1: Total Protein Levels in Different Tissues of *Oziotelphusa senex senex* After Methyl Famesoate Plus Light Treatment

Tissues	*Group A (Normal Untreated)*	*Group B (Normal Light)*	*Group C (Saline)*	*Group D (Methyl Farnesoate)*
Haemolymph	48.299±0.380	***74.980±1.079	***84.095±1.007	*****89.180±0.797
Hepatopancreas	135.723±0.981	***118.505±0.781	***108.301±0.833	***99.255±0.801
Ovary	67.149±1.400	***90.205±0.825	***106.259±0.738	***113.177±0.822
Spermatheca	59.206±0.451	***84.282±0.807	*****89.871±0.716	***99.805±0.959

Each value is mean±SEM of 10 samples, expressed as mg/g wet tissue and mg/ml haemolymph.

: $P < 0.001$; ****: $P < 0.01$; *****: $P < 0.05$; ***: Insignificant Group A vs B vs C vs D.

Table 19.2: Total Carbohydrate Levels in Different Tissues of *Oziotelphusa senex senex* After Methyl Farnesoate Plus Light Treatment

Tissues	*Group A (Normal Untreated)*	*Group B (Normal Light)*	*Group C (Saline)*	*Group D (Methyl Farnesoate)*
Haemolymph	17.282±0.366	***28.260±0.793	*****33.381±0.806	*****38.014±0.930
Hepatopancreas	53.567±0.473	***41.846±0.940	*****36.285±0.800	***30.582±0.773
Ovary	13.637±0.418	****16.460±0.877	****19.530±0.748	*****24.365±0.735
Spermatheca	16.846±0.461	****22.405±0.801	***29.225±0.810	*****33.216±0.8191

Each value is mean±SEM of 10 samples, expressed as mg/g wet tissue and mg/ml haemolymph.

: $P < 0.001$; ****: $P < 0.01$; *****: $P < 0.05$; ***: Insignificant Group A vs B vs C vs D.

shows the carbohydrate content and the Table 19.3 shows the lipid content of the experimental crabs. The mobilization of all the organic contents from the hepatopancreas to ovary and spermatheca via haemolymph in experiment crabs is highly significant ($P<0.001$) than their respective controls.

The changes in biochemical constituents are pronounced in invertebrates, which are cyclic in reproduction, and a great amount of energy must be channeled to the gonads during reproduction. This is reflected in the deposition or depletion of nutrients with the advent or depart of the reproductive period (Lambert and Dehnel, 1974). The shifting of lipid reserves to ovaries during the ovarian development has been reported by Rice and Armitage (1974) in Cray fish, *Orconectes nais* at the time of short and normal photoperiod exposure.

Recently, there are many reports on the accelerated gonadal development in crustaceans through the use of synthetic hormones or neurotransmitters (Diwan, 2005). It has been shown that organic material may be transferred from the midgut gland to the gonads as the animal matures.

Kulkarni and Nagabhushanam (1979) have identified the mobilization of organic reserves during ovarian development in a marine penaeid prawn, *Parapenaeopsis hardwickii,* clearly indicating the remarkable variations in the organic substances.

Teshima and Kanazawa (1983) have reported that the lipid concentration increased during ovarian maturation of the prawn, *P. japonicus* and decreased during spent stage and the hepatopancreatic lipid content decreased as the ovarian lipid content increased and suggested that lipids are important substances for maturation of crustacean ovaries.

Gunamalai and Subramoniam (2002) have reported that haemolymph lipoprotein play a pivotal role in the inter-organ transport of lipids in crustacea. They have reported that moulting and reproduction, being the two major energy demanding physiological processes in the adult decapod crustaceans, requires mobilization of raw materials from the storage organs to the target tissues such as epidermis to ovary, haemolymph lipoprotein help in such mobilization of proteins and lipids.

Table 19.3: Total Lipid Levels in Different Tissues of *Oziotelphusa senex senex* After Tissues Methyl Farnesoate Plus Light Treatment

Tissues	*Group A (Normal Untreated)*	*Group B (Normal Light)*	*Group C (Saline)*	*Group D (Methyl Farnesoate)*
Haemolymph	27.965±0.334	***36.257±0.822	*****40.247±0.820	***46.163±0.782
Hepatopancreas	108.369±0.757	***110.223±0.760	***102.279±0.840	***88.187±0.810
Ovary	35.523±0.579	***90.375±0.380	*****95.362±0.270	***103.415±0.390
Spennatheca	30.823±0.620	***53.260±0.810	*****57.250±0.490	***67.167±0.810

Each value is mean±SEM of 10 samples, expressed as mg/g wet tissue and mg/ml haemolymph.

: $P < 0.001$; ****: $P < 0.01$; *****: $P < 0.05$; ***: Insignificant Group A vs B vs C vs D.

Arcos *et al.* (2003) have reported that the pacific white shrimp, *Litopenaeus vannamei* showed higher vitellogenin levels in the females of low maturation capability and suggested that vitellogenin levels in haemolymph could be used as possible predictive criteria of maturation capability, possibly because they reflect the degree of ovarian development at the time of eyestalk ablation.

Boucard *et al.* (2002) have further reported that in *F. indicus* during the early stages of vitellogenesis the cells of hepatopancreas to synthesize vitellogenin, which would be conveyed by the haemolymph before entering into the oocytes by endocytosis.

Meera (2005) and Meera *et al.* (2006a) have reported the increase in the organic contents in ovary and spermatheca and depletion in hepatopancreas and it is mobilized via haemolymph during gonadal maturation due to Fluoxetine in non ablated, eyestalk ablated and also in light treated *Uannulipes.*

Maitra *et al.* (2007) have reported in the Indian carp, *Catla catla* that the artificial long-days (16L:08D) for 30 days induced precocious ovarian development in the prespawning phase resulting in maturation of gonad atleast two months ahead of tree-living fish, while retardation of gonadal growth was noted in the corresponding group of short photoperiodic (08L : 16D) fish.

In the present investigation though there is a large body of information available for the role of MF in crustacean reproduction, relatively few data, available on light plus MF treatment and their effects on the ovary, spermatheca and on the biochemical changes. Biochemical analysis of different tissues such as haemolymph, hepatopancreas, ovary and spermatheca revealed interesting results. The organic contents such as protein, carbohydrate and lipids and their mobilization via haemolymph to ovaries or spermatheca due to MF treatment are highly significant. The mobilization trend in the MF plus light treated crabs are similar suggesting the deposition of organic contents trom hepatopancreas to ovary or spermatheca via haemolymph.

Acknowledgements

The author J. Jayanthi wholeheartedly thanks the University Grants Commission, Southern Region, Hyderabad for the financial assistance for this study (Funding No: L1705).

References

Barnes, H. and Blackstock, J., 1973. Estimation of lipids in marine animals and tissues: Detailed investigation of the Phosphovanillin method for total lipids. *J. Exp. Mar. Biol. Ecol.*, 12: 103–118.

Lowry, O.H., Rosebourgh, N.K., Farr, A.L. and Randall, R.J., 1951. Protein measurement with folin-phenol reagent. *J. Biol. Chem.*, 193: 265–275.

Roe, J.R., 1955. The determination of sugar in blood and spinal fluid with anthrone reagent. *J. Biol. Chem.*, 20: 335–343.

Chapter 20

Correlation of Quantitative Depletion of Protein Fractions with Anti-ovipositional Activity due to Penultimate Oocyte Resorption in *Periplaneta americana* Induced by Cerberin Glycoside

Mangla Bhide, Rekha Rai,*
Shubhra Agarwal and Deepti Tomar
Department of Zoology, Dr. H.S. Gour Vishwavidyalaya,
Sagar – 470 003, M.P.

ABSTRACT

Periplaneta americana is an omnivorous and significant pest of food, fodder and households. It is prolific breeder and transmitting a number of diseases in human beings and also having high rate of fertility. It is therefore, very significant to control the fertility, the cockroaches were treated with sub-lethal concentrations of cerberin glycoside and protein fractions were detected out from the ovaries of 6, 12, 18, 24 and 30 days control and experimental female insects by paper

* Corresponding author: E-mails: manglabhide@rediffmail.com; manglabhide@yahoo.com

electrophoresis. The results summarized in electrophoretograms (Figure 20.1–20.2) showed the gradual increase in the number of protein fractions in the ovaries of control insects indicate the phase of vitellogenesis while depletion in the number as well as the decrease in the intensity of protein fractions due to the intoxication effect of the treated glycoside was correlated with the degeneration of penultimate oocyte in the ovaries which never attain such maturity, is required for oviposition (Figures 20.3–20.10). So in the treated groups the cerberin glycoside acts as an anti-ovipositional agent and in this way it is quite significant to control the fertility of other insect pests also.

Keywords: *Protein depletion, Anti-ovipositional activity, Cockroach.*

Introduction

Study of protein fractions in the ovarian tissue of control and treated insects is an important aspect of this investigation which is evident from the continuous increase (in control groups) as reported by Bhide and Kharya (2002), Bhide (2003), Bhide *et al.* (2003), Ahi (2004), Bhide and Kumar (2004), Gupta and Rathore (2004), Bhide and Rai (2005) and Rai (2006) in some insects, or regular depletion (in experimental groups) in the number and intensities of protein fractions by electrophoresis. On the basis of these observations one can correlate this increase (in control groups) or decrease (in experimental groups) on molecular level in the ovarian architecture as reported in *Atractomorpha crenulata* by Murugan *et al.* (1996) and Babu *et al.* (1997) after neem extract treatment, Rathore (1999) in *Poekilocerus pictus* after *Annona squamosa* seed extract treatment, in some species of *Dysdercus* by Mugdam and Banerjee (1999), Yadu *et al.* (1999), Mugdam (2000) after plumbagin and gossypol exposure, Gupta and Rathore (2002, 2004) in *Poekilocerus pictus* after neem extract treatment, Bhide (2003) in *Musca domestica* after *Annona squamosa* seed extract treatment, and Bhide and Kumar (2004) in *Periplaneta americana* after *Cassia fistula* seed extract treatment respectively. The plant seed extract intoxication causes large number of degenerated, disintegrated pathological penultimate oocytes which never attain such maturity required for oviposition as observed by Bhide and Kharya (2002) in large number of insects, Bhide and Rai (2005) and Rai (2006) in *Bagrada cruciferarum* respectively. In this way it is quite significant to control the fertility of other pest insects also. So the present investigation was done to know more on these significant aspects and correlated the results with the anti-ovipositional activity of cerberin glycoside.

Materials and Methods

Nymphs of cockroaches were collected locally and reared in the laboratory on normal laboratory condition and provided the balanced diet and water *ad libitum* and the adult female emerged from the last nymphal instars were used for experimental purposes.

Procurement of Seed Kernel of *Cerbera thevetia* and Extraction of Glycoside Cerberin

Cerbera thevetia belonging to family-Apocynaceae is found every where in India. The seeds of these plants (shrub) were collected locally, washed thoroughly, shadow dried and kernel of these seeds (after removing their seed coats), mechanically grounded; in coarse pieces and the glycoside cerberin was extracted by direct total alcohol extraction method (Sharma, 1988). The extract was concentrated and evaporated to dryness in water-bath.

Thin Layer Chromatography Method for the Confirmation of Glycoside in the Extract

The glycoside fraction was separated by TLC (Sharma, 1988). The alcoholic extract samples applied on Silica gel "G" coated glass plates, were developed in solvent system–Chloroform : Ethanol : Acetone : Glacial acetic acid (60 : 40 : 15 : 6).

After development the dried plates were exposed to vapors of concentrated HCl (Hydrochloric acid) by kept the plates over porcelain basin containing concentrated HCl, which was heated over a Bunsen flame and cerberin glycoside was detected out as follows:

One blue colour spot of cerberin, Rf. value 0.05 was detected out.

1 per cent stock solution of cerberin glycoside fraction was prepared and further desired concentrations were made to decide the lethal doses and the detected values were 3.0 per cent LD_{100}, 1.5 per cent LD_{50} (Finney, 1971), 0.75 per cent LD_0 and 0.6 per cent sub lethal concentration used during experimentation in the present investigation.

Preparation of Protein Samples of the Ovaries

6, 12, 18, 24 and 30 days old adult female cockroaches were treated with sub-lethal concentration of cerberin glycoside by hypodermic syringe in the haemolymph. The control and treated groups of cockroaches were dissected out and the ovaries were processed for the extraction of protein samples (method adapted after Jairaman, 1985). The samples were applied in the form of streak on the 1 cm width × 39 cm long Whatman no. 1 chromatography paper strips presoaked in 8.6 pH borate buffer. Six samples were applied at a time.

Electrophoresis of Extracted Samples

Samples were applied at +ve or –ve pole for the separation of +vely and –vely charged protein fractions respectively. VMC (Vertical migration chamber) Systronics model no. 604 V and digital power supply (Systronics model EPA 610) were used for electrophoresis (method applied as provided in the manual supplied by Systronics Co. Ltd., Ahmedabad).

Staining of Paper Strips for the Detection of Protein Fractions

Mercuric bromophenol blue staining method was used for the detection of protein bands on the electrophoresis strips after clearing the background with 1 per cent acetic acid solution and fixed bands in methanol. The data of detected protein fractions were collected and summarized in the form of electrophoretograms (Figures 20.1–20.2).

Histochemical Studies on the Penultimate Oocytes Resorption

For the histochemical studies of the penultimate oocyte resorption, the 6, 12, 18, 24 and 30 days old female cockroaches emerged from the last nymphal instars were treated with sub lethal concentration of cerberin glycoside and were dissected out. The ovaries of control and treated groups were fixed in Carnoy's fluid (6 : 3 : 1) fixative and processed (Pearse, 1960). Paraffin blocks of the ovaries were prepared in usual manner and sections of the ovaries were cut at 6 µm.

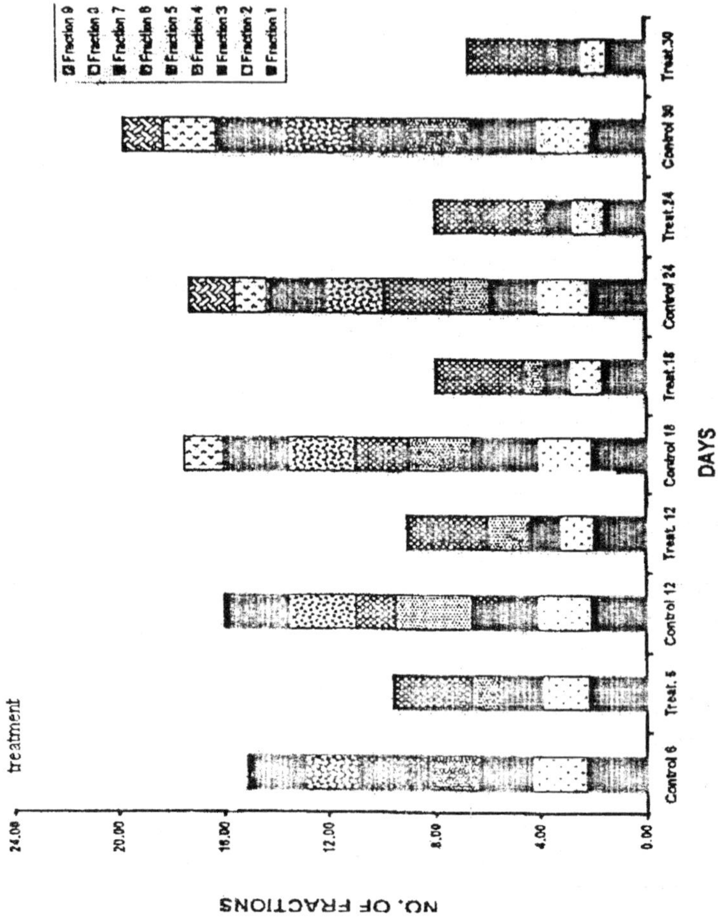

Figure 20.1: Electrophoretogram Showing Detection of Positively Charged Protein Fractions in the Ovaries of Control and Experimental Groups of *Periplaneta americana* after Cerberin Glycoside Treatment

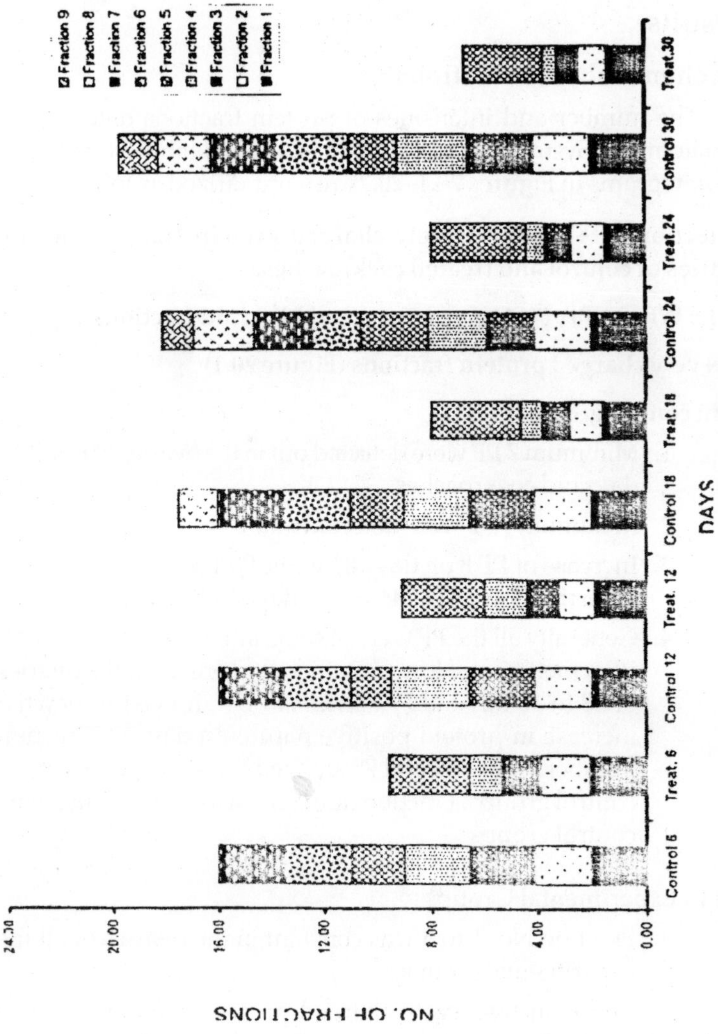

Figure 20.2: Electrophoretogram Showing Detection of Negatively Charged Protein Fractions in the Ovaries of Control and Experimental Groups of *Periplaneta americana* after Cerberin Glycoside Treatment

Detection of Carbohydrate in Ovarian Tissue

For detection of carbohydrate, periodic acid Schiffs reagent method after Pearse (1960) was applied and results expressed in photomicrographs (Figures 20.3–20.10).

Results

Biochemical Observations

The number and intensities of protein fractions detected by densitometer (Systronics Co. Ltd., Ahmedabad) exhibited by electrophoretograms in Figures 20.1–20.2 are summarized as follows:

Detection of +vely and –vely charged protein fractions in the ovaries of control and treated cockroaches:

Note: **Where, PF: Protein fractions and LF: Lipid fractions.**

[1] +vely charged protein fractions (Figure 20.1)

(*i*) In control groups

1. Minimum 7 PF were detected out in the ovaries of 6 and 12 days old cockroaches.
2. Maximum 9 PF were detected out.
3. Increase of PF 8 on day 18th while 9 PF were detected out from the ovaries of 24th to 30th day old *P. americana.*
4. Generally all the PF were of same intensities except the 8th PF which was of low protein +ve intensity in the ovaries of 18 to 30 day old *P. americana* PF 9 showed somewhat increase in protein positive nature on day 30th Fraction number 1st, 2nd, 4th and 7th showed the same pattern in all control groups. Fraction number 1 was constant in all the control groups.

(*ii*) In Experimental Groups

1. Fraction No. 1 to 5 was constant in the ovaries of all the experimental groups.
2. These all five +vely charged protein fractions showed gradual decrease from day 6th onwards while ovaries of 30 days treated insects showed overall depletion in the protein +ve intensity.
3. Fraction No. 5 was of high protein intensity in all the treated groups of *P. americana.*

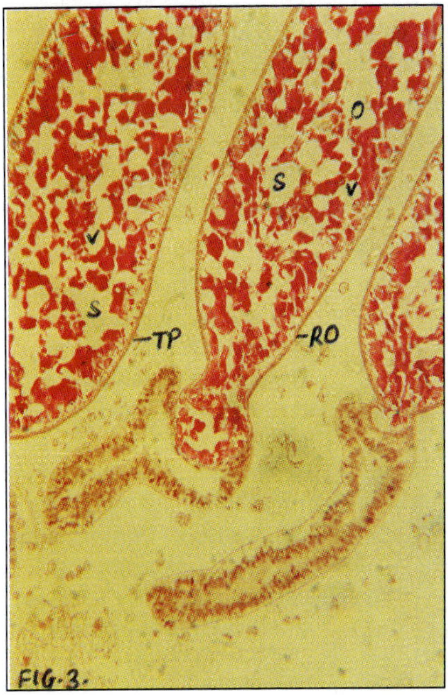

Control (Periodic Acid Schiff's reagent)

Figure 20.3: Section of Ovary of 12 Days Old Cockroach Showing Deposition of PAS +ve Material in the Ooplasm, Less PAS +ve Oocyte Nucleus, Regular Highly PAS +ve Follicular Epithelium (X 100)

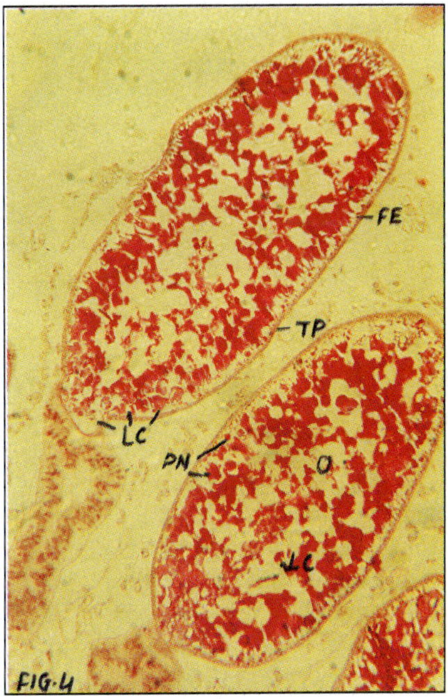

Figure 20.4: Section of Ovary of 24 Days Old Cockroach Showing Fully Grown Big Sized Penultimate Oocytes with Large Quantity of Highly PAS +ve Material and a Number of *Corpora lutea* with Highly PAS +ve *Tunica propria*, Less PAS +ve Cytoplasmic Syncytium and a Central Gap which is PAS –ve in Nature (X 100)

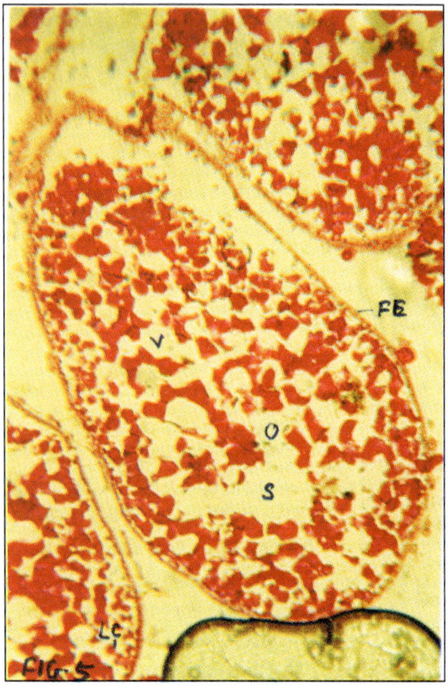

Figure 20.5: Section of Ovary of 30 Days Old Cockroach Showing the Reduction in the Size of Corpora Lutea Indicated by the Folding and Refolding of *Tunica propria*. Cytoplasmic syncytium is very less PAS +ve in nature. Penultimate oocytes showed the presence of highly PAS +ve material in the ooplasm (X 100).

Cerberin Glycoside Treated Groups

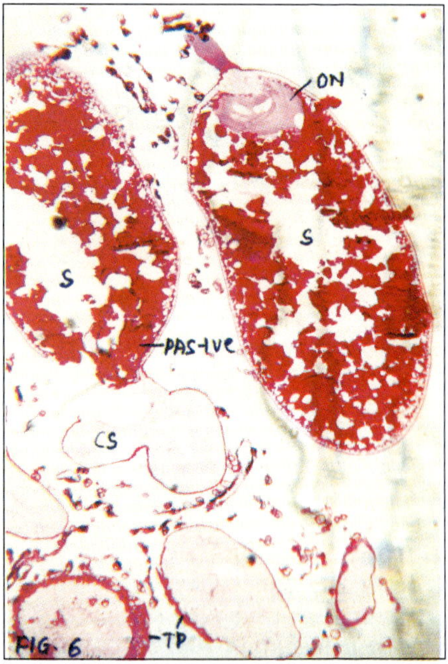

Figure 20.6: Section of Ovary of Six Days Treated Cockroach Showing Deposition of Highly PAS +ve Material in the *Corpora lutea* with PAS +ve Folded *Tunica propria* (X 100)

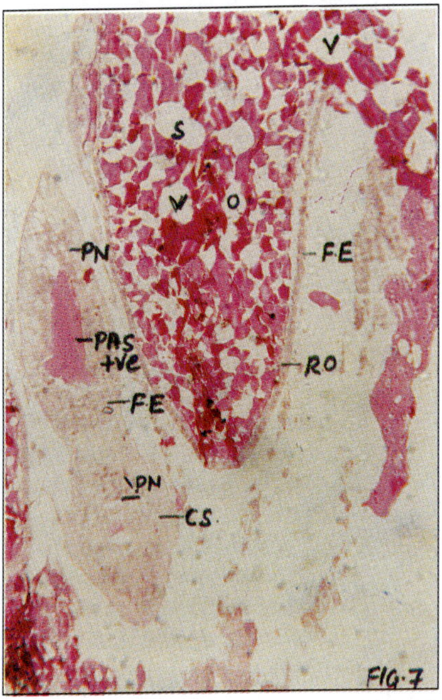

Figure 20.7: Section of Ovary of Twelve Days Treated Cockroach Showing Vacuolization in Ooplasm due to the Resorption of PAS +ve Material from the Corresponding Sites. Detachment of follicular epithelium has observed in one oocyte while other resorbed oocytes exhibited the presence of very less PAS +ve material (X 100).

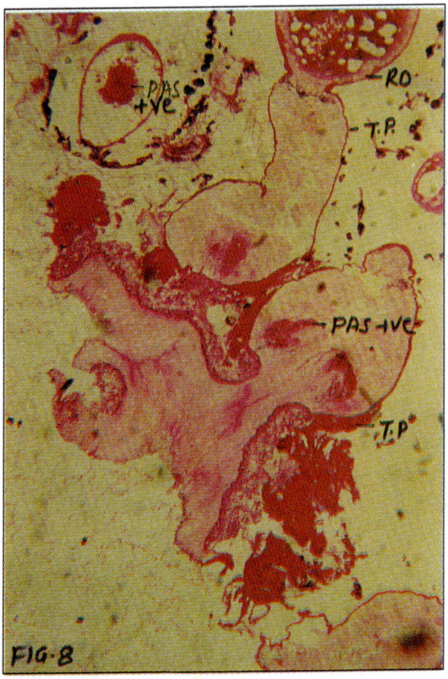

Figure 20.8: Section of Ovary of eighteen Days Treated Cockroach Showing the Resorption of PAS +ve Material from Degenerated Penultimate Oocyte (X 100).

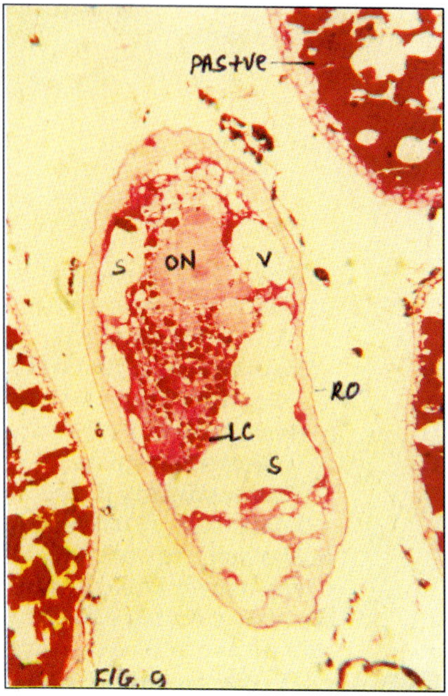

Figure 20.9: Section of Ovary of Twenty-Four Days Treated Cockroach Showing the Shifting of PAS +ve Material at Distal Side Due to Resorption of Most of the Quantity of Ooplasm. Large number of spaces and vacuoles are observed at the periphery of oocyte. Follicular epithelium is less PAS +ve in nature (X 100).

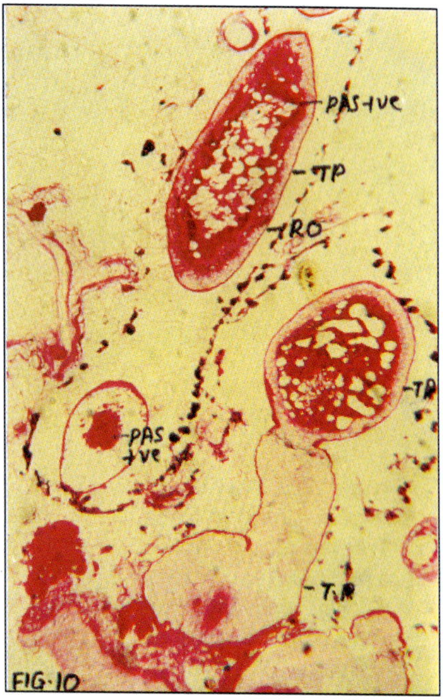

Figure 20.10: Section of Ovary of Thirty Days Treated Cockroach Showing the Reduction in the Size of Penultimate Oocytes. Large number of variable sized vacuoles and spaces are seen in highly PAS +ve ooplasm indicated the site of active resorption. Follicular epithelium is multicellular and surrounded by highly PAS +ve *Tunica propria* (X 100).

Keys to Abbreviations

F.E.: Follicular epithelium; C.G.: Central gap; C.S.: Cytoplasmic syncytium; L.C.: Lecitholytic cell; O.: Ooplasm; O.N.: Oocyte nucleus; PAS +ve.: Carbohydrate rich material; P.N.: Pycnotic nuclei; R.O.: Resorbed penultimate oocyte; S.: Space; T.P.: *Tunica propria* and V.: Vacuoles.

[2] –vely charged protein fractions (Figure 20.2)

(*i*) In Control Groups

1. Minimum 7 PF were detected out.
2. Maximum 9 PF and PF 8 was added in the ovaries of 18 days old cockroaches and 2 PF (PF 8 and 9) were added in the ovaries of 24th and 30th days old insects. Addition of fractions were evident the phase of vitellogenesis.
3. PF 5 showed somewhat decrease in protein intensities from day 6th to 18th day old insects while the pattern of most of the PF from day 6th to 30th was almost the same.

(*ii*) In Treated Groups

1. Fraction No. 1 to 5 was constant in the ovaries of all the experimental groups.
2. Gradual depletion of fraction No. 1, 2, 3 and 4 was observed from day 12th to 30th.
3. On day 30th all the 5 fractions showed very less protein intensity.

Histological Observations on the Ovaries

In Control Groups

Panoistic types of ovaries are present in American cockroaches. The development of oocytes and oviposition were normal as observed in Figures 20.3–20.5, where penultimate oocytes were fully grown size with regular PAS +ve follicular epithelium. Ooplasm exhibited the presence of highly PAS +ve material, while oocyte nucleus is less PAS +ve in nature (Figure 20.4). A number of corpora lutea were formed after oviposition of ootheca at the junction of corresponding ovariole and the peduncle and exhibited a PAS –ve central gap indicated that the oocyte has been laid down from the corresponding follicle. *Tunica propria* was highly PAS +ve in nature while cytoplasmic syncytium and pycnotic nuclei were less PAS +ve (Figures 20.3–20.5).

In Treated Groups

In treated groups the ovaries were less developed and showed the depletion of PAS +ve material in the affected oocytes. The histopathological changes in the penultimate oocytes are as follows:

In Six Days Treated Cockroaches

Pathological penultimate oocytes showed vacuolization and depletion of PAS +ve nature (Figure 20.6). *Tunica propria* was moderately PAS +ve in nature while follicular epithelium was multilayered with PAS +ve nuclei and it became invaded in ooplasm (Figure 20.6).

In Twelve Days Treated Cockroaches

Pathological penultimate oocytes showed the depletion of moderately PAS +ve ooplasm by the presence of large number of variable sized vacuoles and detachment of less PAS +ve follicular epithelium showed the presence of a gap in between it and ooplasm indicated the total resorption of PAS +ve material from the corresponding sites. The resorbed oocytes showed the reduction in size and encircled by multicellular follicular epithelium (Figure 20.7).

In Eighteen Days Treated Cockroaches

Pathological penultimate oocytes showed irregular less PAS +ve multicellular follicular epithelium. Large number of variable sized vacuoles and spaces were exhibited in the ooplasm. Large number of lecitholytic cells found in side the ooplasm, which were migrated from the follicular epithelium (Figure 20.8).

In Twenty-four Days Treated Cockroaches:

Ovaries showed the presence of a number of reduced sized resorbed penultimate oocytes which never attain such maturity required for oviposition due to intoxication of cerberin glycoside

(Figure 20.9). Ovaries also showed the presence of less number of small sized corpora lutea formed by the ovulation of terminal oocytes.

In Thirty Days Treated Cockroaches

Ovaries showing presence of much reduced sized resorbed oocytes, exhibited the complete depletion of PAS +ve material. Resorption of corpus luteum was indicated by the presence of foldings and refoldings of *Tunica propria* (Figure 20.10).

In the present investigation depletion in the protein bands in the ovaries of treated cockroaches due to intoxication of cerberin glycoside and their correlation with deformities of penultimate oocytes resulted into an antiovipositional activity suggested its insecticidal property and this glycoside should be used to control the other insect pests of valuable crops also.

Discussion

Cockroaches are the prolific breeders, transmitting a number of diseases and also having high rate of fertility. In order to control the fertility, cerberin glycoside, extracted from the seed kernel of plant *Cerbera thevetia* was used in the present investigation.

Adult female cockroaches displayed a and duration of treatment dependent (partial to total) sterility which was evident by histopathological and biochemical studies on the ovaries of experimental groups. Ovarian architecture showed the presence of large number of pathological resorbed penultimate oocytes which were reduced in size, deformed with gradually decreased amount of essential nutrients (required for the progressive development of healthy oocytes in control groups) as observed by Bhide (2003), Bhide *et al.* (2003) in *Musca domestica* and Bhide and Rai (2005) and Rai (2006) in *Bagrada cruciferarum* after treatment with some natural plant products.

In treated groups degenerative histopathological changes have been observed in the penultimate oocytes of the treated groups as observed by Bhide and Kharya (2002) in a number of insects due to certain adverse environmental conditions, Bhide (2003) and Bhide *et al.* (2003) in *Musca domestica* by *Annona squamosa* and *Delonix regia* seed extracts treatment respectively.

In the present investigation in *Periplaneta americana* the degenerative changes and resorption of PAS +ve materials from penultimate oocytes resulted into the formation of large number of resorptive bodies in the ovaries as observed by Bhide and Rai (2005) and Rai (2006) in *Bagrada cruciferarum* by *Cerbera thevetia* seed kernel extract respectively.

These resorbed oocytes never attain such maturity which is required for oviposition, so in these treated females of *P. americana* no ootheca formation has been observed in the present investigation hence it could be concluded that the glycoside extracted from *Cerbera thevetia* seed kernel acts as an antiovipositional agent and after formulation should be synthesized in large amount for spraying in the fields to protect our valuable crops from pest insects and in this way it could be possible to control the fertility of pest insects and to increase the economy of farmers of developing country like India.

The biochemical study is the integral aspect of the present investigation and the results should be correlated with the histopathological aspect of ovaries of treated cockroaches. It could be concluded that the increase in the number of PF from day 18th to 30th in the ovaries of control groups were evident from phase of vitellogenesis in *P. americana* in the present investigation as reported by Kumar *et al.* (2003), Bhide (2003) and Bhide *et al.* (2004) in the control groups of *P. americana* and *M. domestica,* Ahi (2004) and Gupta and Rathore (2004) in *Poekilocerus pictus* while drastic demarcating depletion in the number as well as in the intensities of PF in the ovaries of *P. americana* in the present investigation showed the dose and duration of treatment dependent toxicity of cerberin glycoside seed extract as reported in *Atractomorpha crenulata* by Murugan *et al.* (1996) and Babu *et al.* (1997) after neem extract treatment, in some species of *Dysdercus* by Mugdam and Banerjee (1999), Yadu *et al.* (1999), Mugdam (2000) after plumbagin and gossypol exposure, Rathore (1999) in *Poekilocerus pictus* with *Annona squamosa* seed extract treatment, Gupta and Rathore (2002) in *Poekilocerus pictus* after neem extract treatment, Bhide *et al.* (2003) in *Musca domestica* after *Delonix regia* seed extract treatment respectively due to destruction of fat bodies in the haemolymph which synthesizing the vitellogenin protein and other nutrients and their up take by developing oocytes by interfollicular cell spaces, pinocytosis and reverse pinocytosis were made in the phase of vitellogenesis but due to cerberin glycoside intoxication the process of vitellogenesis became arrested resulting into the production of large number of penultimate pathological oocytes (Figures 20.3–20.10) in severely effected cockroaches and exhibited the depletion of protein metabolites and these oocytes became degenerated to form large number of resorbed oocytes which never attain such maturity which is required for oviposition as observed by Bhide and Kharya (2002) in large number of insects of Order–Orthoptera, Diptera, Heteroptera.

In the present investigation depletion in the number of protein fractions could be correlated with the depletion of PAS +ve material in the deformed penultimate oocytes as reported by Bhide and Kumar (2004) in *Periplaneta americana* after *Cassia fistula* seed extract treatment, Bhide and Rai (2005) and Rai (2006) in *Bagrada cruciferarum* due to intoxication of some plant extracts. The ultimate consequences were the total inhibition of ootheca oviposition in the treated cockroaches and in this way fertility control programme of other insect pests is also possible.

Acknowledgements

We are very much thankful to Prof. Smita Banerjee, HOD, Department of Zoology, Dr. H.S. Gour Vishwavidyalaya, Sagar (M.P.) for providing the laboratory facilities and one of the authors (M. Bhide) is grateful to UGC for the financial assistance during the tenure of this research investigation.

References

Ahi, J.D., 2004. Effect of benzedine on the stress proteins of adult *Poekilocerus pictus* (Fabr.) Orthoptera: Acrididae. *Modern Trends in Physiological and Ecological Entomology*, 18: 9–15.

Babu, B., Murugan, K. and. Kavitha, R., 1997. Impact of azadirachtin on quantitative protein and lipid profiles during gonadotropic period of *Atractomorpha crenulata* Feb. *Ind. J. Exp. Biol.*, 35: 998–1001.

Bhide, M., 2003. Control of *Musca domestica* by *Annona squamosa* seed extract. *Him. J. Env. Zool.*, 17(1): 65–73.

Bhide, M. and Kharya, V., 2002. Potentials of resorptive bodies in insects. In: *Biological and Biotechnological Resources*, (Ed.) G. Tripathi and Y.C. Tripathi. Campus Book International, New Delhi, p. 324–350.

Bhide, M. and Kumar, S., 2004. Detection of protein and lipid fractions in the ovaries of *Periplaneta americana* treated with *Cassia fistula* seed extract. *J. Exp. Zool., India,* 7(2): 253–259.

Bhide, M. and Rai, R., 2005. Knock down toxicity of *Cerbera thevetia* seed kernel extract in different nymphal stages and oocyte resorption in *Bagrada cruciferarum* (Kirk). *J. Exp. Zool., India,* 8(1): 29–36.

Bhide, M., Kumar, S. and. Kharya, V., 2003. Control of fertility of *Musca domestica* by *Delonix regia* seed extract. *J. Ecophysiol. Occup. Hlth.,* 3: 219–232.

Finney, D.J., 1971. *Probit Analysis,* 3rd Edn., Cambridge University Press, London.

Gupta, U.S. and Rathore, C.S., 2002. Effect of neem seed oil on the physiology of reproduction in *Poekilocerus pictus. Modern Trends in Entomological Researches in India,* p. 166–201.

Gupta, U.S. and Rathore, C.S., 2004. Effect of *Annona squamosa* seed extract on the physiology of reproduction in *Poekilocerus pictus. Modern Trends in Physiological and Ecological Entomology,* 11: 6–14.

Jairaman, J., 1985. *Laboratory Manual in Biochemistry.* Wiley Eastern Ltd., New Delhi.

Mugdam, S., 2000. Investigation on bioactivity of plumbagin and some related compounds in *Dysdercus koenigii. Ph.D. Thesis,* Dr. H.S. Gour Vishwavidyalaya, Sagar.

Mugdam, S. and Banerjee, S., 1999. Histopathological alteration induced by two naturally occurring quinines in the ovary of *Dysdercus koenigii. Modern Trends in Entomological Researches in India,* 45: 36–39.

Murugan, K., Jahanmohni, P. and Babu, R., 1996. In: *Neem and Environment,* Vol. 1, (Eds.) R.P. Singh, M.S. Chari, A.K. Raheja and W. Kraus. Oxford and IBH, New Delhi.

Pearse, A.G.E., 1960. *Histochemistry: Theoretical and Applied.* J.&A. Churchill, London.

Rai, R., 2006. Control of fertility in *Bagrada cruciferarum* (Kirk) by some plant products. *Ph.D. Thesis,* Dr. H.S. Gour Vishwavidyalaya, Sagar (M.P.).

Rathore, C.S., 1999. Studies on the effect of some plant products on reproduction and haemolymph proteins in some insects. *Ph.D. Thesis*, Dr. H.S. Gour Vishwavidyalaya, Sagar (M.P.).

Sharma, J.D., 1988. *Forensic Sciences and Toxicology*. Suvidha Law House, Bhopal.

Yadu, N., Mugdam, S. and Banerjee, S., 1999. Chemosterilant activity of a natural product gossypol in the red cotton bug *Dysdercus similis*, Freeman (Phrrhocoridae, Heteroptera). *Modern Trends in Entomological Researches in India*, 49: 40–48.

Chapter 21

KD Toxicity of *Dhatura alba* Seed Extract in Different Nymphal Instar Stages and Antioviposional Activity Due to Oocyte Resorption in *Dysdercus similis*, Freeman (Phrrhocoridae : Heteroptera)

Mangla Bhide, Shuhbra Agarwal, Rekha Rai and Deepti Tomar*

Department of Zoology, Dr. H.S. Gour University, Sagar – 470 003, M.P.

ABSTRACT

In the present investigation methanol extract of *Dhatura alba* seeds was evaluated for its nymphicidal activity against all the five nymphal instars stages of *Dysdercus similis* at 1.5 per cent experimental sub lethal concentration. The extract showed nymphicidal activity at 1.5 per cent concentration and the KD_{50} (knockdown 50 per cent) was detected as 15 hrs in I[st] nymphal instars, 22 hrs in IInd nymphal instars, 25 hrs in IIIrd nymphal instars, 29 hrs in IVth nymphal instars and 45 hrs in V[th] nymphal

* Corresponding author: E-mails: manglabhide@rediffmail.com; manglabhide@yahoo.com

instars respectively by the method adapted after Finney (1971). The experimental findings proved the efficacy of the extract in control the fertility of *Dysdercus similis* because the last nymphal instar stages never became hatched into healthy adults and most of the nymphs emerged into malformed adults with less developed gonads and the oocytes of such adúlts emerged from these treated nymphal instar never attained such maturity which is required for oviposition, thus exhibited the antiovipositional activity of *Dhatura alba* in *Dysdercus similis.*

Keywords: *Dhatura, Nymphicidal, Antiovipositional action, Dysdercus similis.*

Introduction

Most of the plant products are used as folk medicines. Tribes also used them to kill or repel the insects and save their crops and vegetables from the damage of insect pests. Secondary metabolites of some plant products participated as defense barriers against insects (Luthria *et al.*, 1993). *Dhatura alba* is a common shrub belongs to family Solanaceae and known as *Kala Dhatura* in Hindi, abundant every where in M.P. The regular survey for several years suggested its insecticidal properties because no insect pest has been found on this plant. All the parts of this plants *e.g.* seed, leaves, stem and flowers are used to control the pests (Anonymous, 1985) as also reported by Girach *et al.* (1994).Dried aerial parts of *Annona* were used as potent insecticide against *Tribolium castaneum* (Atal *et al.*, 1978). Mellini *et al.* (1994) in *Exorista larvarum*, Patil *et al.* (1997) in *Dectynotus carthamii*, Kumar *et al.* (1999) in *Perigea capensis*, Kumar and Bhide (2001) in *Periplaneta americana*, Bhide (2003a, b) in *Musca domestica*, Bhide *et al.* (2004, 2005) in *Periplaneta americana*, Bhide and Rai (2005) and Rai (2006) in *Bagrada cruciferum* reported the insecticidal activity of some plant products and reported the high degree of mortality at higher dose treatment.

Dysdercus similis is a common pest of family–Malvaceae plants, with high rate of fertility and their nymphal stages causes a lot of damage to the standing crops of cotton (cash crop) and *bhindi*, so the present investigation was done to know more about the KD toxicity of *Dhatura alba* on the different nymphal stages and also investigate the antioviposional activity due to resorption of oocytes in *Dysdercus similis.*

Materials and Methods

The materials and methods used in this investigation are as follows:

Procurement of Seeds of *Dhatura alba* and Extraction

Dhatura alba belongs to family–Solanaceae. It grows wildly all over India. The seeds of these plants (shrub) procured locally, washed thoroughly, shadow dried, grind mechanically into coarse pieces and extracted with methanol (BDH, B.P. 60–80°C) by Soxhlet method (Sharma, 1988).The extract was concentrated by evaporation to dryness in water bath. After evaporation of the solvent, 1 per cent stock solution of extract was prepared with distilled water in dark colour glass bottles and stored in refrigerator. The stock solution was further diluted to decide the lethal doses and the detected values were 4.5 per cent KD_{100}, 3.0per cent KD_{50} (Finney, 1971), 2.5 per cent KD_0 and 1.5per cent sub lethal concentration (Table 21.1).

Table 21.1: KD Toxicity of *Dhatura alba* Seed Extract in 1st Nymphal Instar Stages of *Dysdercus similis* (Freeman)

Sl.No.	*Name of Plant*	*Concentration of Extract*	*Duration of Application (in hrs)*	*Per cent Mortality*	*KD Toxicity Value*
1.	*Dhatura alba*	1.5 per cent	25 hrs	100 per cent	KD_{100}
2.		experimental	15 hrs	50 per cent	KD_{50}
3.		concentration	10 hrs	Nil	KD_0
4.			06 hrs	Nil	Experimental concentration

Procurement and Rearing of *Dysdercus similis*

Adult insects of *Dysdercus similis* were collected from the plants of family Malvaceae and were reared on normal laboratory conditions by provided the fruits and leaves of Malvaceae plants to avoid the stress of starvation and water ad libitum to avoid the stress of dehydration. All the nymphal instar stages and 2, 4, 6 and 8 days old adult female insects emerged from control and treated last nymphal instars were used for experimental purposes.

Contact Bioassay

A contact bioassay modified method of Broussalis *et al.* (1999)

was used to test the toxicity of extract. 1.5 per cent concentration of the extract was applied on the glass petri dishes (10 cm diameter). 20 nymphal instars of each stage of *Dysdercus similis* were introduced into the petri dishes separately. Four replicates for each nymphal instar stages were used for the knock down test (*i.e.* those that no longer maintained normal posture and were unable to move or were on their backs) were recorded at 1 min intervals up to the total mortality was achieved. The nymphs were observed under dissecting microscope and mortality was determined when they did not respond to mechanical stimulation. Control groups nymphal instars were released separately on the petri dishes containing only solvent up to 96 hrs. Knock down (KD_{50}) as the hrs needed to produce 50 per cent mortality in experimental stages was determined by Probit analysis method (Finney, 1971).

Histochemical Studies on the Oocytes Resorption

For the histochemical studies on the oocyte resorption the 2, 4 and 6 days old female insects emerged from the treated last nymphal instars were treated with sub lethal concentration of extract (used in nymphicida1 toxicity) and were dissected out. Their ovaries were fixed for lipid detection in Calcium formyl fixative and for the detection of protein and carbohydrate in Carnoy's fluid (6 : 3 : 1) fixative and processed according to Pearse (1966). Paraffin blocks of the ovaries were prepared in usual manner and sections of the ovaries were cut at 6 µm.

Detection of Lipid, Protein and Carbohydrate in Ovarian Tissue

Sections of the ovaries were stained with "Sudan black B" for the detection of lipid, "Mercuric bromophenol blue stain" for the detection of protein and "Periodic acid Schiffs reagent" method after Pearse (1966) for the detection of carbohydrate. The results were expressed in the form of photomicrographs (Figure 21.1–21.4).

Results

The methanolic extract of *Dhatura alba* was tested for "Knock down" toxicity and the time in hrs for KD_{100}, KD_{50} and KD_0 for each nymphal instar stages of *Dysdercus similis* was recorded and summarized in Tables 21.1–21.5.

Table 21.2: KD Toxicity of *Dhatura alba* Seed Extract in IInd Nymphal Instar Stages of *Dysdercus similis* (Freeman)

Sl.No.	*Name of Plant*	*Concentration of Extract*	*Duration of Application (in hrs)*	*Per cent Mortality*	*KD Toxicity Value*
1.	*Dhatura alba*	1.5 per cent	30 hrs	100 per cent	KD_{100}
2.		experimental	22 hrs	50 per cent	KD_{50}
3.		concentration	15 hrs	Nil	KD_{0}
4.			07 hrs	Nil	Experimental concentration

Table 21.3: KD Toxicity of *Dhatura alba* Seed Extract in IIIrd Nymphal Instar Stages of *Dysdercus similis* (Freeman)

Sl.No.	*Name of Plant*	*Concentration of Extract*	*Duration of Application (in hrs)*	*Per cent Mortality*	*KD Toxicity Value*
1.	*Dhatura alba*	1.5 per cent	40 hrs	100 per cent	KD_{100}
2.		experimental	25 hrs	50 per cent	KD_{50}
3.		concentration	18 hrs	Nil	KD_{0}
4.			10 hrs	Nil	Experimental concentration

Table 21.4: KD Toxicity of *Dhatura alba* Seed Extract in IVth Nymphal Instar Stages of *Dysdercus similis* (Freeman)

Sl.No.	*Name of Plant*	*Concentration of Extract*	*Duration of Application (in hrs)*	*Per cent Mortality*	*KD Toxicity Value*
1.	*Dhatura alba*	0.1 per cent	50 hrs	100 per cent	KD_{100}
2.		experimental	29 hrs	50 per cent	KD_{50}
3.		concentration	20 hrs	Nil	KD_{0}
4.			12 hrs	Nil	Experimental concentration

The experiments suggested that 1st nymphal instars were more susceptible to the test solution in comparison to other nymphal stages while last nymphal instar stages were less affected but they developed into small sized, morphological malformed weakened adults with less developed gonads in female insects and it was exhibited by the gradual histopathological and histochemical changes in the ovaries as follows:

In Two Days Old Female

The acrotrophic ovaries of two days old *Dysdercus similes* exhibited the presence of number of pathological oocytes showed the resorption of lipid bodies by irregular follicular epithelium and vacuolization has seen nearby the cortical ooplasm. Intrafollicular lipid spherules were observed. Depletion of L_1, L_2 and L_3 lipid bodies was prominent in irregular reduced sized oocytes. Degeneration of nutritive cords was prominent (Figure 21.1).

Table 21.5: KD Toxicity of *Dhatura alba* Seed Extract in Vth Nymphal Instar Stages of *Dysdercus similis* (Freeman)

Sl.No.	*Name of Plant*	*Concentration of Extract*	*Duration of Application (in hrs)*	*Per cent Mortality*	*KD Toxicity Value*
1.	*Dhatura alba*	1.5 per cent	65 hrs	100 per cent	KD_{100}
2.		experimental	45 hrs	50 per cent	KD_{50}
3.		concentration	30 hrs	Nil	KD_0
4.			15 hrs	Nil	Experimental concentration

In Four Days Old Female

The rate of vacuolization in the highly PAS +ve ooplasm was increased in comparison to two days old female. The variable sized vacuoles were totally PAS –ve in nature. The less PAS +ve follicular epithelium became irregular and thin walled and *Tunica propria* showed the folding and refolding resulted into reduced size of pathological oocytes (Figure 21.2).

In Six Days Old Female

The resorbed oocytes showed the presence of less PAS +ve, irregular and pycnotic follicular epithelium invading in side the less PAS +ve and highly vacuolated ooplasm. Some lecitholytic cells were seen inside the ooplasm at the site of active resorption. The vacuoles were totally PAS –ve in nature and were of variable sized. Some of the vacuoles were fused to form large sized vacuoles. The area of vacuolization denoted the active resorption of ooplasm from the corresponding site (Figure 21.3).

In Eight Days Old Female

The ovarian architecture showed the presence of large number of pathological resorbed oocytes filled with large number of less protein +ve lecitholytic cells. Nutritive cords were protein +ve in nature but showed interruption at some places. Follicular epithelium and *Tunica propria* were intense protein +ve became thin and showed invading inside the protein +ve ooplasm (Figure 21.4).

Discussion

A lot of research work has been done on the insecticidal property of plant products by Luthria *et al.* (1993), Girach *et al.* (1994), Chang *et al.* (1995), Atal *et al.* (1998), Kumar *et al.* (1999), Jaswanth *et al.* (2002), Kumar (2003) and Bhide *et al.* (2004, 2005) and reported high percentage of mortality in nymphal instars and adults at higher concentration.

In the present study 1.5 per cent solution of *Dhatura alba* was tested for KD toxicity in all nymphal instar stages and adult female insects emerged from the treated Vth nymphal instars exhibited the high degree of mortality in nymphs at the time of moulting and the 1st nymphal instars were more susceptible to the test concentration in comparison to other nymphal stages as reported by Atal *et al.* (1978) in *Tribolium castaneum*, Mellini *et al.* (1994) in *Exorista larvarum*, Patil *et al.* (1997) in *Dactynotus carthamii* and Kumar *et al.* (1999) in *Perigea capensis* after the application of some plant extracts, Bhide and Rai (2005) and Rai (2006) in *Bagrada cruciferarum* after treatment with some glycosides of plant origin respectively.

The *Dhatura alba* seed extract being a contact poison for insects which penetrate the body wall and tracheal system resulted into death probably due to the insecticidal activity against *Dysdercus similis*. The methanolic extract of *Dhatura alba* seeds acts as nymphicidal agent in the present investigation as reported by Kumar and Bhide (2001), Kumar (2003), Bhide *et al.* (20m a, b, 2004) in *Musca domestica* after *Cassia fistula* and *Delonix regia* seed extract treatment.

Adult female insects showed dose and duration of treatment dependent partial to total sterility which was evident by the

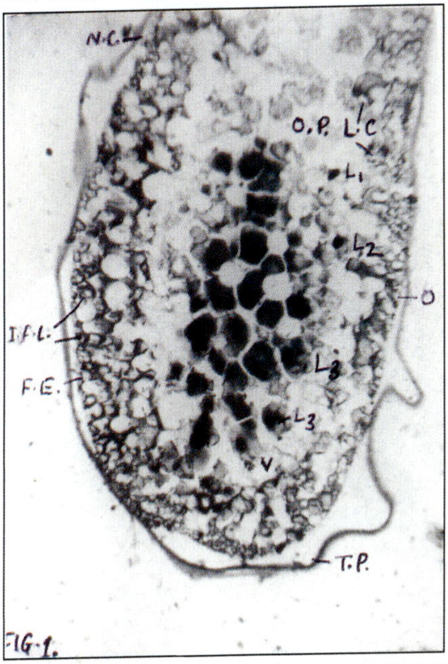

Figure 21.1: Section of Ovary of Two Days Old *Dysdercus similis* Stained with Sudan Black B Stain Showing the Depletion of L_1, L_2 and L_3 Lipid Bodies and Intra Follicular Lipid in Pathological Resorbed Oocytes and Degenerated Nutritive Cords (X 450)

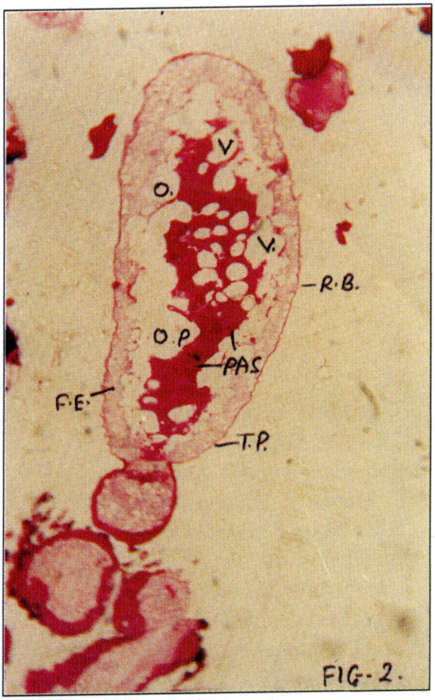

Figure 21.2: Section of Ovary of Four Days Old *Dysdercus similis* Stained with Periodic Acid Schiffs Reagent Showing the Resorption of PAS +ve Material from the Site of Vacuolization. Some lecitholytic cells are seen in the ooplasm (X 450).

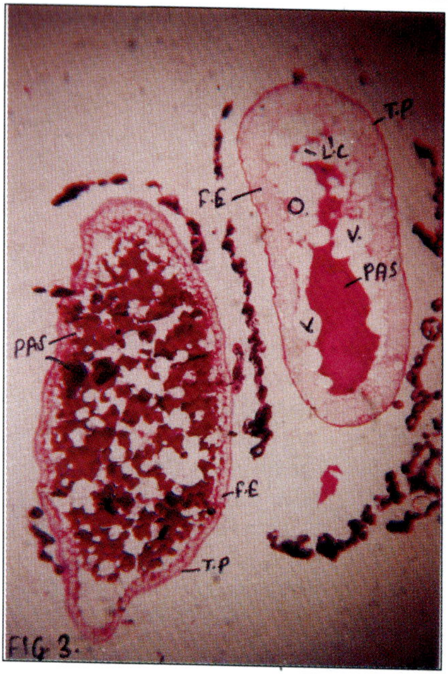

Figure 21.3: Section of Ovary of Six Days Old *Dysdercus similis* Stained with PAS Reagent Showing the Resorption of PAS +ve Material from the Site of Vacuolization. Follicular epithelium showing less PAS +ve material (X 450).

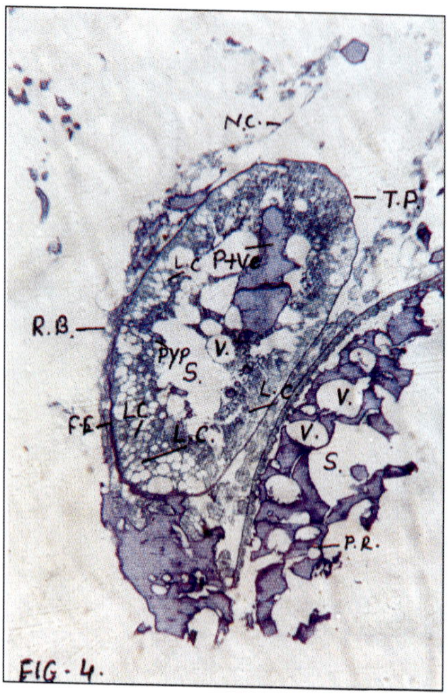

Figure 21.4: Section of Ovary of Eight Days Old *Dysdercus similis* Stained with Mercuric Bromophenol Blue Stain Showing the Resorption of Protein +ve Material from Most of Resorbed Oocytes which are Indicated by Hyper Vacuolization (X 450).

Keys to Abbreviations

F.E.: Follicular epithelium; L_1, L_2, L_3: Lipid bodies; I.F.L.: Intrafollicular lipid; L.C.: Lecitholytic cells; N.C.: Nutritive cords; O.P.: Ooplasm; P.A.S: PAS +ve material; P.R.: Protein +ve material; O.: Oocyte; R.B.: Resorbed oocyte; P.Y.P.: Protein yolk precursors; T.P.: *Tunica propria*; V.: Vacuoles.

morphological, histopathological and histochemical studies on the gonads of the experimental groups. Ovarian architecture showed the presence of large number of pathological resorbed oocytes which were reduced in size, deformed with gradually decreased amount of essential nutrients (required for the progressive development of the healthy oocytes in control groups) resulted into resorption of these pathological oocytes which never attain such maturity required for ovulation and oviposition so the seed extract of *Dhatura alba* acts as antiovipositional agent and no oviposition was observed in treated female insects of *Dysdercus similes* as observed by Bhide (2003a, b) in *Musca domestica* after *Cassia fistula* and *Annona quamosa* seed extract treatment.

The results exhibited that this insecticide is of great economic importance from the agronomic point of view. The reason for using new natural pesticides is that these are active at highly acceptable level as well as these plant products are non-persistent type, biodegradable and their residues not accumulated in the food chains, non toxic to higher animals and showed significant toxic effects on small pest insects and it could be quite significant after formulation in large scale to use these chemicals of plant origin to control the fertility of pests of other crops and to increase the productivity of the developing country like India.

The ovarian architecture showed the presence of large number of resorbed oocytes which never attain such maturity which is required for the oviposition and was the main cause of nil or low rate of fecundity in *Dysdercus similis.* No oviposition has been observed in the treated groups in the present investigation and in this way it is quite significant to control the fertility of *Dysdercus similis,* which is the serious pest of family–Malvaceae and save our valuable products of high nutritive value and increase the economy of Indian farmers.

Acknowledgements

We are very much thankful to Prof. Smita Banerjee, HOD, Department of Zoology, Dr. H.S. Gour Vishwavidhyalaya, Sagar (M.P.) for providing the laboratory facilities and one of the authors (M. Bhide) is grateful to UGC for the financial assistance during the tenure of this research investigation.

References

Anonyomus, A., 1985. The wealth of India, Raw materials. *J. Council of Scientific and Industrial Research,* New Delhi, India, 1: 284–286.

Atal, C.K., Srivastava, J.B., Wali, B.K., Chakravarty, R.B., Dhawan, B.N. and Rastogi, R.P., 1978. Screening of Indian plants for biological activity. *Ind. J. Exp. Biol.,* 16: 330–349.

Bhide, M., 2003a. Control of *Musca domestica* by *Annona squamosa* seed extract. *Himalayan J. Environ. Zool.,* 17(1): 65–73.

Bhide, M., 2003b. Control of fertility of *Musca domestica* by *Delonix regia* seed extract. *J. Ecophysiol. Occup. Hlth.,* 3: 219–232.

Bhide, M., Kumar, S. and Kharya, V., 2004. Control of *Musca domestica* by *Cassia fistula* seed extract. In: *Perspectives in Animal Ecology and Reproduction,* (Eds.) V.K. Gupta and Anil K. Verma, 2: 214–240.

Bhide, M. and Rai, R., 2005. Knock down toxicity of *Cerbera thevetia* seed kernel extract in different nymphal stages and oocyte resorption in *Bagrada cruciferarum* (Kirk). *J. Exp. Zool.,* 8(1): 29–36.

Broussalis, A.M., Ferrao, G.E., Martino, V.S., Coussio, J.D. and Alvarez, J.C., 1999. Argentina plants as potential source of insecticidal compounds. *J. Ethanopharmacol.,* 67: 219–223.

Chang, F.R., Chen, K.S., Ko, F.N., Teng, C.M. and Wu, Y.C., 1995. Bioactive alkaloids from *Annona squamosa. Chin. J. Pharma.,* 47: 10–15.

Finney, D.J., 1971. *Probit Analysis,* 3rd Edn. Cambridge University Press, London.

Girach, R., Aminnudin, D., Siddioni, P.A. and Khan, S.A., 1994. Traditional plant remedies among the Kondh district of Dhenkanal (Orissa). *Int. J. Pharmacogn.,* 32: 274–283.

Jaswanth, A., Ramanatran, M., Krishnaraj, V. and Ruckmani, K., 2002. Insecticidal activity of leaf extract of *Annona reticulata*. *Adv. Pharmacol. Toxicol.*, 3(2): 13–16.

Kumar, S., 2003. Studies on the insecticidal properties of some plant extracts on the female reproductive organs of some insects. *Ph.D. thesis*, Dr. H.S. Gour V.V., Sagar.

Kumar, S. and Bhide, M., 2001. Effect of *Cassia fistula* seed extract on the ovarian tissue of *Musca domestica*. In: *Proc. 88 Session, Ind. Sci., Congo*, Abst. 46, 49.

Kumar, S., Kulkarni, S.R., Basavana, G. and Mallapur, C.P., 1999. Insecticidal property of some indigenous plant extracts against Lepidopteran pests of Safflower. *Annals of Bioresearches*, 4(1): 49–52.

Luthria, D., Ramkrishnan, V. and Banerji, A., 1993. Insect antifeedent activity of furochromes: Structure activity relationship. *J. Nat. Prod.*, 56: 671–675.

Mellini, E., Gardenghi, G. and Coulbaly, A.K., 1994. Anatomical and histological characteristics of the female genital apparatus of *Exorista larvarum*, L. (Diptera: Tachinidae), a parasitoid that lays macro type eggs on the host. *Bollettino Institute di Entomologia*. Guido Grand della Universita degali Studi-di-Biologia, 48: 35–48.

Patil, R.K., Rayar, S.G., Bassapa, H., Hiremath, L.G. and Patil, B.R., 1997. Insecticidal property of indigenous plants against *Dactynotus carthamii*, H.R.L. and its predator *Chrysoperlacamea*. *J. Oilseeds Research*, 14(1): 71–73.

Pearse, A.G.E., 1966. *Histochemistry: Theoretical and Applied*. J.&A., Churchill, London.

Rai, R., 2006. Control of fertility in *Bagrada cruciferarum* (Kirk) by some plant product. *Ph.D. Thesis*, Dr. H.S. Gour Vishwavidhyalaya, Sagar (M.P.).

Sharma, J.D., 1988. *Forensic Sciences and Toxicology*. Suvidha Law House, Bhopal.

Chapter 22

Studies on the Lifecycle of *Moina micrura* Kurz, 1874 (Crustacea: Branchiopoda: Cladocera)

K.K. Subhash Babu and S. Bijoy Nandan***

Department of Marine Biology, Biochemistry and Microbiology,
School of Marine Sciences,
Cochin University of Science and Technology,
Cochin – 16, Kerala

ABSTRACT

The life-cycle of Cladocera *Moina micrura* Kurz, was studied on the basis of laboratory culture techniques and presented. The female *M. micrura* showed single pre-adult instar and 11 adult instars. The maximum number of eggs (18.4±2.2) was observed in the 7th instar, but with the progress in development the number of eggs tended to decrease. The neonates produced from the same brood may be of any sex. The life span of female, with 301.3 hours produced 122.8±11.06 eggs. In males, six instars were observed but after the first instar, one pair of long testis was clearly visible on both sides of the alimentary canal. The

* E-mail: kallikadavil@yahoo.com; **bijoynandan@yahoo.co.in.

males had a higher life span than the females, with 417 hours. This study reports on the life cycle of both parthenogenetic and gamogenetic phase of *Miona micrura* Kurz.

***Keywords:** Cladocera, Moina micrura, Ephippium, Instar, Male.*

Introduction

Understanding the life cycle of cladocerans and other micro crustaceans would help in assessing the secondary productivity of aquatic ecosystems. Moreover, cladocerans are considered as an ideal live feed for nursery ponds and hatcheries of all cultivable fishes (Alikunhi *et al.*, 1955; Parabrahman *et al.*, 1967). Therefore, knowledge on the life cycle and biology of individual species of Cladocera is a prerequisite for the successful culture of these micro-crustaceans. In this context, Micheal (1962), made the first report on the life cycle of female cladocera, *Ceriodaphnia cornuta* Sars from Madurai, India. The other related studies on the life cycle of female cladocerans were by Navaneethakrishnan and Michael (1971) on *Daphnia carinata* King; Murugan and Sivaramakrishnan (1976) on *Scapholeberis kingi* Sars; Murugan (1975) on *Moina micrura* Kurz; Murugan and Venkataraman (1977) on *Daphnia carinata*; Kanaujia (1982) on *Simocephalus vetulus* (O. F. MÜller); Thresiama *et al.* (1991) on *Moina micrura* Kurz and Subhash Babu and Nayar (1993) on *Ceriodaphnia cornuta* Sars. However, available reports on the biology of male *Daphnia lumholtzi* Sars was reported by Sharma *et al.* (1981) and that of *Simocephalus serrulatus* Koch by Subhash Babu and Nayar (1997).

Thus, most of the studies available are on the life cycle of female, with very little information on the male cladocerans. In view of this, the present contribution discusses in detail the lifecycle of both male (gametogenetic) and female (parthenogentic) of *Moina micrura* Kurz.

Materials and Methods

The specimens were collected from a tank in Irinjalakuda (10° 20'42"N, 76°12'44"E), Kerala, India and brought live to the laboratory. Ten ovigerous females were sorted under a stereomicroscope and transferred into an earthen pot of 5 litre capacity containing culture medium. This was maintained as a stock culture and kept outside the laboratory. The culture medium was prepared by green pond water and filtered through a net of 50mm bolting silk. The medium contained mainly unicellular algae (*Chlorella* sp) with a density about 10×10^4 cells/litre. Further 50 mg/litre finely powdered groundnut cake was also added as supplementary food for animals and growing algal cells (modified methodology by Green, 1956).

One ovigerous healthy female isolated from stock culture with the help of a pipette was inoculated into a glass beaker containing 250 ml of the culture medium. The female was kept under constant observation so as to isolate the newly hatched young ones. Young ones (neonates) were individually reared in 10 different petri-dishes containing the culture medium and kept in laboratory under the fluorescent light (40 watts). The time of hatching, moulting, number of neonates hatched in each brood and the life span of all individuals were regularly recorded until their death. Newly released young ones from the brood were subjected to careful observation under low power microscope to verify their sex and isolate the males for observing their life cycle. All observations were made at room temperature 28±2°C. The medium was changed every 24 hours in order to prevent any possible starvation and accumulation of fecal matters.

Ephippial females were also collected from the stock culture and isolated under the microscope. Another set of 5 individuals were simultaneously reared in the same medium for dissections. Dissections were made by using micro-tungsten needles. The identification of the specimens was carried out using the standard monographs (Michael and Sharma, 1988; Dodson and Frey, 1999). The measurements were made with calibrated micrometer and diagrams drawn by camera lucida.

Results

In mature females, the ovaries were seen as a pair of elongated sacs on each sides of the alimentary canal. The content of the ovary was discharged through a small opening at the posterior end of the brood pouch. The discharged mass later became spherical and formed the eggs. The relationship between mean size and the number of eggs produced from each brood and instar duration are given in the Table 22.1.

Table 22.1: Number of Instars, Size, Number of Eggs/Brood and Duration of Instars in *Moina micrura* Kurz at 28±2°C (n=10)

Instar No.	*Mean Length (mm)*	*Mean Height (mm)*	*Eggs/ Brood (X±SD)*	*Instar Duration (in hours)*	*Cumulative Duration of Instars (hours)*
1	0.49	0.24	0	20.0	20.0
2	0.75	0.45	6.0±0	22.2	42.2
3	0.87	0.54	16.0±0.5	22.3	64.5
4	0.98	0.67	16.0±0.6	24.1	88.6
5	1.03	0.71	16.0±1.2	25.0	113.6
6	1.09	0.76	16.0±1.2	28.4	142.0
7	1.12	0.80	18.4±2.2	26.5	168.5
8	1.16	0.80	16.6±1.6	28.2	196.7
9	1.18	0.82	11.3±1.9	27.0	223.7
10	1.20	0.84	10.6±2.3	28.1	251.8
11	1.20	0.84	7.5±4.5	29.5	281.3
12	1.20	0.84	0	20.0	301.3

The morphological features of the embryonic stages of *Moina micrura* Kurz are given below:

1. Eggs are spherical, slightly greenish with transparent edge, encased by a thin egg membrane (Figure 22.1a)
2. Slightly elongated embryo with granulated central region surrounded by dividing cells (Figure 22.1b).
3. Elongated embryo without distinct head lobe but rudiments of antenna was developed. Some fat globules are present in the central region of the developing embryo (Figure 22.1c).
4. Embryo with distinct head lobe and rudiments of antennae and antennules. The egg membrane cast off and two pinkish eyespots were present (Figure 22.1d).
5. The two pink eyespots modified to large dark eyespots, closer to each other. The segmented endopodite and exopodite of the antennae clearly visible with two rudiments of antennule (Figure 22.1e).

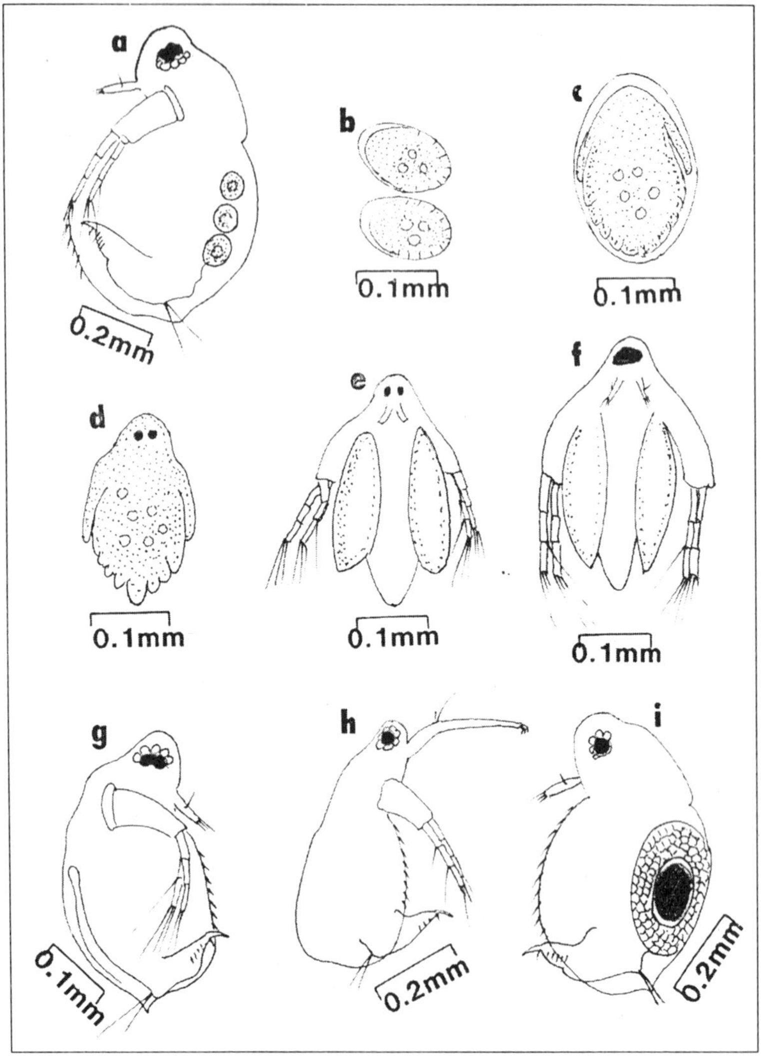

Figure 22.1: ***Moina micrura***

6. Two dark eye spots fused and formed a large single median eye. Towards the ends of this stage all appendages were well developed and the young ones started exhibiting movements (Figure 22.1f).

7. The neonates hatched out from brood chamber resembled the adults in their morphological features (Figure 22.1g).

The young ones were released from the brood pouch by jerking movements of the post abdomen. Female neonates had a mean length of 0.49 mm and width of 0.24 mm. Then it develops into a single pre-adult instar and eleven adult instars in a life span of 301.3 hours with an average instar duration of 25.1±3.2 hours (n=10).

Maximum length of 1.2 mm and width of 0.84 mm were attained on its 10th instar and maximum growth rates were noted in the pre-adult instar. After the first adult instar the animal produced 6 eggs however, the maximum number of eggs (18.4±2.2) was observed in its 7th instar. After the 7th instar there was a decline in eggs production and no eggs were observed at its 12th instar (Figure 22.2). A total of 122.8±11.06 eggs were produced during the life span of 301.3 hours (Table 22.I). The number of eggs in the brood showed pronounced variations in the later instars (7-11th) as compared to earlier periods (Figure 22.2). Moulting occurred any time immediately after the release of young ones from the brood pouch. Other sets of eggs were discharged into the brood pouch immediately after moulting and the parthenogenetic cycle is repeated.

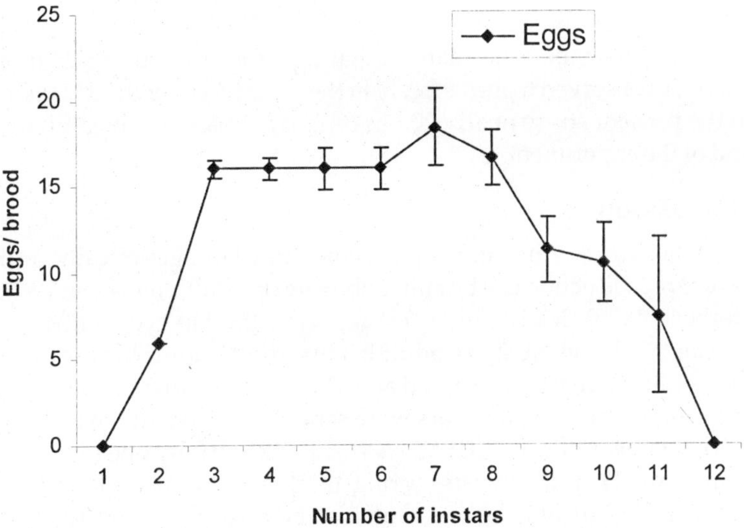

Figure 22.2: Eggs/Brood in *Moina micrura* Kurz (n=10)

Ephippial Female (Sexual Female)

There were numerous ephippial females in the stock culture. They could be distinguished from the pathenogenetic females by its smaller size and presence of dark ephippium on the dorsal side of the carapace (Figure 22.1i). The ephippium was saddle shaped (mean size 0.38mm) which is a modified form of brood pouch and its outer surface ornamented with honeycomb like pattern. Ephippium contained only a single egg, which required fertilization by male and after fertilization the female produces resting eggs. Free ephippia were found in large numbers floating on the surface of the culture medium as well as sticking on the sidewall of the container.

Male

Males were small and oblong with a pair of long prehensile antennules (Figure 22.1h). Newly released males measured 0.49 mm in length and 0.24 mm height with the antennules measuring 0.15 mm. One pair of long testis was clearly visible on the both sides of the alimentary canal after its 1st instar. Six instars were observed in the life span of male for completing the life cycle with a period 417 hours (Table 22.2). The average size of male during the last instar (6th) was 0.89 mm in length and 0.39 mm width after which the animals stopped it's moulting and maintained their size until death. The testis in the adult males opened on the dorsal side of the post abdomen through a pair of sperm ducts. The spermatozoa were spherical in shape with many radiating axons measuring 25 mm and were observed to move freely in the testis just after the 3rd instar. In the present study nearly 20 per cent of animals survived till the end of the experiment.

Discussion

Cladocerans normally reproduce parthenogenetically and sexual reproduction can be either obligate or facultative (Han, 1969; Herbert, 1978). Murugan (1975) reported the life-cycle of *Moina micrura* Kurz, where 2 pre adult instars and 11 adult instars of 13 days of life span were observed and also indicated that the duration of pre-adult and adult instars were same (24 hours). In the present study in it was noticed that *M. micrura* Kurz had only one pre adult instar and 11 adult instars during the life span of the 12.6 days (Table 22.1). Further, the pre-adult instar duration was shorter than the average adult instar duration which varied from 22.2-28.4 hours.

Similar observations were also made in other tropical cldocerans (Venkataraman, 1981; Kanaujia, 1987; Subhash Babu and Nayar, 1993).

Table 22.2: Numbers of Instars, Duration of Instars and Size in *Moina micrura* Kurz (Male) at 28±2°C (n=10)

Instar No	*Mean Length (mm)*	*Mean Height (mm)*	*Duration of Instars*	*Cumulative Duration of Instars (in hours)*
1	0.49	0.24	42.1	42.1
2	0.69	0.32	48.0	90.1
3	0.81	0.36	48.5	138.6
4	0.83	0.38	55.1	193.7
5	0.86	0.38	58.5	152.2
6	0.89	0.39	164.8	417.0

In the present study, the clutch size in each female during their lifespan was 122.8±11.06, whereas Murugan (1975) reported only 61.8 eggs. The number of eggs in each brood and duration of instar are directly influenced by many factors such as amount of the food available to animal (Anderon and Jenkins, 1943), temperature of the culture medium (Kanaujia, 1988) and genetic constitution of the species (Banta and Wood, 1939). Recent studies on the life history of Cladocera has revealed that the instar duration, number of instars for primiparous stage, time to primiparous stage and clutch size were influenced not only by temperature and food but also by chemical signals produced by predators (Han, 1984; Schwtz, 1984; Havel and Dodson, 1989), which however, needs further study.

Increase in the size of individuals in each instar was found to be more rapid during the pre-adult phase in this study (Table 22.1). Similar observations were reported by Green, (1956) on studies in *Daphnids*. In the present study, number of eggs in the first primiparous instar was 6, whereas maximum number of eggs was observed in the 7th instar, after which there was a decline and no eggs were observed at its last instar (Figure 22.2). Similar observations were made by Murugan, (1975) and Venaktaraman (1981) in *Moina micrura* and *Daphnia carinata*. However, Navaneethakrishnan and Michael (1971) reported that the eggs in each brood gradually increased from the first instar to the last instar in *Daphnia carinata* king, whereas Murugan and Job (1982) observed only two eggs in all instars of *Ledigia acanthocercoids* (Fischer) without increase or decrease in the rate of egg production.

It is known that, cladocerans also reproduce sexually but is not obligatory (Shan, 1969; Herbert, 1978). The embryonic stages in *Moina micrura* Kurz revealed a close similarity with general pattern of development in cladocerans with the only exception in duration of embryonic period (Hutchinson, 1967). Studies by Dodson and Frey, (1991) reported that males of *Moina* sp. and *Daphniopsis* sp. developed from sexually derived resting eggs. But, in the present study, males

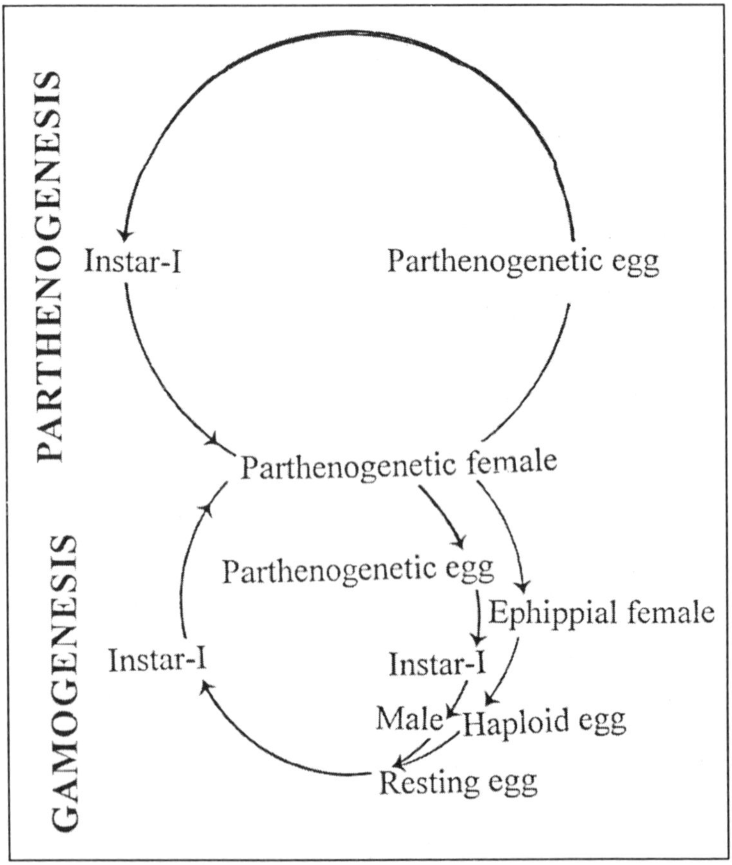

Figure 22.3: Life-Cycle Pattern of *Moina micrura* Kurz

hatched out from parthenogenetic eggs. They also reported that asexual formation of males could be induced in certain cladocerans due to deterioration in environmental factors such as, poor food concentration, crowding and decreasing photo- periods.

In the present investigation, the life span of male was longer than that of female and moulting also occurred in the adult phase (Table 22.2). On the contrary, Sharma *et al.* (1981) and Subhash Babu and Nayar (1997) reported a shorter life span of male than that of female in both *D. lumholtzi Sars* and *Simocephalus serrulatus* Koch. However, in the genus *Scapholebris* and *Daphnia,* males continue to moult even after attaining maturity (Dodson and Frey, 1991).

As observed in the present study, the spermatozoa were spherical with radiating axons. Spherical spermatozoa with radiating axons are probably a generic character, in *M. micrura* (Golden, 1968). In natural systems, males appear in the population before the appearance of ephippial females (Shan, 1969). In this study also, large number of ephippial females appeared along with the males in the stock culture due to over crowding. Similar observations were made by Pennak (1991); Michael (1962) and Hutchinson (1967). Therefore, over crowding could be a regulating factor for the formation of males and subsequent appearance of ephippial females in the natural population of Cladocera.

Conclusion

Cladocera *Moina micrura* showing parthenogenetic mode of reproduction in the favorable environmental condition. Whereas, under the unfavorable environmental condition the animal will switchover its reproductive strategies from normal parthenogenetic to gamogenetic phase (Figure 22.3).

Acknowledgements

Authors are very much thankful to Dr. C. K. G. Nayar, Director, Kerala Limnological Research Institute (KLRI), Christ College Campus, Irinjalakuda, Kerala, for his constructive comments on the manuscript.

References

Alikunhi, K.H., Chaudhuri, H. and Ramachandran, V., 1955. On the mortality of carp fry in nursery ponds and the role of plankton in their survival and growth. *Ind. J. Fish*, 2: 257–313.

Banta, A.M. and Wood, T.R., 1993. General studies in sexual reproduction. *In Banta*, p. 131–181.

Dodson, S. I., 1989. Predator induced reaction noms. *Bioscience*, 39: 447–452.

Dodson, S.L. and Frey, D.G., 1991. Cladocera and other branchiopoda. In: *Ecology and Classification of North American Freshwater Invertebrates*, pp. 723–786, (Eds.) James H. Throp and Alan P. Covich. Academic Press, Inc. San Diego, 911 pp.

Goulden, C.E., 1968. The systematic and evelution of Moinidae. *Trns. Amer. Phil. Soc.*, 58: 1–101.

Green, J., 1956. Growth size and reproduction in *Daphnia* (Crustacea: Cladocera). *Proc. Zool. Soc. Lond*, 126: 173–204.

Herbert, P.D.N., 1978. The population biology of *Daphnia* (Crustacea: Daphniidae). *Biological Review*, 53: 387–426.

Havel, J.E. and Dodson, S.I., 1987. Reproductive coasts of Chaoborus induced polymorphism in *Daphnia pulex*. *Hydrobiologia*, 150: 273–281.

Han, B.J., 1984. Influence of temperature on life history characteristics of two sibling species of *Eurycercus* (Cladocera: Chydoridae). *Canadian Journal of Zoology*, 63: 891–898.

Hutchinson, C.E., 1967. *A Treatise on Limnology, Vol. VII: Introduction to Lake Biology and Limnoplankton*. John Wiley and Sons, New York, 1115 pp.

Kanaujia, D.R., 1988. Preliminary observations on the life history and culture of *Ceriodaphnia cornuta* Sars (Cladocera: Daphniidae). *Ind. J. Anim. Sci.*, 58(1): 150–154.

Kanaujia, D.R., 1987. Biology and development in *Simocephalus vetulus* (O. F. Muller, 1976) (Cladocera: Daphniidae). *Ind. Animal Sci.*, 57(1): 1153–1160.

Kanaujia, D.R., 1984. Life history, ephippia development, cyclomorphosis and temperature effects on the life-cycle in *Daphnia lumholtzi* Sars. (Cladocera: Daphniidae). *J. Bombay Nat. Hist. Soc.*, 80(2): 442–448.

Kanaujia, D.R., 1982. Instar duration, instar number, egg production and longevity in *Ceriodaphnia cornuta* Sars at two temperature ranges. *J. Bombay Nat. Hist. Soc.*, 79(2): 441–445.

Michael, R.G. and Sharma, B.K., 1988. Fauna of India and adjacent countries (Indian Cladocera). *Zoological Survey of India*, 262 pp.

Michael, R.G., 1962. Seasonal events in a natural population of the Cladoceran *Ceriodaphnia cornuta* Sars and observations on its life cycle. *J. Zool. Soc., Ind.*, 14: 211–218.

Murugan, N. and Job, S.V., 1982. Laboratory studies on the life cycle of *Ledigia acanthocercoides* Fisher (1854) (Cladocera: Chydoridae). *Hydrobiologia*, 89: 6–16.

Murugan, N. and Venkataraman, K., 1977. Study of the *in vitro* development of the parthenogenetic egg of *Daphnia carrinata* king (Cladocera: Daphniidae). *Hydrobiologia*, 52(2–3): 129–134.

Murugan, N. and Sivaramakrishnan, K.G., 1976. Longivity, instar duration, growth, reproduction and embryonic development in *Scapholebris kingii* Sras (1903) (Cladocera: Daphniidae). *Hydrobiologia*, 50: 75–80.

Murugan, N., 1975. Egg production, development and growth in *Moina micrura* Kurz (1874) (Cladocera: Moinidae). *Freshwat. Biol.*, 5: 245–250.

Navaneethakrishnan, P. and Michael, R.G., 1971. Egg production and growth in *Daphnia carinata* King. *Proc. Ind. Acad. Sci.*, 72: 117–123.

Pennak, R.W., 1991. *Freshwater Invertebrates of the United States*, 3rd Edn. John Wiley and Sons Inc., New York.

Parabrahman, M., Khan, A.N. and Lakshminarayana, J.S.S., 1967. Occurrence, growth and feeding habits of *Moina dubia* Gurney and Richard and its role in the stabilization of sewage. *J. Mar. Biol. Asian India*, Symposium on Crustacea Part–II: 586–594.

Shan, R.K., 1969. Life cycle of a chydorid cladoceran *Pleuroxus denticulatus* Birge. *Hydrobiologia*, 31: 203–238.

Schwartz, S., 1984. Life history strategies in *Daphnia*: A review and predictions. *Oikos*, 42: 114–122.

Sharma, S., Sharma, B.K. and Michael, R.G., 1981. Laboratory studies on the longevity, instar duration and growth of *Daphnia lumholtzi* Sars (Cladocera: Daphniidae). *Curr. Sci.*, 50: 200–206.

Subhash Babu, K.K. and Nayar, C.K.G., 1997. Laboratory studies on the life cycle of *Simocephalus serrulatus* Koch 1881 (Cladocera: Crustacea). *J. Bombay Nat. Hist. Soc.*, 94(2): 317–321.

Subhash Babu, K.K. and Nayar, C.K.G., 1993. Observations on the life cycle of *Ceriodaphnia cornuta* Sars. *J. Zool. Soc., Kerala*, 3(2): 13–17.

Thresiama, J., Mercy, T.V.A. and Thampy, D.M., 1991. Production and population density of *Moina micrura* Kurz culture in different media. *J. Zool. Soc., Kerala*, 1(1): 21–31.

Venkataraman, K., 1981. Field and laboratory studies on *Daphnia carinata* King (Cladocera: Daphniidae) from a seasonal tropical pond. *Hydrobiologia*, 78: 221–225.

Chapter 23

Life Cycle Studies and Growth Dynamics of *Moina macrocopa*

Satinder Kour
Department of Zoology, Government College for Women, Gandhi Nagar, Jammu

ABSTRACT

The *in situ* development of *Moina macrocopa* has constantly been observed, during present investigations. The different developmental stages along with the details of time taken to complete different stages of development were recorded. The length, breadth brood and fecundity were tabulated in regular time slots.

Keywords: Life cycle, Crustacean, Cladoceran, Moina macrocopa.

Introduction

Moina macrocopa are small mostly freshwater crustaceans. Some people know them as a "water flea". They move in short jerky motions, this is how they got their nickname. There are different types of daphniidae all over the world. The two types that are the most varied and the biggest food source for young and adult freshwater fish are *Moina* and *Daphnia*. Moina are the species of daphniidae used in culture the most often. *Moina* have both reproductive and non-reproductive stages. Groups of daphniidae mainly consist of females that are asexual. This however, is only in mist environments. The result is laying resting egg triggered by the lack of food, low oxygen supply, high population density, or low temperatures.

The *Moina's* nutritional needs are different depending on age and what they've eaten in the past. Adults have higher fat content than that of a juvenile. A live *Moina* is made up of 95 per cent water, 4 per cent protein, 0.54 per cent fat, 0.67 per cent carbohydrates, and 0.15 per cent ash. Fish fry need the fatty acid composition for survival and growth. Although all daphnids have needs, each species may have different ones. Freshwater daphnia obviously need freshwater, although some daphnia can be found in slightly salty water. Daphnids are usually able to live in poor water quality and dissolved oxygen levels range from close to zero super saturation. The temperature variance is convenient for temperature changes throughout the year.

Materials and Methods

The laboratory reared newly hatched young ones (10 in number)were segregated form the stock and each of these reared in a small petridish filled with nutrient/culture medium (containing 50 ml of aqueous solution of lg^{-1} of poultry droppings). Normal growth (length, breadth), reproduction till completion of life cycle were made regarding the length increment, member of eggs per brood and life span. Studies were carried out at different temperature ranges *viz.*, 9–15°C, 18–23°C, 23–26°C, 28–33°C.

The eggs were allowed to develop in the brood chamber of the mother. Carapace being a transparent see through structure, various stages of development within the brood pouch were studied under the microscope, 12 hourly. Observations regarding gross morphological changes were made at intervals from the appearance of eggs within the brood pouch till the completion of development and subsequent release. Only distinct stages were recorded (Figures 23.1a–g).

Results and Discussion

Life Cycle Studies and Growth Dynamics

The *in situ* development of *Moina macrocopa* has constantly been

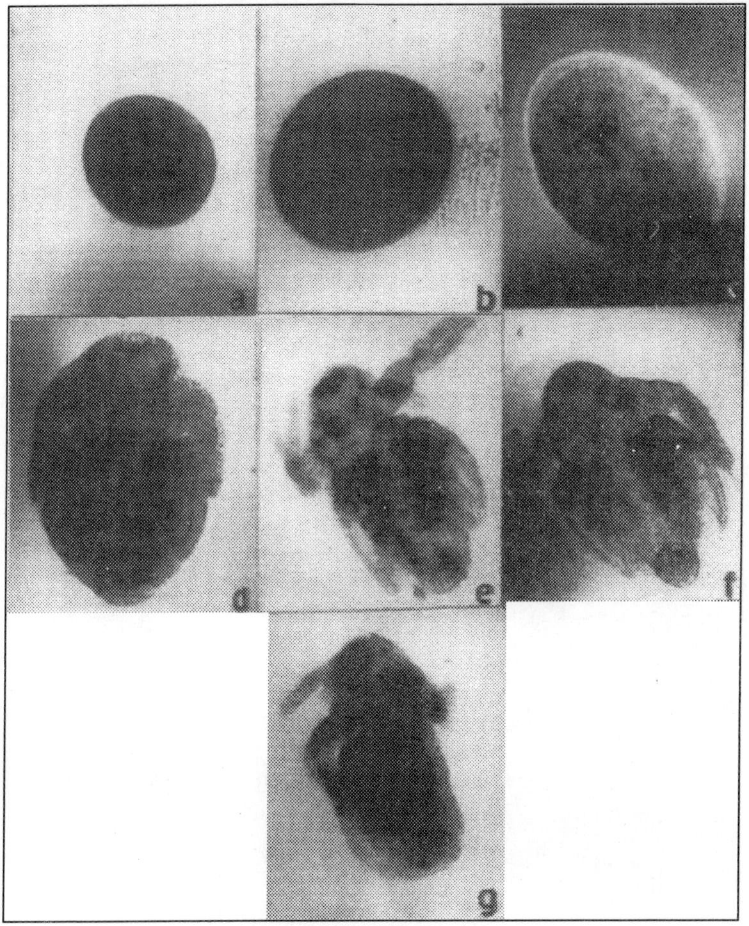

Figure 23.1a–g: Embryonic Developmental Stages of *Moina macrocopa*

a: Round egg stage; b: Initiation of fat cell formation; c: Embryo elongation stage; d: Appearance of rudiments for paired eyes and appendages; e: Formation of distinct paired eyes and appendages; f: Fusion of paired eyes is evident, development progressing; g: Formation of single compound eye, alimentary canal and appendages is complete.

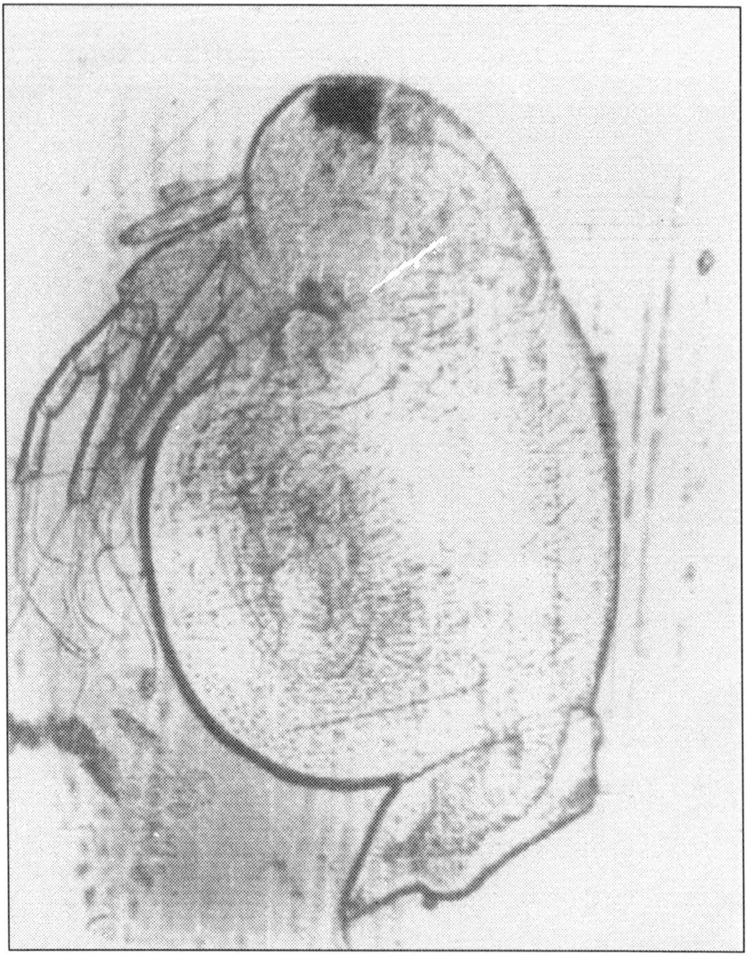

Figure 23.2: Adult of *Moina macrocopa*

observed, during present investigations. The different developmental stages along with the details of time taken to complete different stages of development are presented in Table 23.1–23.4; Figures 23.1a–g and 23.2.

The stages embryonic development of *M. macrocopa* follow those defined by Green (1956) and Murugan and Sivaramakrishnan (1973).

Tables 23.1a–e: Showing a Complete Picture of Life Cycle Studies of *Moina macrocopa* (Temp. 9–15°C)

(a)

Life Span (in days)	*Length (in mm)*									
	Number of Individuals									
	1	*2*	*3*	*4*	*5*	*6*	*7*	*8*	*9*	*10*
1	0.43	0.46	0.45	0.48	0.51	0.53	0.54	0.56	0.57	0.59
2	0.50	0.51	0.55	0.54	0.53	0.57	0.57	0.59	0.59	0.58
3	0.63	0.60	0.65	0.61	0.68	0.60	0.66	0.68	0.67	0.69
4	0.74	0.70	0.71	0.73	0.79	0.71	0.75	0.80	0.79	0.81
5	0.85	0.84	0.85	0.81	0.90	0.85	0.86	0.98	1.01	1.02
6	0.97	0.95	0.95	0.92	1.01	0.95	0.97	1.08	1.08	1.08
7	1.05	1.04	1.04	1.02	1.14	1.65	1.07	1.09	1.09	1.09
8	1.15	1.14	1.17	1.14	1.78	121	120	1.19	120	1.24
9	1.23	1.25	126	1.34	128	125	1.32	125	125	129
10	1.30	M	1.33	1.31	1.30	1.29	M	1.32	1.31	1.32
11	1.35	M	1.38	M	1.35	1.34	M	1.38	1.35	1.35
12	1.38	M	1.40	M	1.37	1.35	M	1.39	M	1.37
13	1.40	M	1.42	M	1.39	1.37	M	1.41	M	1.40
14	1.61	M	1.43	M	M	M	M	1.42	M	1.43
15	1.63	M	M	M	M	M	M	1.44	M	1.45
16	M	M	M	M	M	M	M	M	M	M

(b)

Life Span (in days)	Breadth (in mm)									
	Number of Individuals									
	1	*2*	*3*	*4*	*5*	*6*	*7*	*8*	*9*	*10*
1	0.20	0.23	0.21	0.25	0.27	0.25	0.24	0.23	0.24	0.21
2	0.33	0.35	0.34	0.35	0.37	0.35	0.30	0.33	0.34	0.31
3	0.41	0.44	0.42	0.43	0.46	0.44	0.42	0.41	0.44	0.42
4	0.43	0.45	0.44	0.45	0.47	0.45	0.42	0.43	0.45	0.43
5	0.45	0.47	0.46	0.47	0.50	0.51	0.47	0.48	0.45	0.45
6	0.62	0.66	0.64	0.63	0.70	0.62	0.56	0.56	0.65	0.68
7	0.77	0.77	0.75	0.73	0.80	0.74	0.78	0.79	0.77	0.80
8	0.82	0.87	0.85	0.87	0.90	0.85	0.88	0.90	0.87	0.82
9	0.89	0.92	0.89	0.90	0.91	0.90	0.91	0.92	0.90	0.91
10	0.90	M	0.90	0.90	0.91	0.91	0.90	0.91	0.91	0.92
11	0.91	M	0.91	M	0.92	0.93	M	0.92	0.94	0.91
12	0.95	M	0.91	M	0.93	0.94	M	0.93	M	0.94
13	0.94	M	0.92	M	0.94	0.94	M	0.94	M	0.95
14	0.94	M	0.93	M	M	M	M	0.91	M	0.91
15	0.94	M	M	M	M	M	M	0.91	M	0.92
16	M	M	M	M	M	M	M	M	M	M

(c)

Life Span (in days)	Moult/Instar (Number)									
	Number of Individuals									
	1	2	3	4	5	6	7	8	9	10
1	–/1	–/1	–/1	–/1	–/1	–/1	–/1	–/1	–/1	–/1
2	–/1	–/1	–/1	–/1	–/1	–/1	–/1	–/1	–/1	–/1
3	1Mo/2	1Mo/2	1Mo/2	1Mo/2	1Mo/2	1Mo/2	1Mo/2	1Mo/2	1Mo/2	1Mo/2
4	1Mo/2	1Mo/2	1Mo/2	1Mo/2	1Mo/2	1Mo/2	1Mo/2	1Mo/2	1Mo/2	1Mo/2
5	2Mo/3	2Mo/3	2Mo/3	2Mo/3	2Mo/3	2Mo/3	2Mo/3	2Mo/3	2Mo/3	2Mo/3
6	3Mo/4	3Mo/4	3Mo/4	3Mo/4	3Mo/4	3Mo/4	3Mo/4	3Mo/4	3Mo/4	3Mo/4
7	4Mo/5	4Mo/5	4Mo/5	4Mo/5	4Mo/5	4Mo/5	4Mo/5	4Mo/5	4Mo/5	4Mo/5
8	5Mo/6	5Mo/6	5Mo/6	5Mo/6	5Mo/6	5Mo/6	5Mo/6	5Mo/6	5Mo/6	5Mo/6
9	6Mo/7	6Mo/7	6Mo/7	6Mo/7	6Mo/7	6Mo/7	6Mo/7	6Mo/7	6Mo/7	6Mo/7
10	7Mo/8	M	7Mo/8	7Mo/8	7Mo/8	7Mo/8	7Mo/8	7Mo/8	7Mo/8	7Mo/8
11	8Mo/9	M	8Mo/9	8Mo/9	8Mo/9	8Mo/9	M	8Mo/9	8Mo/9	8Mo/9
12	9Mo/10	M	9Mo/10	9Mo/10	9Mo/10	9Mo/10	M	9Mo/10	M	9Mo/10
13	10Mo/11	M	10Mo/11	10Mo/11	10Mo/11	10Mo/11	M	10Mo/11	M	10Mo/11
14	11Mo/12	M	11Mo/12	M	M	M	M	11Mo/12	M	11Mo/12
15	12Mo/13	M	12Mo/13	M	M	M	M	12Mo/13	M	12Mo/13
16	M	M	M	M	M	M	M	M	M	M

(d)

Life Span (in days)	Hatching									
	Number of Individuals									
	1	2	3	4	5	6	7	8	9	10
1	–	–	–	–	–	–	–	–	–	–
2	–	–	–	–	–	–	–	–	–	–
3	–	–	–	–	–	–	–	–	–	–
4	–	–	–	–	–	–	–	–	–	–
5	–	–	–	–	–	–	–	–	–	–
6	–	–	–	–	–	–	–	–	–	–
7	–	–	–	–	–	–	–	–	–	–
8	7	8	9	12	10	9	9	7	10	11
9	9	12	10	13	13	10	11	12	10	10
10	10	M	10	13	12	12	13	12	10	10
11	12	M	10	M	8	8	M	9	8	8
12	11	M	9	M	8	9	M	9	M	8
13	10	M	8	M	8	M	M	9	M	10
14	10	M	9	M	M	M	M	10	M	10
15	10	M	M	M	M	M	M	9	M	9
16	M	M	M	M	M	M	M	M	M	M

(e)

Life Span (in days)	*Cumulative Number of Hatchings*									
	Number of Individuals									
	1	*2*	*3*	*4*	*5*	*6*	*7*	*8*	*9*	*10*
1	–	–	–	–	–	–	–	–	–	–
2	–	–	–	–	–	–	–	–	–	–
3	–	–	–	–	–	–	–	–	–	–
4	–	–	–	–	–	–	–	–	–	–
5	–	–	–	–	–	–	–	–	–	–
6	–	–	–	–	–	–	–	–	–	–
7	–	–	–	–	–	–	–	–	–	–
8	7	8	9	12	10	9	9	7	10	11
9	16	20	19	25	23	19	20	19	20	21
10	26	M	29	38	35	31	33	31	30	31
11	38	M	39	M	43	39	M	40	38	35
12	49	M	48	M	51	48	M	49	M	47
13	59	M	56	M	58	M	M	58	M	57
14	69	M	65	M	M	M	M	68	M	67
15	79	M	M	M	M	M	M	77	M	76
16	M	M	M	M	M	M	M	M	M	M

Table 23.2a–e: Showing a Complete Picture of Life Cycle Studies of *Moina macrocopa* (Temp. 18–23°C)

(a)

Life Span (in days)	*Length (in mm)*									
	Number of Individuals									
	1	*2*	*3*	*4*	*5*	*6*	*7*	*8*	*9*	*10*
1	0.55	0.46	0.45	0.54	0.57	0.53	0.57	0.50	0.54	0.53
2	0.79	0.70	0.73	0.81	0.80	0.74	0.85	0.76	0.82	0.80
3	1.0	0.91	0.96	0.97	1.0	0.97	1.05	0.98	1.0	0.94
4	1.13	1.03	1.07	1.08	1.07	1.10	1.17	1.08	1.0	1.0
5	1.29	1.12	1.23	1.10	1.15	1.10	1.20	1.05	1.18	1.18
6	1.40	1.38	1.35	1.20	1.20	1.23	1.23	1.25	1.25	1.25
7	1.45	1.39	1.25	1.30	1.30	1.30	1.34	1.34	1.35	1.36
8	1.45	1.39	M	1.30	1.30	1.30	M	1.34	1.40	M
9	1.48	1.49	M	1.32	1.40	M	M	1.40	1.40	M
10	1.48	1.40	M	1.40	1.40	M	M	1.40	1.40	M
11	1.40	1.43	M	1.42	M	M	M	1.42	1.42	M
12	1.43	M	M	1.32	M	M	M	M	1.40	M
13	1.45	M	M	1.40	M	M	M	M	M	M
14	M	M	M	1.43	M	M	M	M	M	M
15	M	M	M	M	M	M	M	M	M	M

(b)

Life Span (in days)	Breadth (in mm)									
	Number of Individuals									
	1	*2*	*3*	*4*	*5*	*6*	*7*	*8*	*9*	*10*
1	0.25	0.20	0.21	0.27	0.24	0.23	0.26	0.23	0.24	0.23
2	0.38	0.35	0.35	0.44	0.44	0.35	0.44	0.32	0.35	0.39
3	0.81	0.63	0.67	0.71	0.74	0.62	0.82	0.65	0.76	0.67
4	0.83	0.70	0.70	0.76	0.80	0.77	0.80	0.70	0.67	0.68
5	0.84	0.78	0.82	0.70	0.75	0.71	0.86	0.78	0.79	0.79
6	0.85	0.81	0.80	0.75	0.77	0.76	0.88	0.88	0.75	0.80
7	0.90	0.81	0.90	0.80	0.80	1.0	0.89	0.84	0.86	0.84
8	0.90	0.82	M	1.06	0.79	0.98	M	0.90	0.86	M
9	0.83	0.88	M	0.95	0.85	0.81	M	0.95	0.80	M
10	1.0	1.0	M	0.82	0.85	M	M	0.89	0.84	M
11	0.85	0.85	M	0.80	M	M	M	0.84	0.82	M
12	0.95	M	M	0.82	M	M	M	M	0.92	M
13	0.90	M	M	0.82	M	M	M	M	M	M
14	M	M	M	0.85	M	M	M	M	M	M
15	M	M	M	M	M	M	M	M	M	M

(c)

Life Span (in days)	(Moult/Instar Number)									
	Number of Individuals									
	1	*2*	*3*	*4*	*5*	*6*	*7*	*8*	*9*	*10*
I	–/1	–/1	–/1	–/1	–/1	–/1	–/1	–/1	–/1	–/1
2	1Mo/2	1Mo/2	1Mo/2	1Mo/2	1Mo/2	1Mo/2	1Mo/2	1Mo/2	1 Mo/2	1 Mo/2
3	2Mo/3	2Mo/3	2Mo/3	2Mo/3	2Mo/3	2Mo/3	2Mo/3	2Mo/3	2Mo/3	2Mo/3
4	3Mo/4	3Mo/4	3Mo/4	3Mo/4	3Mo/4	3Mo/4	3Mo/4	3Mo/4	3Mo/4	3Mo/4
5	4Mo/5	4Mo/5	4Mo/5	4Mo/5	4Mo/5	4Mo/5	4Mo/5	4Mo/5	4Mo/5	4Mo/5
6	5Mo/6	5Mo/6	5Mo/6	5Mo/6	5Mo/6	5Mo/6	5Mo/6	5Mo/6	5Mo/6	5Mo/6
7	6Mo/7	6Mo/7	6Mo/7	6Mo/7	6Mo/7	6Mo/7	6Mo/7	6Mo/7	6Mo/7	6Mo/7
8	7Mo/8	7Mo/8	M	7Mo/8	7Mo/8	7Mo/8	M	7Mo/8	7Mo/8	M
9	8Mo/9	8Mo/9	M	8Mo/9	8Mo/9	8Mo/9	M	8Mo/9	8Mo/9	M
10	9Mo/10	9Mo/1 0	M	9Mo/10	9Mo/10	M	M	9Mo/10	9Mo/10	M
11	10Mo/11	10Mo/1l	M	10Mo/11	M	M	M	10Mo/11	10Mo/11	M
12	11Mo/12	M	M	11Mo/12	M	M	M	M	11Mo/12	M
13	12Mo/13	M	M	12Mo/13	M	M	M	M	M	M
14	M	M	M	M	M	M	M	M	M	M
15	M	M	M	M	M	M	M	M	M	M

(d)

Life Span (in days)	Hatching									
	Number of Individuals									
	1	2	3	4	5	6	7	8	9	10
1	–	–	–	–	–	–	–	–	–	–
2	–	–	–	–	–	–	–	–	–	–
3	–	–	–	–	–	–	–	–	–	–
4	20	18	20	20	20	18	18	19	19	20
5	20	18	18	18	19	19	18	19	18	19
6	20	21	20	18	18	19	18	20	18	19
7	25	22	20	15	18	19	23	22	20	20
8	17	22	19	18	18	19	19	19	20	23
9	22	18	M	24	15	20	M	18	23	M
10	18	20	M	20	17	18	M	24	20	M
11	21	19	M	20	18	M	M	18	19	M
12	18	17	M	18	M	M	M	18	19	M
13	17	M	M	20	M	M	M	M	18	M
14	18	M	M	23	M	M	M	M	M	M
15	18	M	M	19	M	M	M	M	M	M

(e)

Life Span (in days)	Cumulative Number of Hatchings									
	Number of Individuals									
	1	*2*	*3*	*4*	*5*	*6*	*7*	*8*	*9*	*10*
1	–	–	–	–	–	–	–	–	–	–
2	–	–	–	–	–	–	–	–	–	–
3	–	–	–	–	–	–	–	–	–	–
4	20	18	20	20	20	18	18	19	19	20
5	40	36	38	38	39	37	36	38	37	39
6	60	57	58	56	57	56	54	58	55	58
7	85	79	78	71	75	75	77	80	75	78
8	102	101	97	89	93	94	96	99	95	101
9	124	119	M	113	108	114	M	117	118	M
10	142	139	M	133	125	132	M	135	138	M
11	163	158	M	153	143	M	M	153	157	M
12	181	165	M	171	M	M	M	M	176	M
13	198	M	M	191	M	M	M	M	194	M
14	216	M	M	214	M	M	M	M	M	M
15	234	M	M	233	M	M	M	M	M	M

Table 23.3a–e: Showing a Complete Picture of Life Cycle Studies of *Moina macrocopa* (Temp. 23–26°C)

(a)

Life Span (in days)	*Length (in mm)*									
	Number of Individuals									
	1	*2*	*3*	*4*	*5*	*6*	*7*	*8*	*9*	*10*
1	0.47	0.45	0.48	0.45	0.49	0.49	0.48	0.48	0.47	0.48
2	0.56	0.59	0.57	0.55	0.60	0.59	0.58	0.57	0.55	0.59
3	0.78	0.79	0.74	0.76	0.79	0.78	0.79	0.76	0.76	0.79
4	0.86	0.89	0.85	0.87	0.89	0.85	0.88	0.90	0.89	0.90
5	1.05	1.05	1.03	1.07	1.08	1.03	1.04	1.07	1.09	1.08
6	1.13	1.15	1.11	1.14	1.14	1.10	1.11	1.15	1.17	1.15
7	1.17	1.20	1.16	1.18	1.17	1.16	1.15	1.20	1.21	1.19
8	1.45	1.48	1.46	1.46	1.44	1.43	1.48	1.49	1.49	1.46
9	1.42	1.49	1.47	1.49	1.47	1.45	1.50	1.51	1.52	1.49
10	1.50	1.53	1.50	1.57	1.54	1.57	1.58	M	1.59	1.54
11	1.52	M	1.58	1.59	M	1.57	1.60	M	1.60	1.59
12	1.63	M	1.67	1.68	M	M	1.68	M	1.67	M
13	M	M	1.68	M	M	M	1.69	M	M	M
14	M	M	1.69	M	M	M	M	M	M	M
15	M	M	M	M	M	M	M	M	M	M

(b)

Life Span (in days)	Breadth (in mm)									
	Number of Individuals									
	1	*2*	*3*	*4*	*5*	*6*	*7*	*8*	*9*	*10*
1	0.26	0.24	0.27	0.24	0.26	0.27	0.25	0.26	0.25	0.28
2	0.34	0.33	0.36	0.34	0.35	0.36	0.34	0.33	0.34	0.37
3	0.56	0.54	0.56	0.53	0.53	0.56	0.52	0.54	0.56	0.58
4	0.71	0.73	0.75	0.72	0.74	0.75	0.74	0.71	0.74	0.58
5	0.87	0.89	0.85	0.83	0.85	0.87	0.85	0.83	0.82	0.77
6	0.96	0.98	0.97	0.92	0.95	0.94	0.93	0.93	0.95	0.86
7	1.01	1.03	1.00	1.02	1.04	1.02	1.01	1.00	1.03	0.97
8	1.03	1.06	1.02	1.08	1.07	1.08	1.04	1.03	1.05	1.01
9	1.05	1.07	1.04	1.07	1.08	1.08	1.06	1.05	1.07	1.06
10	1.07	1.08	1.06	1.09	1.09	1.09	1.07	M	1.08	1.07
11	1.08	M	1.06	1.09	M	1.09	1.09	M	1.08	1.07
12	1.09	M	1.08	1.09	M	M	1.17	M	1.08	M
13	M	M	1.15	M	M	M	M	M	M	M
14	M	M	1.18	M	M	M	M	M	M	M
15	M	M	M	M	M	M	M	M	M	M

(c)

Life Span (in days)	(Moult/Instar Number)									
	Number of Individuals									
	1	2	3	4	5	6	7	8	9	10
1	–/1	–/1	–/1	–/1	–/1	–/1	–/1	–/1	–/1	–/1
2	–/1	–/1	–/1	–/1	–/1	–/l	–/1	–/1	–/1	–/1
3	IMo/2	IMo/2	IMo/2	IMo/2	IMo/2	IMo/2	IMo/2	IMo/2	1 Mo/2	1Mo/2
4	2Mo/3	2Mo/3	2Mo/3	2Mo/3	2Mo/3	2Mo/3	2Mo/3	2Mo/3	2Mo/3	2Mo/3
5	3Mo/4	3Mo/4	3Mo/4	3Mo/4	3Mo/4	3Mo/4	3Mo/4	3Mo/4	3Mo/4	3Mo/4
6	4Mo/5	4Mo/5	4Mo/5	4Mo/5	4Mo/5	4Mo/5	4Mo/5	4Mo/5	4Mo/5	4Mo/5
7	5Mo/6	5Mo/6	5Mo/6	5Mo/6	5Mo/6	5Mo/6	5Mo/6	5Mo/6	5Mo/6	5Mo/6
8	6Mo/7	6Mo/7	6Mo/7	6Mo/7	6Mo/7	6Mo/7	6Mo/7	6Mo/7	6Mo/7	6Mo/7
9	7Mo/8	7Mo/8	7Mo/8	7Mo/8	7Mo/8	7Mo/8	7Mo/8	7Mo/8	7Mo/8	7Mo/8
10	8Mo/9	8Mo/9	8Mo/9	8Mo/9	8Mo/9	8Mo/9	8Mo/9	M	8Mo/9	8Mo/9
11	9Mo/10	M	9Mo/10	9Mo/10	M	9Mo/10	9Mo/10	M	9Mo/10	9Mo/10
12	10Mo/11	M	10Mo/11	10Mo/11	M	M	10Mo/11	M	10Mo/11	M
13	M	M	11Mo/12	M	M	M	11Mo/12	M	M	M
14	M	M	12Mo/13	M	M	M	M	M	M	M
15	M	M	M	M	M	M	M	M	M	M

(d)

Life Span (in days)	Hatching									
	Number of Individuals									
	1	*2*	*3*	*4*	*5*	*6*	*7*	*8*	*9*	*10*
1	–	–	–	–	–	–	–	–	–	–
2	–	–	–	–	–	–	–	–	–	–
3	–	–	–	–	–	–	–	–	–	–
4	18	20	20	19	21	19	18	17	20	21
5	20	22	19	17	18	20	21	20	20	20
6	21	21	19	20	18	19	19	18	19	21
7	23	20	24	22	22	23	20	24	21	23
8	22	20	23	22	23	20	23	23	23	22
9	18	19	20	20	20	17	17	19	20	24
10	17	20	18	20	19	18	19	18	20	18
11	19	17	20	21	17	20	20	M	22	17
12	18	M	19	22	M	18	17	M	19	19
13	18	M	17	M	M	M	M	M	18	M
14	M	M	18	M	M	M	M	M	M	M
15	M	M	M	M	M	M	M	M	M	M

(e)

Life Span (in days)	Cumulative Number of Hatchings									
	Number of Individuals									
	1	*2*	*3*	*4*	*5*	*6*	*7*	*8*	*9*	*10*
1	–	–	–	–	–	–	–	–	–	–
2	–	–	–	–	–	–	–	–	–	–
3	–	–	–	–	–	–	–	–	–	–
4	18	20	20	19	21	19	18	17	20	21
5	38	42	39	36	39	39	39	37	40	41
6	59	63	58	56	57	58	58	55	59	62
7	82	83	82	78	79	81	78	79	80	85
8	104	103	105	100	102	101	101	102	103	107
9	122	122	125	120	123	118	118	121	123	131
10	139	142	143	140	142	136	137	139	143	149
11	158	159	163	161	159	156	157	M	165	165
12	176	M	182	183	M	174	174	M	184	185
13	194	M	199	M	M	M	190	M	202	M
14	M	M	117	M	M	M	M	M	M	M
15	M	M	M	M	M	M	M	M	M	M

Table 23.4a–e: Showing a Complete Picture of Life Cycle Studies of *Moina macrocopa* (Temp. 28–33°C)

(a)

Life Span (in days)	*Length (in mm)*									
	Number of Individuals									
	1	*2*	*3*	*4*	*5*	*6*	*7*	*8*	*9*	*10*
1	0.50	0.50	0.48	0.49	0.51	0.48	0.50	0.49	0.50	0.49
2	0.67	0.66	0.63	0.57	0.64	0.68	0.63	0.65	0.62	0.64
3	0.96	0.95	0.90	0.97	0.91	0.90	0.93	0.91	0.97	0.92
4	1.10	1.10	1.08	1.09	1.12	1.06	1.21	1.06	1.13	1.08
5	1.21	1.20	1.10	1.12	1.10	1.20	1.23	1.12	1.22	1.13
6	1.28	1.27	1.20	1.21	1.22	1.29	M	1.23	1.28	1.21
7	1.30	1.31	M	1.28	1.29	1.30	M	1.29	1.34	1.30
8	1.38	1.39	M	1.34	M	1.39	M	1.38	M	1.34
9	1.42	1.43	M	1.40	M	1.43	M	1.41	M	M
10	M	1.46	M	M	M	1.45	M	1.44	M	M
11	M	1.48	M	M	M	1.48	M	M	M	M
12	M	M	M	M	M	M	M	M	M	M

(b)

Life Span (in days)	Breadth (in mm)									
	Number of Individuals									
	1	2	3	4	5	6	7	8	9	10
1	0.23	0.21	0.20	0.20	0.24	0.21	0.22	0.22	0.25	0.23
2	0.38	0.36	0.34	0.33	0.35	0.35	0.39	0.35	0.39	0.36
3	0.85	0.84	0.82	0.83	0.89	0.85	0.85	0.83	0.89	0.87
4	0.88	0.87	0.85	0.86	0.90	0.89	0.89	0.88	0.91	0.89
5	0.89	0.89	0.88	0.89	0.91	0.90	0.90	0.89	0.93	0.91
6	0.95	0.92	0.93	0.93	0.95	0.96	M	0.94	0.96	0.95
7	0.98	0.97	M	0.97	0.98	0.99	M	0.95	0.97	0.98
8	0.99	0.98	M	0.99	M	0.99	M	0.98	M	1.00
9	1.01	1.01	M	1.01	M	1.01	M	0.99	M	M
10	M	1.02	M	M	M	1.02	M	1.01	M	M
11	M	1.02	M	M	M	1.02	M	M	M	M
12	M	M	M	M	M	M	M	M	M	M

(c)

Life Span (in days)	(Moult/Instar Number)									
	Number of Individuals									
	1	2	3	4	5	6	7	8	9	10
1	–/1	–/1	–/1	–/1	–/1	–/1	–/1	–/1	–/1	–/1
2	–/1	–/1	–/1	–/1	–/1	–/l	–/1	–/1	–/1	–/l
3	1Mo/2	1Mo/2	1Mo/2	1 Mo/2	1Mo/2	1Mo/2	1Mo/2	1Mo/2	1Mo/2	1Mo/2
4	2Mo/3	2Mo/3	2Mo/3	2Mo/3	2Mo/3	2Mo/3	2Mo/3	2Mo/3	2Mo/3	2Mo/3
5	3Mo/4	3Mo/4	3Mo/4	3Mo/4	3Mo/4	3Mo/4	3Mo/4	3Mo/4	3Mo/4	3Mo/4
6	4Mo/5	4Mo/5	4Mo/5	4Mo/5	4Mo/5	4Mo/5	M	4Mo/5	4Mo/5	4Mo/5
7	5Mo/6	5Mo/6	M	5Mo/6	5Mo/6	5Mo/6	M	5Mo/6	5Mo/6	5Mo/6
8	6Mo/7	6Mo/7	M	6Mo/7	M	6Mo/7	M	6Mo/7	M	6Mo/7
9	7Mo/8	7Mo/8	M	7Mo/8	M	7Mo/8	M	7Mo/8	M	M
10	M	8Mo/9	M	M	M	8Mo/9	M	8Mo/9	M	M
11	M	9Mo/l0	M	M	M	9Mo/10	M	9Mo/10	M	M
12	M	M	M	M	M	M	M	M	M	M

(d)

Life Span (in days)	Hatching									
	Number of Individuals									
	1	*2*	*3*	*4*	*5*	*6*	*7*	*8*	*9*	*10*
1	–	–	–	–	–	–	–	–	–	–
2	–	–	–	–	–	–	–	–	–	–
3	–	–	–	–	–	–	–	–	–	–
4	10	12	10	9	13	9	10	9	10	9
5	15	13	12	12	16	11	13	12	14	13
6	14	15	14	14	15	13	M	13	12	14
7	13	15	M	14	13	14	M	14	16	16
8	15	17	M	14	M	16	M	17	M	18
9	14	14	M	15	M	14	M	15	M	M
10	M	12	M	M	M	10	M	12	M	M
II	M	13	M	M	M	10	M	M	M	M
12	M	M	M	M	M	M	M	M	M	M

(e)

Life Span (in days)	Cumulative Number of Hatchings									
	Number of Individuals									
	1	*2*	*3*	*4*	*5*	*6*	*7*	*8*	*9*	*10*
1	–	–	–	–	–	–	–	–	–	–
2	–	–	–	–	–	–	–	–	–	–
3	–	–	–	–	–	–	–	–	–	–
4	10	12	10	9	13	9	10	9	10	9
5	25	25	22	21	29	20	23	21	24	22
6	39	40	26	35	44	33	M	34	36	36
7	52	55	M	49	57	47	M	48	52	52
8	67	72	M	63	M	63	M	65	M	70
9	81	86	M	78	M	77	M	80	M	M
10	M	98	M	M	M	87	M	92	M	M
11	M	111	M	M	M	97	M	M	M	M
12	M	M	M	M	M	M	M	M	M	M

Various stages of embryonic development of *M. macrocopa* are classified as:

Stage I

Egg is a translucent spherical structure with a transparent peripheral region (Figure 23.1a). It has been observed that the stage-I lasts for 15–16 hrs. in laboratory.

Stage II

During stage-II, the translucent area of developing egg fat cell start forming (Murugan and Sivaramakrishana, 1973). There is a gradual increase in the peripheral transparent zone of the egg. In the centre of the egg there is present a well marked clear area (Figure 23.1b). In *M. macrocopa* these changes are completed in 6–8 hours after stage-I.

Stage III

In this stage there is elongation of egg both anteriorly as well as posteriorly along with the initiation of cleavage. The duration of such, changes takes place in 10–13 hours (Figure 23.1c).

Stage IV

In the next 4–8 hours the embryo elongates. The rudiments of head, antennary and trunk are formed. Eyes in this stage are still rudimentary (Figure 23.1d).

Stage V

It has been observed that further elongation of embryo is continuous during present studies. Head has become distinct. The alimentary canal is distinctly formed. Rudiments of paired eyes, ocellus and limbs are also formed (Figure 23.1e). It takes 7–13 hrs to complete this stage and then the embryo enters the next stage.

Stage VI

In this stage the antennae are well marked and clearly visible. The paired eyes are in proximity with one another (Figure 23.1f). This stage lasts for 7–9 hrs.

Stage VII

In this stage the embryo attains maximum length along with proper setting of rudimentary paired eyes. These eyes attain dark colour in this stage. The development of alimentary canal is complete by now. Thoracic legs become distinctly visible. After acquiring their last stage of development, juveniles emerge from the brood pouch. This stage takes 10–16 hrs to complete (Figure 23.1g).

The miniature adult called as neonates are just like adult in appearance except being smaller in size. After being released from the brood pouch of adult, it passes through various number of pre-adult and adult instars (Figure 23.2).

Number of Pre-adult Instars

The observations on the life history of *M. macrocopa* have revealed that the number of pre-adult instars is 2 at the temperature range of 23–26°C. Langer (1991 unpublished) has also reported 2–3 pre-adult instars in case of *M. macrocopa* at optimum range of temperature at 25–27°C.

A perusal of Tables 23.5–23.8; indicate an inverse relationship between temperature and the number of pre-adult instars. A lower number of pre-adult instars (2–3) is characteristic of higher temperature range during present studies and a higher number of pre-adult instars (5) is characteristic of lower temperature range.

These results are in partial agreement with those reported by Murugan and Sivaramakrishnan (1976) in case of *M. micrura* in which low number of pre-adult instars (2) at the higher temperature range of 28–31°C have been observed.

Number of Adult Instars

The number of adult instars in *M. macrocopa* vary at different temperature regime as observed in present studies (Tables 23.5–23.8). At 9–15°C temperature regime, the number of adult instars is 10; at 18–23°C, the number of adult instar is 13; at 23–25°C the number of adult instar is 12 and at 28–33°C, the numbers of adult instars is 9. These results are in line with those of Langer (1991 unpublished) who is case of *M. macrocopa* has also recorded similar observations. Similar results with regards to the variations in the number of instars in case of *Daphnia carinata* has also been reported by Venkataraman and Job (1980).

Murugan and Sivaramakrishnan (1976), while investigation life cycle of *M. micrura* at temperature range of 28–31°C also reported that the number of adult instars is 11.

Table 25.5: Mean Length, Mean Breadth, Average Number of Eggs/Brood, Moult/Instar Number and Cumulative Frequency of Eggs in each Instars of *Moina macrocopa* at 9–15°C

Instar No./ Age (in days)	*Mean Length (mm)*	*S.D (±)*	*Mean Breadth (mm)*	*S.D (±)*	*Moult/ Instar No.*	*Average Brood Size*	*Cumulative No. of Eggs*
1	0.512	0.052	0.23	0.021	–/1	–	–
2	0.545	0.032	0.34	0.015	–/1	–	–
3	0.647	0.032	0.43	0.015	1 M/2	–	–
4	0.753	0.039	0.44	0.014	1 M/2	–	–
5	0.897	0.073	0.48	0.017	1 M/2	–	–
6	0.998	0.059	0.65	0.024	2 M/2	–	–
7	1.068	0.033	0.76	0.027	2 M/2	10	10
8	1.182	0.030	0.86	0.026	3 M/2	13.1	23.1
9	1.259	0.019	0.90	0.011	4 M/2	13.3	36.4
10	1.312	0.011	0.90	0.403	5 M/2	11.22	47.62
11	1.359	0.013	0.92	0.010	6 M/2	10.28	57.80
12	1.375	0.013	0.93	0.016	7 M/2	10.5	68.30
13	1.397	0.012	0.93	0.016	8 M/2	11.0	79.30
14	1.412	0.014	0.92	0.017	9 M/2	11.0	90.30
15	1.429	0.015	0.93	0.015	10 M/2	10.33	100.63
16	1.473	0.024	0.93	0.014	11 M/2	9.0	109.63

Duration of Instars

Perusal of Tables 23.5–23.8; reveals that the duration of pre-adult instar is more at lower temperature regime (9–15°C) as compared to the higher temperature regime (28–33°C). Such longer duration of pre-adult instars observed in *M. macrocopa* during present investigations may be attributed to slow growth at low temperature. At 9–15°C temperature range the first preadult instar duration is prolonged to 2 days and 2nd one lasted for 3 days and 3rd one for again 2 days, but adult instar duration has been observed to be approximately constant. At other temperature ranges presently investigated the duration of instars is somewhat constant lasting one day each (Tables 23.5–23.8).

Table 23.6: Mean Length, Mean Breadth, Average Number of Eggs/Brood, Moult/Instar Number and Cumulative Frequency of Eggs in Each Instars of *Moina macrocopa* at 18–23°C

Instar No./ Age (in days)	*Mean Length (mm)*	*S.D (±)*	*Mean Breadth (mm)*	*S.D (±)*	*Moult/ Instar No.*	*Average Brood Size*	*Cumulative No. of Eggs*
1	0.524	0.039	0.23	0.020	–/1	0	0
2	0.78	0.043	0.38	0.042	–/1	0	0
3	0.97	0.036	0.70	0.068	–/1	18.8	18.8
4	1.07	0.051	0.74	0.054	1 M/2	9.0	27.0
5	1.16	0.067	0.78	0.049	2 M/2	8.6	35.6
6	1.27	0.066	0.80	0.048	3 M/2	9.8	45.4
7	1.33	0.055	0.86	0.058	4 M/2	7.7	53.1
8	1.35	0.055	0.89	0.089	5 M/2	10.4	63.4
9	1.39	0.060	0.86	0.057	6 M/2	9.0	72.4
10	1.39	0.05	0.9	0.473	7 M/2	6.6	79.0
11	1.41	0.73	0.83	0.019	8 M/2	7.0	86.0
12	1.38	0.046	0.90	0.061	9 M/2	6.0	92.0
13	1.42	0.025	0.86	0.04	10 M/2	0	0
14	1.43	0	0.85	0	11 M/2	0	0

The observation made by Langer (1991, unpublished) in *M. macrocopa* and in some other cladoceran species *viz., D. similes,* and *S. vetulus,* indicates that first primiparous (egg bearing) instar is distinctly of longer duration than the pre-adult instars. A similar conclusions have previously been drawn in case of *D. pulex* by Anderson *et al.* (1937); in case of *D. magna* by Anderson and Jenkins (1942); in case of *D. carinata* by Navaneethakrishnan and Michael (1971) and in case of *S. acutirostratus* by Murugan and Sivaramakrishnan (1973).

Many workers have also supported the present viewpoint, that the preadult instars may be of same or longer duration than that of the first primiparous instar (Murugan, 1975 in *M. micrura;* Murugan and Sivaramakrishnan, 1976; in *Scapholebris kingi*).

Table 23.7: Mean Length, Mean Breadth, Average Number of Eggs/Brood, Moult/Instar Number and Cumulative Frequency of Eggs in Each Instars of *Moina macrocopa* at 23–26°C

Instar No./ Age (in days)	*Mean Length (mm)*	*S.D (±)*	*Mean Breadth (mm)*	*S.D (±)*	*Moult/ Instar No.*	*Average Brood Size*	*Cumulative No. of Eggs*
1	0.475	0.012	0.258	0.012	–/1	–	–
2	0.576	0.015	0.346	0.012	–/1	–	–
3	0.774	0.016	0.548	0.017	1 M/2	12.3	12.1
4	0.878	0.018	0.736	0.018	2 M/2	14.7	27
5	1.059	0.020	0.853	0.020	3 M/2	19.1	46.1
6	1.136	0.022	0.948	0.018	4 M/2	21.1	67.2
7	1.179	0.019	1.017	0.012	5 M/2	22.8	90
8	1.464	0.019	1.045	0.015	6 M/2	23.1	113.1
9	1.486	0.020	1.063	0.012	7 M/2	19.2	132.3
10	1.553	0.026	1.077	0.010	8 M/2	17.1	149.4
11	1.581	0.025	1.081	0.010	9 M/2	16.4	165.8
12	1.666	0.018	1.086	0.015	10 M/2	14.6	180.4
13	1.685	0.005	1.16	0.01	11 M/2	13	193.4
14	1.69	0	1.18	0	12 M/2	13	206.4

Extent of Growth during Instars

The measurement of growth has recorded a progressive increase in the individual in each instar. A generalized observation on growth pattern during present studies indicate that the growth rate gradually slows down during the reproductive phase, but is rapid in the pre-adult phase. The present study regarding the growth pattern also supports the view point expressed by Murugan and Sivaramakrishnan, (1973, 1976) that the rapid pre-adult growth is a common feature for cladocerans irrespective of climatic difference. Greater growth in pre-adult instars may be attributed to diversion of energy towards gonadal maturation in the latter, and extra energy is utilized for the propagation as reported by Crabatree (1975) and Paloheimo *et al.* (1982) in case of *Daphnia.* The extent of growth per instar in cladocerans, is said to be positively correlated with the food supplement (Hutchinson, 1967) and temperature variations (Malhotra and Langer, 1990b) and Sharma and Sahai (1990).

Table 23.8: Mean Length, Mean Breadth, Average Number of Eggs/Brood, Moultl/Instar Number and Cumulative Frequency of Eggs in Each Instars of *Moina macrocopa* at 28–33°C

Instar No./ Age (in days)	*Mean Length (mm)*	*S.D (±)*	*Mean Breadth (mm)*	*S.D (±)*	*Moult/ Instar No.*	*Average Brood Size*	*Cumulative No. of Eggs*
1	0.495	0.010	0.22	0.015	–/1	–	–
2	0.649	0.019	0.361	0.019	–/1	–	–
3	0.932	0.029	0.85	0.023	1 M/2	10.20	10.20
4	1.103	0.041	0.88	0.017	2 M/2	13.40	23.60
5	1.163	0.050	0.89	0.013	3 M/2	13.90	37.50
6	1.243	0.033	0.94	0.012	4 M/2	14.55	52.05
7	1.304	0.019	0.87	0.011	5 M/2	17.12	69.17
8	1.37	0.021	0.98	0.006	6 M/2	15.66	84.83
9	1.418	0.011	1.024	0.038	7 M/2	13.60	98.43
10	1.45	0.008	1.13	0.008	8 M/2	13.00	111.43
11	1.48	0	1.2	0	9 M/2	11.50	122.90

Edlen (1937) reported that growth in cladocernas is accompanied by moulting, after which increase in size is rapid, being completed in less than a minute in *D. magna.*

The perusal of Tables 23.5–23.8; Figures 23.3–23.4 indicate that in *M. macrocopa,* growth continues till a specific size is attained after which ceasation or reduced growth rate is evident, as has also been recorded in case of *D. magna* (McArthur and Baillie, 1929) and in *S. vetulusi* (Langer, 1987 unpublished). In contrast many workers have reported that the growth in some cladocerans like *D. magna* (Edlen, 1937) and *D. pulex* (Anderson *et al.,* 1937) continues till the end of life cycle.

Effect of Temperature on Growth

With increase in temperature form low temperature range (9–15°C) to higher temperature range (28–33°C) an increase in growth in *M. macrocopa* has been observed during present study (as a

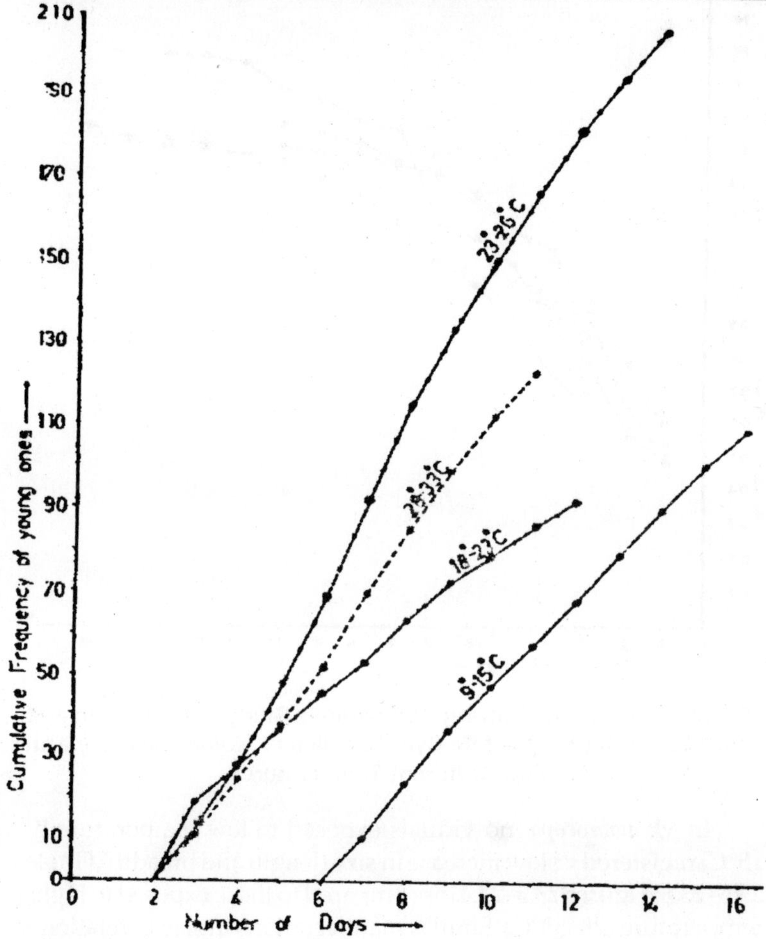

Figure 23.3: Graph Showing Comparative Study of Cumulative Frequency of Young Ones and Number of Days in the Life Cycle Studies of *Moina macrocopa* at Four Different Temperatures

consequence of shortening of duration of instars in present investigations) (Tables 23.5–23.8; Figures 23.3–23.4). These findings are in agreement with the observations made in this regards for *Daphnia* species (Brown, 1927; McArthur and Baillie, 1929; Anderson *et al.*, 1937).

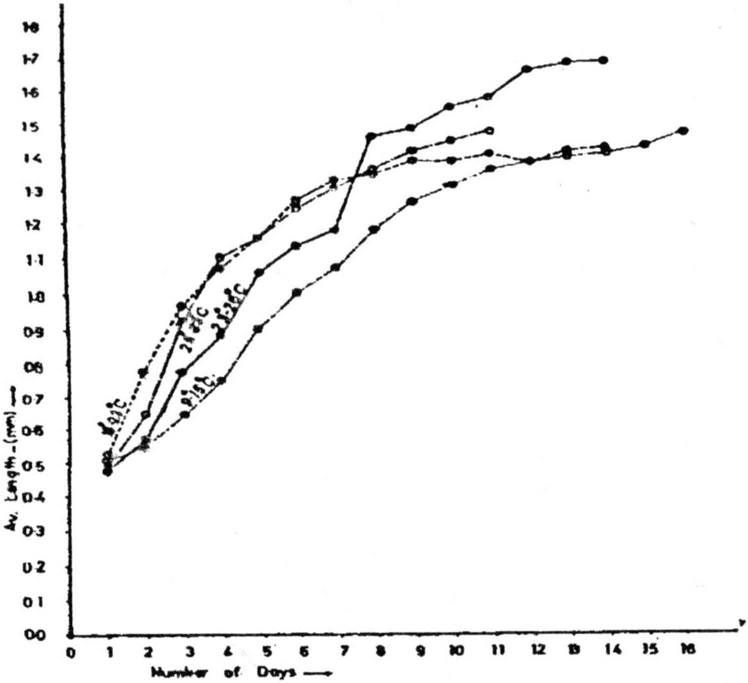

Figure 23.4: Graph Showing Comparative Study of Av. Length and Number of Days in the Life Cycle Studies of *Moina macrocopa* at Four Different Temperatures

In *M. macrocopa* individuals exposed to low temperature (9–15°C) registered a slow increase in size (length and breadth) (Tables 23.5–23.8; Figures 23.3–23.4) as compared to those exposed to higher temperature (28–33°C). Similar manifestations in size in relation to temperature have been reported for *D. magna* (Anderson, 1932) *D. pulex* (Anderson *et al.*, 1937) and several other cladoceran species (Munro and White, 1975; Vijverberg, 1980 and Mitchel and Williams, 1982).

Mode of Egg Production

During the present investigations, a unimodal egg production has been observed in *M. macrocopa* when reared at different temperature ranges (Tables 23.5–23.8; Figure 23.5). Where a peak is recorded between 4–5 adult instar at 18–23°C; around 5th adult instar at 23–26°C and 28–30°C; and whereas at another temperature range

(9–15°C) presently investigated again a single peak around 6th adult instar has been observed (Tables 23.5–23.8; Figures 23.5–23.6). These results are in agreement with the findings reported in many temperate cladocerans *D. magna* (Anderson *et al.*, 1937); *D. longispina* (Ingle *et al.*, 1937); *D. magnai* (Enderson and Jenkins; 1942) and *M. macrocopa* Langer (1991) unpublished) where also single peak in egg production occurs around 5th adult instar.

During present observations on fecundity in *M. macrocopa* it has been observed that the fecundity increases with the increase in body length and body breadth (Tables 23.5–23.8; Figures 23.3–23.4). This phenomenon of increase in fecundity in relation to growth is well known in different species of cladocerans as recorded by several workers (Green, 1956; Hebert, 1978 and Jana and Pal, 1983b).

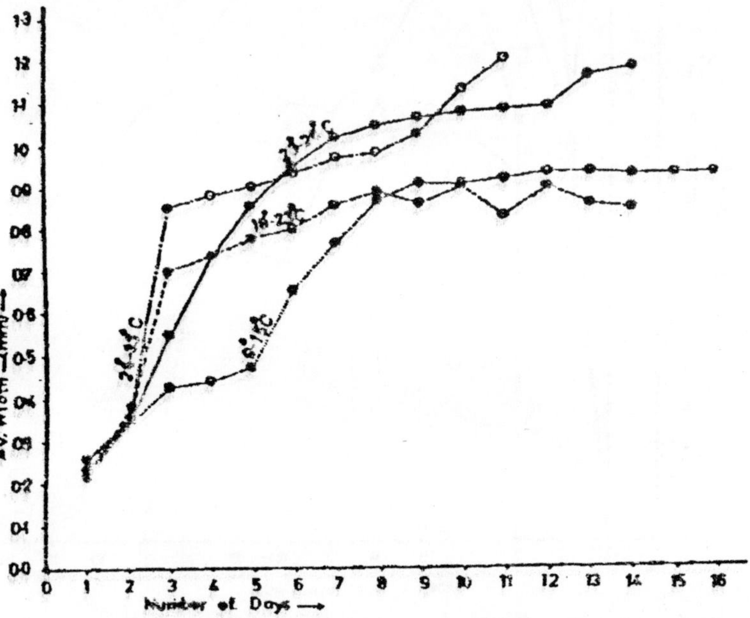

Figure 23.5: Graph Showing Comparative Study of Av. Width and Number of Days in the Life Cycle Studies of *Moina macrocopa* at Four Different Temperatures

During present investigations, an important fish food organism, *Moina macrocopa* (Straus, 1820), a cladoceran species was selected for the study of various biological aspects such as:

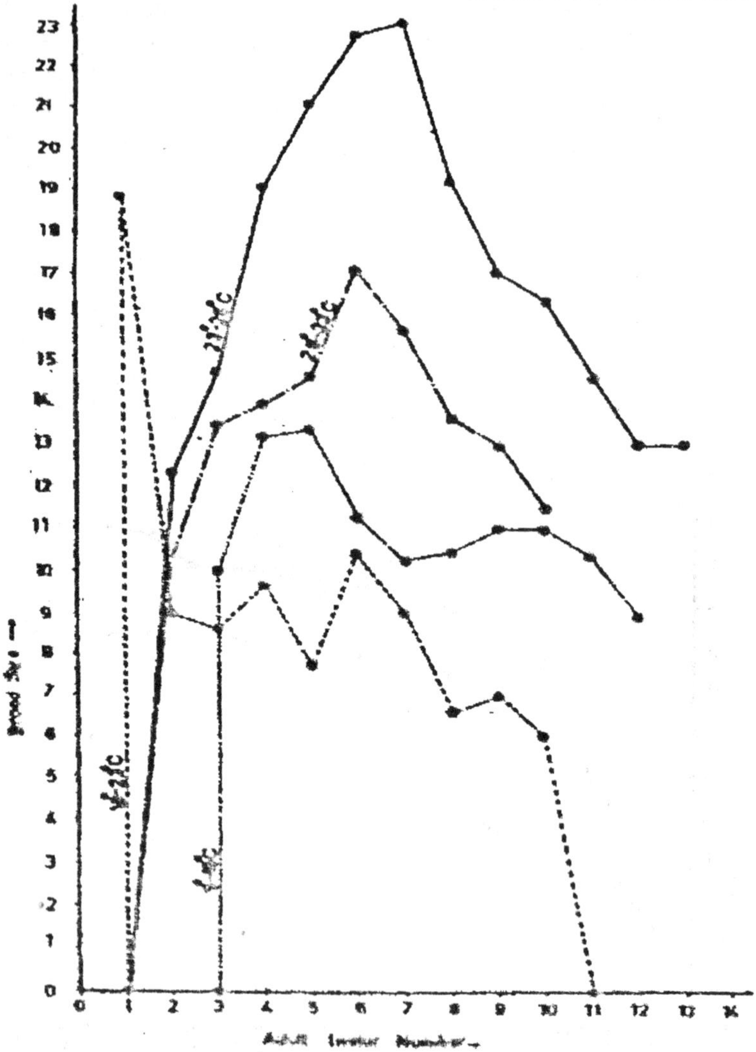

Figure 23.6: Graph Showing Comparative Study of Brood Size and Number of Adult Instar in the Life Cycle Studies of *Moina macrocopa* at Four Different Temperatures

V–1 Life cycle studies and growth dynamics.

Effect of varying temperature on different aspects of life cycle *viz.:*

1. Number and duration of instars

2. Extent of growth during instars.
3. Effect on growth.
4. Mode of egg production and effect on egg production.

V–2 Effect of different nutrient media on growth pattern, brood size and population dynamics of *M. macrocopa.*

V–3 Effect of inoculum density on growth pattern fecundity and population dynamics of *M. macrocopa.*

V–4 Culture of *M. macrocopa* under maintained and static conditions.

V–5 Life cycle studies and growth dynamics:

During the present studies seven embryonic stages have been recorded during the embryonic development of *M. macrocopa.*

Number and Duration of Instars

An inverse relationship between temperature and number of instars has been observed in *M. macrocopa.*

The number of adult instar were observed to be highest at the intermediate temperature regime of 18–23°C (13), a minimum was recorded at higher temperature regime of 28–33°C.

Temperature also bears an inverse relationship with duration of instars.

Extent of Growth during Instars

Growth has been observed to be rapid in pre-adult instar in *M. macrocopa,* but slows down during reproductive phase.

Effect on Growth

With the increase in temperature from low temperature range (9–15°C) to higher temperature range (28–33°C) an increase in growth (in *M. macrocopa*) has been observed as a consequences of shortening of duration of instars at higher temperature. Present studies also reveal that there is an optimum limit within the extended biokinetic range for development of *M. macrocopa,* lying between 18–26°C. Above or below which features negative to growth are manifested.

Mode of Egg Production and Effect on Egg Production

A unimodal pattern of egg production during life span has been observed in *Moina macrocopa*, with peaks occurring around 4th–5th adult instar, at different temperature ranges. An increase in fecundity has also been observed to be associated with increase in body growth.

Summary and Conclusions

Effect on Growth

With the increase in temperature from low temperature range (9–15°C) to higher temperature range (28–33°C) an increase in growth (in *M. macrocopa*) has been observed as a consequences of shortening of duration of instars at higher temperature. Present studies also reveal that there is an optimum limit within the extended biokinetic range for development of *M. macrocopa*, lying between 18–26°C. Above or below which features negative to growth are manifested.

Mode of Egg Production and Effect on Egg Production

A unimodal pattern of egg production during life span has been observed in *Moina macrocopa*, with peaks occurring around 4th–5th adult instar, at different temperature ranges. An increase in fecundity has also been observed to be associated with increase in body growth.

Effect of Different Nutrient Media on Growth Patterns, Brood Size and Population Dynamics of *M. macrocopa*

Growth Patterns

During present studies on *M. macrocopa* a steady growth has been observed to be followed by a sudden decline

The maximum growth (mean length of *M. macrocopa* has observed in. set fed on rice bran + cow-dung (1.42 mm).

It is thus concluded, from the data on growth, from al the nutrient sources tried presently, the combination of rice bran+cow-dung is the best so far as the linear growth of *M. macrocopa* is concerned.

Brood Size

Egg production in *M. macrocopa* recorded a trimodal pattern in all the nutrient combinations, except for *Chlorella*, where it remained bimodal.

Maximum average number of eggs (22) have been observed in culture raised on rice bran+cow-dung which provide enough energy for growth and reproduction in *M. macrocopa*.

It is indicated that age of culture also effects the egg production in *M. macrocopa* (wherein old cultures have been observed to be unfavourable for egg production).

Population

Maximum average number of *M. macrocopa* individuals have been obtained in rice bran+cow-dung combination (14500 inds/2 lit). It is thus concluded that the best combination, for culture of *M. macrocopa* (out of all the combinations of nutrient sources presently tried) is rice bran+cow-dung, which when applied in combination yield better results.

Effect of Inoculum Density on Growth Patterns, Fecundity and Population Dynamics of *M. macrocopa*

Body length/body breadth recorded variations in *M. macrocopa* with varying inoculum density but no significant relationship between inoculum density and growth (body length/body breadth) has been drawn.

Heavy mortality among populations of *M. macrocopa* has been observed in all the sets due to increased competition for food and space. among individuals. Minimal space appears to be a critical factor for such heavy mortality.

References

Anderson, B.G., 1932. The number of pre-adult instars, growth, relative growth and variation in *D. magna*. *Biol. Bull. Woods. Hole.*, 63: 81–98.

Anderson, B.G. and Jenkins, J.C., 1942. A time study events in the life span of *Daphnia magna*. *Biol. Bull.*, 83: 260–272.

Anderson, B.G., Lumel, H. and Zupanic, L.J., 1937. Growth and variability in *Daphnia pulex*. *Biol. Bull., Wood's Hole*, 73: 433–463.

Benider, A. *et al.*, 2002. Growth of *Moina macrocopa* (Straus 1820) (Gustacea : Cladocera): Influence of trophic conditions, population density and temperature. *Hydrobiologia*, 1–3: 468.

Brown, L.A., 1927. Temperature characteristics for duration of an instar in cladocerans. *J. Gen. Physiol.*, 10: 111–119.

Crabtree, S.J., 1975. Population dynamics in *Daphnia pulex*. *Dissertation*. University of Toronto, Toronto, Ontario, Canada.

Edlen, A., 1937. *Experimentelle'wachstumsstudien an Daphnia magna*. Lunds Universtets Arsskrift. N.F. Ard., 2(24): 1–29.

Green, J., 1956. Growth, size and reproduction in *Daphnia* (Crustacea : Cladocera). *Proc. Zool. Soc. Lond.*, 126: 173–205.

Hebert, P.D.N., 1978. The population biology of *Daphnia* (Crustacea : Daphnidae). *Biol. Rev.*, 53: 387–426.

Hutchinson, G.E., 1967. *A Treatise on Limnology, Vol. 2: Introduction to Lake Biology and the Limnoplankton.* John Wiley and Sons Inc., New York, 1115 pp.

Ingle, L., Wood, T.R. and Banta, A.M., 1937. A study of the longevity, growth, reproduction and heart rate in *Daphnia longispina* as influenced by limitations in quantity of food. *J. Exp. Zool.*, 76: 325–352.

Jana, B.B. and Pal. G.P., 1983b. Population growth in *Daphnia* Indian and carinata King as a direct function of available space. *Indian Journal of Experimental Biology*, 21(21): 153–155.

Langer, S., 1987. Culture and some biological aspects of a cladoceran, *Simocephalus vetulus. M. Phil. Dissertation*, University of Jammu, 115 pp.

Langer, S., 1991. Culture and some biological aspects of some important fish food organisms. *Ph.D. Thesis*, University of Jammu.

Malhotra, Y.R. and Langer, S., 1990b. Biological aspects of *Moina macrocopa* in relation to temperature variations. *J. Freshwater Biol.*, 2(2): 111–115.

McArthur, J.W. and Baillie, W.H.T., 1929. Metabolic activity and duration of life. II. Metabolic rates and their relation to longevity in *Daphnia magna. J. Exptal. Zool.*, 53: 243–268.

Mitchell, B.D. and Williams, W.D., 1982. Population dynamics and production of *Daphnia carinata* King and *Simocephalus expinosus* (Koch) in waste stabilization ponds. *Aust. J. Mar. Freshwater Res.,* 33: 837–864.

Munro, I.G. and White, R.W.G., 1975. Comparison of the influence of temperature on the egg development and growth of *Daphnia longispina* O.F. Muller from two habitats in Southern England. *Ocealogia,* 20: 157–165.

Murugan, N. and Sivaramakrishnan, K.G., 1973. *Acutirostratus* (Daphnidae) laboratory studies: The biology of simocephalus of the life span, instar duration, egg production, growth and stages in embryonic development. *Freshwater Biol.,* 3: 77–83.

Murugan N. and Sivaramakrishnan, K.G., 1976Laboratory studies on the longevity, instar duration, growth, reproduction and embryonic development in Scapholebris Kingi Sars (1903) (Cladocera : Daphnidae). *Hydrobiologia,* 50(1): 75–80.

Murugan, N., 1975. Egg production, development and growth in *Moina micrura* (Kurz.) (1974) (Cladocera : Moinidae). *Freshwater Biol.,* 5: 245–250.

Paloheimo, J.E., Crabtree, S.J. and Taylor, W.I.D., 1982. Growth model of Daphnia. *Canadian Journal of Fisheries and Aquatic Sciences,* 39: 598–606.

Ramirez, E.M., Sarma, S.S.S. and Nandini, S., 2004. Recovery patterns of *Moina macrocopa* exposed previously to different concentrations of cadmium and methyl parathion: Life table demography and population growth studies. *Hydrobiologea,* 1: 526.

Sharma, N. and Sahai, Y.N., 1990. Laboratory culture of fish food *Daphnia cladoceran. Recent Trends in Limnology,* p. 197–203.

Venkataraman, K. and Job, S.V., 1980. Effect of temperature on the development, growth and egg production in *Daphnia carinata* King (Cladocera : Daphnidae). *Hydrobiologia,* 68: 217–224.

Vijverberg, J., 1980. Effect of temperature in laboratory studies on development and growth of cladocera and copepoda from Jheukmeer, The Netherlands. *Freshwater Biol.,* 10: 319–340.

Chapter 24

Recent Advances in Fish Pheromone Research: An Overview

A.K. Pandey[1] *and Saurabh Pandey*[2]
[1]*National Bureau of Fish Genetic Resources (ICAR), Canal Ring Road, Telibagh, P.O. Dilkusha, Lucknow – 226 002*
[2]*Department of Zoology, D.D.U. Gorakhpur University, Gorakhpur – 273 009*

ABSTRACT

There are increasing evidences that most of the vital activities of fish such as alarm communication, sex attraction and synchronization of reproductive processes, individual identification, group cohesion, parent-offspring recognition, territorial markings as well as migration are shaped by pheromones. The involvement of pheromones is known only by the behavioural responses because chemically they are largely unknown. Aspects pertaining to fright reaction and alarm substances, sex pheromones, social structure and individual recognition together with migration have been elaborated in the light of recent advances in the field. Problems associated with chemoreception due to environmental pollution as well as potential of utilizing pheromones in advancing maturation and breeding of cultivable species and management of invasive non-native/pest species are also highlighted.

Keywords: Alarm pheromones, Individual recognition, Sex pheromones, Migration, Aquatic pollution, Fishery management.

Introduction

Pheromones are externally secreted semiochemicals employed in intercommunications among individuals of the same species. There exist convincing evidence that such system is widely operative among the organisms ranging from protozoans to mammals. There are instances of chemical communication in aquatic plants too, especially for the attraction of gametes. Recent studies have shown that most of the vital activities in fish like alarm communication, sex attraction as well as synchronization of the reproductive processes, individual identification, group cohesion, parent-offspring recognition, territorial markings and migration are shaped by pheromones (Liley, 1982; Pfeiffer, 1982; Pandey, 1984, 1998; Chivers and Smith, 1998; Mann *et al.*, 2003). Pheromones of the terrestrial organisms are volatile having low molecular weight ranging from 50–300 while such semiochemicals in the aquatic species appears to be of larger size (high molecular weight) and soluble in water (Smith, 1992; Sorensen *et al.*, 1998; Pandey, 2003, 2005).

Alarm Pheromones

Alarm substance cells (ASC) of the ostariophysan and other groups of fish release alarm pheromone from their skin on mechanical injury, presumably by a predator that elicits alarm reaction among the conspecifics and members of the closely-related species. Depending upon biology of the species, the alann reaction includes dashing, c-turn, immobility, tight schooling, hiding and avoidance of the area (Pfeiffer, 1982; Smith, 1992). There are instances that the predator northern pike (*Esox lucius*) consuming the preys like *Pimephales promelas* and *Culaea incostans* containing the ASCs in skin, releases the alarm pheromone in the fecal matter which are detected by the members of the prey or closely-related species. The active space of alarm pheromone can exceed 58,000 l/cm^2 skin. The chemically induced alarm reaction may be transmitted visually among other members of the school/shoal (Pfeiffer, 1977; Smith, 1977, 1982a). There are reports that experience also plays a role in the alarm pheromone perception (Smith, 1982b; Mathis and Smith, 1993; Brown *et al.*, 1995; Chivers *et al.*, 1995; Wisenden *et al.*, 1995). Stabell and Lwin (1997) observed phenotypic changes like increase in body depth in crucian carp (*Carassius carassius*) exposed to the odour of northern pike fed on crucian carps. This phenotypic change appears to be due to primer effect of the alarm pheromone.

The alarm substance of minnow was suggested to be purine- or pterine-like (double ringed compounds usually associated with pigments), non-volatile and extremely soluble in water. Histamine was also implicated to be alarm substance of certain cyprinids. Pfeiffer and Lemke (1973) reported the alarm substance of the European minnows (*Phoxinus phoxinus*) to be pterine (isoxanthopterine) that might be conjugated with protein for water solubility as well as species specificity. However, Kasumyan and Lebedeva (1977, 1979) showed that the alarm pheromone of *P. phoxinus* possesses molecular weight approximately 1100, behaves in alkaline medium (pH 8-9) as anion and breaks down when heated. They suggested it to be a carbohydrate or related compound containing amino groups. Pfeiffer (1982) found several heterocyclic compounds to induce effective alarm response. Hypoxanthine-3-N-oxide has been reported to be an active component of the ostariophysan alann signaling system (Pfeiffer *et al.*, 1985; Brown *et al.*, 2000, 2003). Laboratory as well as field studies have revealed that the traps labeled with the fathead minnows skin extract, hypoxanthine-3-N-oxide, or pyridine-Noxide caught significantly fewer fish than did those labelled with distilled water (Brown *et al.*, 2000). The central bundles in both tractus olfactorii of the medial olfactory tract of the fish conduct the nervous excitation elicited by alarm substance to the higher center in brain (Pfeiffer *et al.*, 1984) resulting in the elevation of plasma cortisol, glucose, potassium (K^+) and sodium (Na^+) levels. Plasma calcium (Ca^{2+}) recorded an initial increase followed by a decline to the below nonnallevel. These changes suggest that alarm pheromone functions as stressor in fish (Lebedeva *et al.*, 1994, 1999).

Female Sex Pheromones

Salmonid males congregate around the cages/pens of the mature females during breeding seasons. Fishermen of Mississippi usually catch large number of male channel catfish (*Ietalurns punctatus*) by placing ripe' females in cages (Timms and Kleerekopper, 1972). Water from gravid females seems to increase the nesting and other activities associated with reproduction among the males in anabantids and cichlids (Crapon de Caprona, 1974). Tavolga (1956) was the first who systematically studied the sexual behaviour of an estuarine goby (*Bothygobius soporator*) and demonstrated that the spenniated (matured) male discriminates between males and females, and also between non-gravid and gravid females on the basis of chemical cues released through their vent. Ovarian fluid was sufficient to evoke courtship behaviour in males even in the absence of females. A female sex pheromone appears to facilitate the pair formation and copulation in the marine carcharhinid shark. Also, mature female sea-lamprey, *Petromyzon marinus*, releases a pheromone in the ovarian fluid that attracts male conspecifics (Teeter, 1980). Later studies revealed that a definite threshold of circulating levels of androgen is essential to render the male differentially sensitive to the sex attractant released by the gravid females.

Mature females of goldfish (*Carassius auratus*), yellowfin Baical sculpin (*Cottocomephorus grewingki*), Atlantic salmon (*Salmo salar*), rainbow trout (*Salmo gairdeneri = Oncorhynchus mykiss*), Baikal cisco (*Coregonus autumnalis migratorius*), common carp (*Cyprinus carpio communis*) and *Barilius bendelisis* release primer pheromone in the urine (Liley, 1982; Sorensen *et al.*, 1998; Sorensen and Stacey, 1999, 2004; Pandey, 2003) which probably enhances the plasma gonadotrophin (GtH) and testosterone (T) levels (within 15–30 minutes in case of goldfish) and milt (sperm and seminal fluid) volume as well as sperm motility among males after prolonged exposure (6 hours in case of goldfish); Besides urine, the priming pheromone has also been detected recently in bile of the female rainbow trout. There exists report showing a reduction in pituitary dopamine turnover resulting in increased gonadotropin secretion in the male goldfish exposed to the female sex pheromone. Interestingly, mature female *Oreochromis niloticus* becomes attractive to males after 17, 20-dihydroxy-4-pregnene-3-one administration. Though the exact chemical identity of the primer pheromone(s) is not known but the maturation-inducing hormone, 17, 20-dihydroxy-4pregnene-3-one and its sulphate or glucuronide metabolites are the likely candidates because they induce electro-olfactogram (EOG) recordings in the olfactory tract even at the low concentration ranging between 10^{-9} –10^{-13} mM. The primer pheromone released by the Atlantic salmon (*Salmo salar*) has been suggested to be F-type prostaglandins. Urine of the female evokes maximum response in the conspecific males emphasizing species-specificity of the pheromone.

A number of externally fertilizing fish release chemical coupling (s) during ovulation which elicits attraction (locomotor behaviour), courtship and spawning in male conspecifics (Murphy *et al.*, 2001; Laberge and Hara, 2003). Such releaser pheromones induce an immediate response (within to-30 minutes) without altering the plasma concentration of gonadotrophin. Amouriq (1965) suggested the female sex attractant (substance dynamogene) of *Lebestes reticulatus* to be an estrogen as hexestrol dipropinoate elicited courtship behaviour among males. Okada *et al.* (1978) showed that the fi-action (FP_1) on DEAE cellulose column that evoked male courtship behaviour in the pond smelt declined in activity after heat (80°C; 5 minutes) or trypsin (30°C; 60 minutes) treatment and presumed the substance in genital fluid was either a protein or a substance involving protein in its structure. Kawabata (1993) reported amino acids as inducer of sexual behaviour in male *Rhodurus occellatus occellatus.* Recent studies have suggested the metabolites of prostaglandin F_{2a} (PGF_{2a}) like 15-keto- and 13,14-dihydro-15-keto-PGF_{2a} to function as releaser pheromones in *Pimephales promelas, Carassius auratus, Salvelinus alpinus, Misgumus anguillicaudatus, Oncorhynchus rhodurus* and *Astyanax mexicanus.* Interestingly, tetrodoxine has been reported to be the sex attractant pheromone to spermiating males in the marine puffer fish, *Fugu niphobles* (Matsumura, 1995). It has been suggested that the releaser pheromone(s) functions as reproductive isolating principle in sympatrically breeding gourami (*Trichogaster trichopterus* and *T. pectoralis*), suckers (*Catastomus catastomus* and *C. commersonii*) and masu salmon/rainbow trout (*Oncorhynchus masu* and *O. mykiss*). As per definition, the pheromones should be species specific but there are instances of closely-related species responding to the sex pheromones in fish.

Male Sex Pheromones

During breeding season, male become attractive to the female conspecifics. In some regions of France, male lampreys nearing sexual maturity are placed by the fishermen in the traps to attract females (Johnson *et al.*, 2006). Attraction on sexually mature female brook trout to mature male conspecifics appears to be mediated by pheromone (Young *et al.*, 2003). The odouriferous secretion of the cutaneous anal gland (rich in mesorchial interstitial cells) of male *Blennius pavo* attracts the females during the breeding season. It has been recorded that appendices of the 1st and 2nd anal fin spines of the mature male *B. pavo* produce a substance that attracts conspecific females. Numerous invaginations are found in the anal appendices of this species and production of pheromone form this organ is dependent on the circulating gonadotropin levels. The gill gland of male swordtail characin, *Corynopoma riisei* appears to be the source of female sex attractant pheromone. Sperm-duct gland of freshwater goby, *Padogobius martensi* appears to secrete sex attractant substance. Jaski (1939) was the first who reported that males of guppy (*P. reticulatus*) secrete a "chemical coupling" which synchronizes the sexual receptivity among the females. Teeter (1980) showed the presence of a pheromone in urinogenital fluid of male sea lamprey (*P. marinus*) that elicits courtship in the females. Out of the nine steroid hormones such as dehydroepiandrosterone, testosterone, dihydrotestosterone, progesterone, androstenedione, estrone, oestradiol, corticosterone and cortisol tested, only testosterone elicited the preference response in spawning-run female sea lamprey (Adams *et al.*, 1987). Testosterone or a closely-related structural derivatives has been suggested to function as sex pheromone in the male cyclostome. The male sexual pheromone of *Gobius jozo* has been identified as etiocholenolone glucuronide. Mature males of *S. alpinus* release metabolites of prostaglandin F_{2a} to attract mature females and induce spawning behaviour.

Seminal vesicles of the male African catfish (*Clarias gariepinus*) secrete sex pheromone that attracts females during the breeding season (Resink, 1988; van Weerd, 1990; van den *Hurk* and Jenkins, 1992). Though 17 free and 8 glucuronide metabolites of sex hormones have been identified in the seminal vesicle fluid (SVF) but Sβ-pregnene-3α, 17α-diol-20-one glucuronide and 5β-androstene-3α, 17α-diol are the two more potent olfactory stimulants with threshold value 10^{-9}. Wild breeding male African catfish recorded 20-folds elevation in level of 5β-pregnene-3α, 17α-diol-20-one glucuronide in the seminal vesicle as compared to those maintained under captivity which may probably be the reason for lack of natural spawning in the catfish kept in ponds (Schoonen, 1987; Resink, 1988).

Mature males of zebrafish (*Brachydanio rerio*) release a pheromone that induced ovulation in females. Gas chromatography-mass spectrometry (GC-MS) studies have revealed the presence of glucuronides of 5α-androstane-3α,17β-diol and cholesterol in male-holding waters. It has been suspected that these compounds, probably originating from liver, may be functioning as ovulation-inducing pheromone.

The most profound effect of males on the females seems to be its influence over the neuroendocrine (hypothalamo-hypophysial-gonad axis) of the latter. Mosher (1954) was the first to observe that introduction of a male in the group of females stimulates egg production and its continued presence led to regular oviposition cycles. In the angelfish (*Pterophyllum sealare*), blue gourami (*T. triehopterus*), Indian catfish (*Heteropneustes fossilis*) and zebrafish (*Brachydanio rerio*), paired females with males possessed ovaries filled with matured ova while ovaries of the grouped females (without males) contain almost no ova but numerous oocytes at various stages of development (Pandey *et al.*, 2000; Pandey, 2005; Gerlach, 2006). There exist reports that the male pheromone(s) of African catfish (*C. gariepinus*) accelerates the ovarian recrudescence and ovulation in females probably by enhancing plasma gonadotropin (GtH) level (van Weerd, 1990; van den Hurk and Resink, 1992). Recently, Degani and Schreibmam (1993) observed that the aquarium water in which male gourami (mature) built nests induced release of immunoreactive gonadotropin (GtH) in the females which, in turn, enhanced plasma 17β-estradiol (E_2), 17α,20β-dihydroxy-4-pregnene-3-one (17,20-P), 5β-pregnene-3α,17α,20β-triol and 11-ketotestosterone (11-KT) levels as well as vitellogenic activities suggesting the primer effects of male pheromone in this species.

Intrasexual Stimulants and School Formation

Though the reports on the odouriferous secretions eliciting male-male and female-female stimulations are rare in vertebrates but the possibilities have been raised from time to time. In fishes, the male three-spined stickleback (*Gasterosteus aeuleatus*) gets stimulated by smelling the odours of its own nest during breeding season. The males of *Hypsoblennius jenkinsi*, *H. robustus* and *H. genitilis* secrete a chemical during "high courting and mating periods" that releases the sexual appetitive behaviour and receptivity in the males. Since unisexual groupings of the females (without males) disrupt (inhibit) the ovarian development, it offers an opportunity to explore the female-female interacting pheromones in fish (Pollack *et al.*, 1978; Pandey, 2005; Gerlach, 2006).

A pheromone is found in the milt of male herring (*Clupea harengus pallasi*) that triggers spawning behaviour (papilla extension) and release of milt in the males. Sherwood *et al.* (1991) reported that this pheromone shows hydrophobic properties similar to those of polar steroids, prostaglandins or other conjugate forms, and at least one form appears to contain a sulphate or glucuronide group. The male pheromone appears to play important role in synchronization of spawning in the school of herrings. Recent studies have shown that the elevated levels of plasma 3α,17α-dihydroxy-5β-pregnen-20-one and 17α-hydroxyprogesterone coincided with responsiveness to the spawning pheromone in this species (Carolsfield, *et al.*, 1997).

McFarland and Moss (1967) speculated possibility of the involvement of pheromones in school formation and attraction of the mature conspecifics towards the spawning grounds. Presence of an aggregating pheromone has been implicated in the schooling Japanese marine catfish, *Plotosus lineatus.* Bloom and Perlmutter (1977) demonstrated the presence of male-male and female-female attractants in the zebra fish (*B. rerio*). Teeter (1980) also hypothesized involvement of pheromone(s) in the aggregation of adult lampreys (*P. marinus*) prior to upstream movement or to keep them together in night during the migration. It has been speculated that the sexual-aggregating pheromones of zebrafish are two in number-one attracting male conspecifics and other the females. These attractants appear to be contained in the cholesterol-ester fraction (R*f*-0.94) of the thin-layer chromatogram and the attraction response was not affected by food or diet.

Individual and Species Recognition

Varying degrees of socialism have been recorded in fish, however, such tendency is more marked among the visually-deficient species. The school of roach (*Ritulus ritulus*) does not disintegrate during night because members of the school maintain contact through chemical cues. There exist records that individuals of *Ictalurus natalis, Clupea herrangus pallasi, Caecobarbus geertsi, Haplochromis burtoni, Astyanix jordani, Salmo alpinus, Notropis lutrensis, Anguilla anguilla, Oncorhynchus kisutch, Oncorhynchus nerka, ktalurus nebulosus, Astyanax mexicanus, Phreactichthys andruzzi, Carassius auratus, Heteropneustes fossilis* and *Danio rerio* are attracted towards the waters of the tank previously occupied by their conspcifics. Even broods and youngs are recognizable to their parents by the odours. The parents of Midas cichlid fish (*Cichlasoma citrinellum*) recognize to their youngs by means of pheromone. Barnett (1981) reported that fly of Midas cichlid fish discriminates even between mother and father on the basis of urine odour and suggested that the steroid titers in urine could provide information on sex and that of peptide chains could indicate species. It appears that secretions in the skin mucus, urine as well as metabolites of bile acids (voided in fecal matter) are involved in individual identification and group cohesion in these fishes (Todd, 1971; Selset, 1980; Barnett, 1981; Zhang, 1997). Interestingly, glucurnidation of sex hormones also takes place in the skin of the African catfish (*Clarias gariepinus*) which may probably have certain implication in individual/sex recognition Recently, synthesis of 17α,20α,17α,20β-dihyhroxy4-pregnen-3-one, 11-ketotestosterone and other sex hormone conjugates have been recorded in the gill of goldfish, common carp and trout suggesting that the organ might play role in secretion of pheromones in some species.

Beneficial conditioning of the waters by conspecifics and inhibition of growth under over-crowded condition appear to be mediated through pheromones (Solomon, 1977). Eels (*Anguilla anguilla*) are usually caught in the traps dripped with conspecific skin extracts (Saglio, 1982). Todd (1971) found that fighting as well as stress alter chemistry of the pheromones of the yellow bullhead (*ktalurns natalis*) secreted in the mucus. Diets also appear to modify body odour of the catfish. Interestingly, Colyer and Jenkins (1976) indicated the presence of an aggression-suppressing pheromone(s) in the Siamese fighting fish, *Betta spendens*. Carr and Carr (1986) have observed that the aggression-suppressing pheromone of the brown bullhead (*I. nebulosus*) gets inactivated by heat and protease treatments and suggested it to be a protein with molecular weight less than 10,000.

During reproductive period, the males of some species deposit a layer of mucus on the spawning surface which may serves as chemical signals marking the breeding sites and may perhaps communicate status of the occupants. There exist reports on the substrate marking behaviour in fishes and secretions of skin (mucus), urophysis and urine have been implicated in the production/release of the territorial marking pheromones (Stabell, 1987). Possibly, these marks bear the same biological significance as marking fluid (MF) of the mammals.

Pheromones and Migration

How migratory fishes locate their pathways from sea to freshwater and *vice versa* remains a mystery. Work done on homing of the salmonids has demonstrated that these fishes migrate from the nursery areas to their parental river streams, out to sea and then back to their birth place to spawn (basis of the home stream theory). It was suggested that each stream had unique bouquet of odours and fish has to follow its nose in detection and discrimination of the natal stream. Though the role of metabolic products of a specific population (pheromones) as homing cues for the salmonids was proposed in 1924 and 1934 but this hypothesis was totally eclipsed by the concept of "olfactory imprinting" propounded by Hasler (1983). According to this theory, the youngs lock onto the odours of aquatic vegetations, soil run-off and other organic chemicals present in the parent stream in brain and they retain the olfactory memory during the interim period of homeward migration.

Nordeng (1971) realized the pheromone odour tracks help guide homing of the Arctic char (*Salmo alpinus*). Solomon (1973) demonstrated that the migrating Atlantic salmon (*S. salar*) prefers the river in which youngs of its own populations were living. After exhaustive studies on the migratory behaviour of *S. alpinus, S. trutta* and *S. safar*, Nordeng (1977) rejuvenated the pheromone hypothesis by emphasizing that homeward navigation is an inherent response to population-specific pheromone trails released from the descending smolts. The role of pheromones in upstream orientations has been confirmed in the anadromous rainbow trout (*O. mykiss*) and sea lamprey (*P. marinus*). Miles (1968) has demonstrated that elvers of the catadromous European eel (*A. rostrata*) are attracted towards the water of the streams containing the conspecific adults. Sorensen (1984) also confirmed this finding during his study at Rhode Island. Recently, Stabell (1992) remarked that the specific homing to a native site is under genetic influence and the possible genetic contamination of pheromones due to hatchery escapes or random stocking programmes might seriously interfere with the homing performance and population structure. He emphasized for the concept of chemical ecology in management of salmonid fisheries.

Our knowledge regarding the chemical nature of these navigational cues is still fragmentary. Miles (1968) found that the compound(s) involved with attraction of the elvers in *A. rostrata* to be biodegradable, unaffected by autoclaving and non-volatile. Atema *et al.* (1973) reported that the alwife (*Alosa pseudoherengu*) homing pheromones are heat-stable, non-volatile, polar and having molecular weight less than 1000. Doving *et al.* (1980) demonstrated the presence of migratory cues of *S. alpinus* in the fecal matter of the smolts and suggested it to be derivatives of bile salts like taurolithocholate, taurolithocholic sulphate, sulphotaurolithocholate or taurochenodeoxycholate which may convey the information to the brain through medial portion of the olfactory tract. By employing electrophysiological, biochemical and behavioural approaches, Zhang (1997) also confirmed the chemostimulatory roles of 9 out of 38 authentic bile acids in lake char (*Salvelinus namaychush*) and remarked that the olfactory sensitivity and specificity of bile acids are affected by (*i*) position and orientation of hydroxyls, (*ii*) hydrosulphation, (*iii*) side-chain length and (*iv*) side-chain substituents of the molecules. Li and Sorensen (1997) and Li *et al.* (2002) have identified peteromyzonol sulphate and allocholic acids as migratory pheromone in the sea lamprey, *P. marinus*.

Chemoreception and Water Pollution

Four groups of chemicals like (*i*) amino acids, (*ii*) steroids, (*iii*) prostaglandins and (*iv*) bile acids (and their derivatives) have been found to be detected by the fish even at minute (10^{-7}–10^{-13} mM) concentration and each group of chemicals have different receptors in the olfactory system. Hara (1992) has given a hypothetical three-dimensional model to show the mode of interaction of amino acids, which serve as feeding stimulants in fish, with olfactory receptor sites. It has been observed that detergents even at the concentration of 0.5 ppm (much lower value that inflict lethal damage) impaired functioning of the receptor by causing erosion in the chemosensory organ. Bloom *et al.* (1978) showed that sublethal (0.5 ppm) exposure of zinc affected the behaviour to the sexual aggregating pheromones in the zebrafish. Linear alkylbenzene sulphate (LAS) significantly depressed chemo-attraction of water conditioned by conspecifics to juvenile *Salvelinus fontinalis*. Laboratory studies have revealed that the freshwater spawning migration of salmons is being impaired by acidification of rivers. Electrophysiological recordings of the olfactory epithelium of the adult male *S. salar* to testosterone and urine of the ovulated females significantly reduced at pH 5.5–4.5

and abolished at pH 3.5. Exposures to sublethal concentrations of diazinon and carbofuran significantly reduced the ability of mature Atlantic salmon parr to priming pheromone of ovulated females. The pernicious effects of an anionic detergent (sodium lauryl sulphate), pH, mercury (Hg), silver (Ag), cadmium (Cd), copper (Cu), nickle (Ni) and zinc (Zn) on the olfactory system are now well established. By calculating IC_{50} value (concentration which inhibits electro-olfactographic response by 50 per cent), Ag, Hg, Cu and Cd are found to elicit most depressive effects on the chemoreception in salmonids. The whitefish (*Coregonus clupeaformis*) and rainbow trout (*O. mykiss*) exposed to 2.4 µm copper sulphate ($CuSO_4$) solution for two weeks exhibited loss in the phospholipid stainable granules in the receptor neurons of the olfactory mucosa, however, recovery in the granules were observed when the fishes were transferred to freshwater for 12 weeks. Phospholipids have been reported to be involved in the electrophysiological (electro-olfactogram, EOG) activities of the olfactory epithelium. There are reports that olfactory epithelial cytochrome P-450 and mono oxygenase (mixed function oxidase, MFO) are activated by the pollutants like hydrocarbons and heavy metals. Since the present trends of rapid industrialization and increase in population have drastically changed the aquatic environment, it is imperative to undertake the studies regarding the interactions of various pollutants with the pheromonal communication system in fishes. Barnett (1981) remarked that a chemically polluted environment might interfere with the success of parent care.

Conclusions

Earlier fishes were viewed as leaky bags that slowly release various chemicals which can serve as a chemical picture of the animal that others can smell and identify as to species, sex, stress level and perhaps size as well as individuality (Atema, 1980, 1996). Though the chemical nature of pheromones and sites of their biosynthesis (except alarm substances) are not clearly identified, there are growing evidences that these chemicals do have communicative roles among this class of vertebrates. Since sex pheromones are widely distributed among fish, a better understanding of their role is *sine qua non* to understand their role in reproductive physiology (Deftaipont and Sorensen, 1993). Interestingly, GnRH has been identified as a potent olfactory stimulant in fish but its role in synchronization of reproductive process is not known. Since these signals are species-specific and operate at molecular levels, it is imperative to utilize them for management and conservation of fisheries (Korkum, 2004; Sorensen and Stacey, 2004; Klassen *et al.*, 2005; Wagner *et al.*, 2006). In aquacultural operations, they may be employed to advance puberty and accelerate ovarian recrudescence as well as selective stimulants to induce spawning at a time and place convenient to the managers (von Weerd, 1990; van den Hurk and Resink, 1992; Zhao *et al.*, 2003; Gerlach, 2006; Hong *et al.*, 2006; Johnson *et al.*, 2006). Since they are externally active, their delivery would eliminate the stress associated with the hormone or drug administration (Stacey, 2003). They may also be employed as artificial baits, selective attractants, growth stimulators as well as inhibitors of aggression and cannibalism. Recent studies point to the potential exploitation of semiochemicals in propagating the migratory fishes in new rivers and streams.

Acknowledgements

Senior author is grateful to Prof. A.G. Sathyanesan, FNA and Prof. C.I. Dominic, FNA for their constant encouragement and inspiration. Help extended by Prof. R.J.F. Smith, Prof. N.R. Liley, Prof. T.J. Hara, Prof. N.E. Stacey, Prof. P.W. Sorensen, Prof. A.C. Rossi, Dr. S.A. Ali and Ms. Rosalie N. Schaffer are thankfully acknowledged.

References

Adams, M.A., Teeter, J.H., Katz, Y. and Johnson, P.B., 1987. Sex pheromones of the sea lamprey (*Petromyzon marinus*) steroid studies. *J. Chem. Ecol.*, 13: 387–395.

Amouriq, L., 1965. Origine de la substance dynamo gene emise par *Lebestes reticulatus* female (Poisson : Poeciliidae : Cyprinodontiformes). *C.R. Acad. Sci.* (*Paris*), 260: 2334–2335.

Atema, J., 1980. Smelling and tasting underwater. *Oceanus*, 23: 2–18.

Atema, J., 1996. Eddy chemotaxis and odour landscape: Exploration of nature with animal sensor. *Biol. Bull.*, 191: 219–238.

Atema, J., Jacobson, S., Todd, J.H. and Baylan, D., 1973. The importance of chemicals in stimulating behaviour of marine organisms: Effects of altered environmental chemistry on animal communication. In: *Bioassay Techniques and*

Environmental Chemistry, (Ed.) G.E. Glass. Ann Arbor Science Pub., New York, pp. 177–197.

Barnett, C., 1981. The role of urine in parent-offspring communication in a cichlid fish. *Z. Tierpsychol.*, 52: 173–182.

Bloom, H.D. and Perlmutter, A., 1977. A sexual aggregating pheromone system in zebra fish, *Brachydanio rerio* (Hamilton-Buchanan). *J. Exp. Zool.*, 199: 215–226.

Bloom, H.D., Perlmutter, A. and Seeley, R.J., 1978. Effect of sublethal concentration of zinc on an aggregating pheromone system in the zebra fish, *Brachydanio rerio* (Hamilton-Buchanan). *Environ. Pollut.*, 17: 127–131.

Brown, E.G., Chivers, D.P. and Smith, R.J.F, 1995. Fathead minnows avoid conspecific and heterospecific alarm pheromones in the feces of northern pike. *J. Fish Biol.*, 47: 387–393.

Brown, G.E., Adrian, J.C., Smyth, E., Leet, H. and Brennan, S., 2000. Ostariophysan alarm pheromones: Laboratory and field tests of the functional significance of nitrogen oxides. *J. Chem. Ecol.*, 26: 139–154.

Brown, G.E., Adrian, J.C., Naderi, T., Harvey, M.C. and Kelly, J.M., 2003. Nitrogen oxides elicit antipredator responses in juvenile channel catfish, but not in convict cichlids of rainbow trout: Conservation of the ostariophysan alarm pheromone. *J. Chem. Ecol.*, 29: 1781–1796.

Carolsfield, J.M., Scott, A.P. and Sherwood, N.M., 1997. Pheromone induced spawning in pacific herring. 2. Plasma steroids distinctive to fish responsive to spawning. *Horm. Behav.*, 31: 269–276.

Carr, M.G. and Carr, J.E., 1986. Characterization of an aggression-suppressing pheromone in the juvenile brown bullhead (*Ictalurus nebulosus*). *Copeia*, 1986: 540–545.

Chivers, D.P. and Smith, R.J.F., 1998. Chemical alarm signaling in aquatic predatory-prey systems: A review and prospectus. *Ecoscience*, 5: 338–352.

Chivers, D.P., Wisenden, B.D. and Smith, R.J.F., 1995. The role of experience in the response of fathead minnows (*Pimephales promelas*) to skin extract of Iowa darters (*Etheostoma exile*). *Behaviour*, 132: 665–674.

Colyer, S.W. and Jenkins, C., 1976. Pheromonal control of aggressive display in Siamese fighting fish (*Betta splendens*). *Percept. Motor Skills*, 42: 47–54.

Crap on de Caprona, M.D., 1974. The effect of chemical stimuli from conspecifics on the behaviour of *Haplochromis burtoni* (Cichlidae : Pisces). *Experientia*, 30: 1394–1395.

Defraipont, M. and Sorensen, P.W., 1993. Exposure to the pheromone 17α,20β-dihydroxy-4pregnen-3-one enhances the behavioural spawning success, sperm production and sperm motility of male goldfish. *Anim. Behav.*, 46: 254–256.

Degani, G. and Schreibman, M.P., 1993. Pheromone of male blue gourami and its effect on vitellogenesis, steroidogenesis and gonadotropin cells in pituitary of the female. *J. Fish Biol.*, 43: 475–485.

Doving, K.B., Selset, R. and Thommesen, G., 1980. Olfactory sensitivity to bile acids in salmonid fish. *Acta Physiol. Scand.*, 108: 123–131.

Gerlach, G., 2006. Pheromonal regulation of reproductive success in female zebrafish: Female suppression and male enhancement. *Anim. Behav.*, 72: 1119–1124.

Hara, T.J., 1992. *Fish Chemoreception*. Chapman and Hall, London.

Hasler, A.D., 1983. Synthetic chemicals and pheromones in homing salmon. In: *Control Processes in Fish Physiology*, (Eds.) J.C. Rankin, T.J. Pitcher and R.T. Duggan. Croom-Helm, London, pp. 103–116.

Hong, W.-S., Chen, S.-X., Zhang, Q.-Y. and Zheng, W.-Y., 2006. Sex organ extracts and artificial hormonal compounds as sex pheromones to attract broodfish and to induce spawning of Chinese blacksleeper (*Bostrichthys sinensis* Lacepede). *Aquacult. Res.*, 37: 529–534.

Jaski, C.J., 1939. Ein oestrus-zyklus bei *Lebestes reticulatus* (Peters). *Proc. Koninkl. Ned Akad Wetenschap.*, 42: 201.

Johnson, N.S., Siefkes, M.J. and Li, W., 2005. Capture of ovulating female sea lampreys in traps baited with spermiating male sea lampreys. *North Am. J. Fish. Manage.*, 25: 67–72.

Johnson, N.S., Luehring, M.A., Siefkes, M.J. and Li, W., 2006. Mating pheromone reception and induced behaviour in ovulating female sea lampreys. *North Am. J. Fish. Manage.*, 26: 88–96.

Kasumyan, A.O. and Lebedeva, N.E., 1977. On the chemical nature of the alarm substance from skin of the European minnow (*Phoxinus phoxinus*). *Biol. Nauk.*, 1: 37–41.

Kasumyan, A.O. and Lebedeva, N.E., 1979. New data on the nature of the alarm pheromone in cyprinids. *J. Ichthyol.*, 19: 109–114.

Kawabata, K., 1993. Induction of sexual behaviour in male fish (*Rhodeus ocellatus ocellatus*) by amino acids. *Amino Acids*, 5: 323–327.

Klassen, W., Adams, J.V. and Twohey, M.B., 2005. Modeling the suppression of sea lamprey populations by use of the male sex pheromone. *J. Great Lake Res.*, 31: 166–173.

Korkum, L.D., 2004. Pheromone signaling in conservation. *Aquatic Conserv. Mar. Freshw. Ecosyst.*, 14: 327–331.

Laberge, F. and Hara, T.J., 2003. Behavioural and electrophysiological responses to F-prostaglandins, putative spawning pheromones, in three salmonid fishes. *J. Fish Biol.*, 62: 206–221.

Lebedeva, N.Y., Lebedeva, V.I. and Tgolovkina, V., 1994. Pheromone of the trout stress inducer. *Biophysics*, 39: 527–530.

Lebedeva, N.Y., Tomkevich, M.S., Golovkina, N.A., Vosyliene, M.Z. and Golovkina, T.A., 1999. On hormesis of alarm pheromone of cyprinids. *Pheromones*, 6: 65–68.

Li, W. and Sorensen, P.W., 1997. Highly independent olfactory receptor sites for naturally occurring bile acids in the sea lamprey, *Petromyzon marinus* to pheromonal bile acids. *J. Compo Physiol.*, 180A: 429–438.

Li, W., Scott, A.P., Siefkes, M.J., Yan, H.G., Liu, Q., Yun, S.S. and Gage, D.A., 2002. Bile acids secreted by male sea lamprey that act as pheromone. *Science*, 296: 138–141.

Liley, N.R., 1982. Chemical communication in fish. *Can. J. Fish. Aquat. Sci.*, 39: 22–35.

Losey, G.S., 1969. Sexual pheromone in some fishes of the genus, *Hypsoblennius* Gill. *Science*, 163: 181–183.

McFarland, W.N. and Moss, S.A., 1967. Internal behavour of fish in schools. *Science*, 156: 260–262.

Mann, K., Turnell, E., Atema, J. and Gerlach, G., 2003. Kin recognition in juvenile zebrafish (*Danio rerio*) based on olfactory cues. *Biol. Bull.*, 205: 224–225.

Mathis, A. and Smith, R.J.F., 1993. Intraspecific and cross-superorder responses of chemical alarm signals by brook stickleback. *Ecology*, 74: 2395–2404.

Matsumura, K., 1995. Tetradoxin as a pheromone. *Nature*, 378: 563–564.

Miles, S.G., 1968. Rheotaxis of elevers of the American eel (*Anguilla rostrata*) in the laboratory to water from different streams in Nova Scotia. *J. Fish. Res. Bd. Can.*, 25: 1591–1602.

Mosher, C., 1954. Observations on the spawning behaviour and the early larval development of the sargassum fish, *Histrio histrio*. *Zoologica*, 39: 141–156.

Murphy, C.A., Stacey, N.E. and Corkum, L.D., 2001. Putative steroidal pheromones in the round goby, *Neogobius melanostomus* : Olfactory and behavioural responses. *J. Chem. Ecol.*, 27: 443–470.

Nordeng, H., 1971. Is the local migration of anadromous fishes determined by pheromones? *Nature*, 233: 411–413.

Nordeng, H., 1977. A pheromone hypothesis for homeward migration in anadromous salmonids. *Gikos*, 28: 155–159.

Okada, H., Sakai, D.K. and Sugiwaka, K., 1978. Chemical stimulus on the reproductive behaviour of the pond smelt. *Sci. Rep. Hokkaido Fish Hatch.*, 33: 89–99.

Pandey, A.K., 1984. Chemical signals in fishes: theory and application. *Acta Hydrochim. Hydrobiol.*, 12: 463–478.

Pandey, A.K., 1998. Chemical communication in fishes: An overview. *Fishing Chimes*, 18(2): 15–20.

Pandey, A.K., 2003. Current status and potential applications of fish pheromones in aquaculture and fishery management. In: *Aquaculture Medicine*, (Eds.) I.S. Bright Singh, S.S. Pai, R. Philip and A. Mohandas. School of Environmental Sciences, Cochin University of Science and Technology, Cochin, pp. 271–292.

Pandey, A.K., 2005. Recent advances in fish pheromone research with emphasis on their potential applications in fisheries. *J. Appl. Zool. Res.*, 16: 210–216.

Pandey, A.K., Mitra, A. and Sarkar, B., 2000. An instance of chemical communication in the Indian catfish, *Heteropneustes tossilis* (Bloch). In: *Fifth Indian Fisheries Forum,* January 17–20, Central Institute of Freshwater Aquaculture, Bhubaneswar, p. 83.

Pfeiffer, W., 1977. The distribution of fright reaction and alarm substance cells in fishes. *Copeia,* p. 653–665.

Pfeiffer, W., 1982. Chemical signals in communication. In: *Chemoreception in Fishes,* (Ed.) T.J. Hara. Elsevier Sci. Pub. Co., Amsterdam, pp. 307–336.

Pfeiffer, W. and Lemke, J., 1973. Untersuchungen zur Isolierung und Identifizierung des Schreckstoffes aus der Hault der Elritze, *Phoxinus phoxinus* (L.) (Cyprinidae, Ostariophysi, Pisces). *J. Compo Physiol.*, 82: 407–410.

Pfeiffer, W., Mangold-Wemado, U. and Neusteurer, P., 1984. Identification of nerve bundle in the tractus olfactorius of the tench, *Tinca tinca* L., which conduct the nervous excitation elicited by alarm substance. *Experientia,* 40: 219–220.

Pfeiffer, W., Riegelbauer, G., Meier, G. and Scheibler, B., 1985. Effects of hypothanthine-3(N)oxide and hypothanthine-1(N)-oxide on central nervous excitation of the black. tetra, *Gymnocorymbus ternetzi* (Characidae, Ostariophysi, Pisces) indicated by dorsal light response. *J. Chem. Ecol.*, 11: 507–524.

Pollack, E.I., Becker, L.R. and Haynes, K., 1978. Sensory control of mating in the blue gourami, *Trichogaster trichopterus* (Pisces : Anabantidae). *Behav. Biol.*, 22: 92–105.

Resink, J.W., 1988. Steroid glucuronides as pheromones in the reproduction of the African catfish, *Ciarias gariepinus*. *Ph.D. Thesis,* University of Utrecht, Utrecht, The Netherlands.

Saglio, P., 1982. Use of intraspecific biological extracts to trap eels (*Anguilla anguilla* L.) in the field: Demonstration of the pheromonal attractivity of the skin mucus. *Acta Gecol. Applic.*, 3: 223–231.

Schoonen, W.G.E.I., 1987. Analysis of steroid synthesis in the reproductive organs of the Mrican catfish, *Clarias gariepinus*: A basis for the study of sex pheromones. *Ph.D. Thesis,* University of Utrecht, Utrecht, The Netherlands.

Selset, R., 1980. Chemical methods for fractionation of odorants produced by char smolts and tentative suggestions for pheromone origins. *Acta Physiol. Scand.,* 108: 97–103.

Sherwood, N.M., Kyle, A.L., Kreiberg, H., Warby, C.M., Magnus, T.H., Carolsfield, I. and Price, W.S., 1991. Partial characterization of a spawning pheromone in the herring, *Clupea herengus pallasi. Can. J. Zool.,* 69: 91–97.

Smith, R.J.F., 1977. Chemical communication as adaptation. In: *Chemical Signals in Vertebrates,* (Eds.) D. Muller-Schwarze and M.M. Mozell. Plenum Press, New York, pp. 303–320.

Smith, R.J.F., 1982a. The adaptive significance of alarm substance-night reaction system in fish. In: *Chemoreception in Fishes,* (Ed.) T.J. Hara. Elsevier/North-Holland, Amsterdam, pp. 327–347.

Smith, R.J.F., 1982b. Reaction of *Percina nigrofasciata, Ammocrypta beani* and *Etheostoma swaini* (Percidae, Pisces) to conspecific and intergeneric skin extracts. *Can. J. Zool.,* 60: 1067–1072.

Smith, R.J.F., 1992. Alarm signals of fishes. *Rev. Fish Biol. Fish.,* 2: 33–63.

Solomon, D.J., 1973. Evidence for pheromone influenced homing by migrating Atlantic salmon (*Salmo salar* L.). *Nature,* 244: 231–232.

Solomon, D.J., 1977. A review of chemical communication in freshwater fish. *J Fish Biol.,* 11: 363–376.

Sorensen, P.W., 1984. Juvenile eels rely on odour cues for migration. *Maritimes,* 28: 8–10.

Sorensen, P.W. and Stacey, N.E., 1999. Evolution and specialization of fish hormonal pheromones. In: *Advances in Chemical Signals in Vertebrates,* (Eds.) R.E. Johnston, D. Muller-Schwarze and P.W. Sorensen. Plenum Press, New York, pp. 15–47.

Sorensen, P.W. and Stacey, N.E., 2004. Brief review of fish pheromones and discussion of their possible uses in control on non-indigenous teleost fishes. *New Zealand J. Mar. Freshw. Res.,* 38: 399–417.

Sorensen, P.W., Christensen, T.A. and Stacey, N.E., 1998. Discrimination of pheromonal cues in fish: Emerging parallel with insect. *Curr. Opinion Neurobiol.*, 8: 458–467.

Stabell, O.B., 1987. Intraspecific pheromone discrimination and substrate marking by Atlantic salmon parı. *J. Chem. Ecol.*, 13: 1625–1644.

Stabell, O.B., 1992. Olfactory control of homing behaviour in salmonids. In: *Fish Chemoreception,* (Ed.) T.J. Hara. Chapman and Hall, London, pp. 249–270.

Stabell, O.B. and Lwin, M.S., 1997. Predator-induced phenotypic changes in crucial carp caused by chemical signals from conspecifics. *Environ. Biol. Fish.*, 49: 145–149.

Stacey, N.E., 2003. Hormones pheromones and reproductive behaviour. *Fish Physiol. Biochem.*, 28: 229–235.

Tavolga, W.N., 1956. Visual, chemical and sound stimuli as cues in sex discriminatory behaviour of the gobid fish, *Bothygobio soporator. Zoologica,* 41: 49–64.

Teeter, J.H., 1980. Pheromonal communication in sea lamprey, *Petromyzon marinus. Can. J. Fish. Aquat. Sci.*, 37: 2123–2132.

Timms, A.M. and Kleerekoper, H., 1972. The locomotor responses of male *Ictaturns punctatus,* the channel catfish. *Trans. Am. Fish. Soc.*, 101: 302–310.

Tood, J.H., 1971. The chemical languages of fishes. *Scient. Am.*, 224(5): 98–108.

van den Hurk, R. and Resink, J.W., 1992. Male reproductive system as sex pheromone producer in teleost fish. *J. Exp. Zool.*, 261: 204–213.

Weerd, J.H., 1990. Pheromones and ovarian growth in the African catfish, *Clarias gariepinus. Ph.D. Thesis,* University of Wageningen, Wegeningen, The Netherlands.

Wagner, C.M., Jones, M.L., Twohey, M.B. and Sorensen, P.W., 2006. A field test verifies that pheromones can be useful for sea lamprey (*Petromyzon marinus*) control in the Great Lake. *Can. J. Fish. Aquat. Sci.*, 63: 475–479.

Wisenden, B.D., Chivers, D.P., Brown, G.E. and Smith, R.J.F., 1995. The role of experience in risk assessment: Avoidance of areas chemically labeled with fathead minnow alarm pheromone by conspecifics and heterospecifics. *Ecoscience,* 2: 116–122.

Young, M.K., Micek, B.K. and Rathbun, M., 2003. Probable pheromone attraction on sexually mature brook trout to mature male conspecifics. *North Am. J. Fish. Mange.,* 23: 276–282.

Zhang, C., 1997. Bile acids as potential pheromones in lake char, *Salvelinus namaycush*: An electrophysiological, biochemical and behavioural study. *Ph.D. Thesis.* University of Manitoba, Winnipeg, Canada.

Zhao, W., Hong, W., Zhang, Q., Jiang, X. and Wu, D., 2003. Induction of maturation and spawning by sex pheromones in female skipper (*Boleophthalamus pectinirostris*). *Mar. Sci., Bull.,* 5: 87–95.

Chapter 25

Study on Radioprotective Effect of Melatonin with Reference to Reproductive Behaviour of Mice

A.L. Bhatia, Dhankesh Meena and Shikha Patni*

Radiation Biology Laboratory, University of Rajasthan, Jaipur – 302 004

ABSTRACT

Melatonin is an endogenous hormone produced by the pinealocyte in the pineal gland during the dark hours (night) of the day-night cycle. To decrease or compensate the effect of age and oxidative stress, taking melatonin and other anti oxidative drugs is increased in developed countries. Its effect on different parameters has already been studied. The most important and drastic gender differences in aging are related to the reproductive organs. An investigation has been done to evaluate protective effect of melatonin on the reproductive behavior against radiation induced damage. Melatonin had been reported to inhibit the effects of GnRH in a dose-dependent manner as the inhibitory effects of melatonin on LH and FSH release are age-dependent. The present investigation has been undertaken to evaluate the reproductive efficiency or reproductive performance of melatonin supplemented male mice in the presence and absence of radiation. It was observed

* E-mail: armbha@gmail.com.

that melatonin does not significantly alter the reproductive behavior of male mice whereas a dose of 4 Gy caused such an ill effect. Melatonin supplemented male mice though showed an urge towards female mice, yet in sporadic cases it was not conspicuous due to drowsiness (perhaps) or seemingly an imperfect protocol for sexual urge. However, a positive response to female mice was evident during the maze learning. Female used in the experiment were always found normal. Litter size did not show any significant alteration when melatonin was supplemented to male mice and females were fertilized by them. All these experiments, however, were made using rather high doses of melatonin. In present investigation low chronic administration of melatonin did not cause any significant effect on male sexual as well as reproductive performance just unlike those reported earlier in females.

Keywords: *Melatonin, Mice, Litter size, Neonates, Reproductive performance.*

Introduction

Despite enormous medical progress during the past few decades, the last years of life are still accompanied by increasing ill health and disability. The ability to maintain active and independent living for as long as possible is a crucial factor for ageing healthily and with dignity. To decrease or compensate the effect of age and oxidative stress, taking melatonin and other antioxidative drugs is increased in developed countries. Its effect on different parameters has already been studied. The most important and drastic gender differences in aging are related to the reproductive organs. In distinction to the course of reproductive ageing in women, with the rapid decline in sex hormones expressed by the cessation of menses, men experience a slow and continuous decline. This decline in endocrine function involves: a decrease of testosterone, dehydro epiandrosterone (DHEA), estrogens, thyroid stimulating hormone (TSH), growth hormone (GH), IGF1, and melatonin. Now it is very obvious that melatonin has an effect on sexual urge of organism.

Melatonin is an endogenous hormone produced by the pinealocyte in the pineal gland during the dark hours (night) of the day-night cycle. Light suppresses the production of melatonin. Melatonin has been shown to participate in a number of physiological processes like regulation of reproduction (Reiter, 1998), sleep (Waldheizer *et al.*, 1998) mood and behaviour (Zu danova *et al.*, 1998) and circadian rhythm (Yu *et al.*, 1993) immune function and body temperature. In many species of animals, the pineal gland acts as "biological clock" (Wainwright, 1982). Melatonin alone has been shown to alter the phase of circadian rhythms (Turek *et al.*, 1976). It has been proposed as a "chronobiotic" (Armstrong, 1989). A chronobiotic is defined as substances that can re-entrain short term dissociated and long-term desynchronized circadian rhythm (Armstrong, 1989).

Melatonin production has an impact on every stage of our life. Newborns produce very little of it and get it from mother milk. Then, at about three months of age-which is the stage of development when they start sleeping longer stretches at night and being more alert during the day-melatonin levels rise. From about the age of one, melatonin levels are more or less constant for a decade. Then, just before puberty, they go down sharply. Recent studies have demonstrated that this decline is the body's signal to the sex glands to set sexual maturation in motion. So clear is the signal meant to be, that a child who maintains unusually high levels of melatonin will experience a delay in the onset of pubescence. There have actually been rare cases, in which the melatonin level is so uncharacteristically high in adolescence that sexual maturation simply does not occur. Therefore, melatonin deserves serious attention for its impressive potency in sex drive.

Synthesis of melatonin also occurs in other areas of the body, including the retina, the gastrointestinal tract, skin, bone marrow and in lymphocytes, from which it may influence other physiological functions through paracrine signaling. Bone marrow cells (Tan *et al.*, 1999a) have melatonin concentration orders of magnitude greater than those in the serum; this is also true for bile (Tan *et al.*, 1999 b) and cerebrospinal fluid (Skinner and Malpaux, 1999) where the levels of melatonin are very much higher than in blood (Pang, 1986 and Reiter, 1991). Within individual cells, melatonin may not be evenly distributed within subcellular compartments (Pang, 1986 and Reiter, 1991) and generally rather little is known about the concentrations of melatonin within cells.

Melatonin also occurs in human milk (Illnerova *et al.*, 1993) and in a variety of common plant food (Dubbels *et al.*, 1995). Such as Bananas beet, cucumber and tomato. The occurrence of melatonin in vegetation parallel the occurrence of it precursor compound serotonin, which as long been known to occur in plant material, particularly in tropical food.

Chemically, melatonin is N-acetyl-5-methoxyserotonin, a methylated and acetylated derivative of serotonin, which in turn is derived from tryptophan. Serotonin is first acetylated by N-acetyltransferase (probably the rate-limiting step), and then methylated by hydroxy indole orthomethyltransferase (HIOMT) to form melatonin. The practical significance of the chemistry is that melatonin, in contrast to serotonin, is lipid-soluble, enabling it to readily traverse membranes and enter the brain and other tissues (even tissues without melatonin receptors). Melatonin synthesis depends on intact beta-adrenergic receptor function. Norepinephrine, the prototypical betaagonist (Rothkamm and Lobrich, 2003) activates the N-acetyltransferase. Predictably, beta-receptor blockers depress melatonin secretion (Anne, 2002).

Oral administration of melatonin causes significant decrease in sex hormones like serum estradiol, testosterone, and DHEAS concentrations in adult dogs. (Ashley *et al.*, 1999). Melatonin is known to inhibit male and female sex behaviour, but this effect has been reported only after repeated administration of sustained doses of the hormone. Melatonin may exert opposite effects on male and female sex behaviour depending on the dose and duration of treatment (Drago *et al.*, 1999). Melatonin protects against the effects of chronic stress on sexual behavior in male rats (Brotto *et al.*, 2001).

The enzymes of melatonin synthesis are activated and depressed, respectively, by darkness and light. Release of melatonin follows a circadian rhythm: "circa" (about) "dias" (a day). It rises and falls in a 24-hour pattern that is to some extent controlled by light. The phases of light and darkness act as synchronizers of the pattern, determining the timing of the rise and fall (Sminia *et al.*, 1996). Thus, during the night (or in darkness) pineal activity and melatonin synthesis and release get increased (Rothkamm and Lobrich, 2003) and during the day (or upon exposure to bright light) they get depressed and sometimes barely measurable (Kuo *et al.*, 1993), thus melatonin has been described as the "hormone of darkness" (Ushakova *et al.*, 1999).

Apart from being influenced by environmental light/dark cycles, the pineal, through melatonin, conditions the internal environment by setting and maintaining the internal clocks governing the natural rhythms of body function. This apparent clock setting property of melatonin has led to the suggestion that it is a "chronobiotic"–a substance that alters and potentially normalizes biological rhythms.

The influence of melatonin on biochemical and physiological processes is so broad that it seems unlikely that it is exerting direct effects. Rather, it appears that melatonin manages and adjusts the timing of other critical processes and biomolecules (hormones, neurotransmitters, etc.), which in turn exert numerous peripheral actions (Konopacka *et al.*, 1998). Studies in our laboratory (Manda and Bhatia, 2003a,b; Bhatia and Manda, 2004) indicate the protective role of melatonin against age, cyclophosphamide and radiation induced oxidative stress.

An investigation has been done to evaluate the radiopathology of the testis and protective effect of melatonin on testicular weight, tubular diameter and germ cell population, histology, and biochemistry against radiation induced damage (Meena, 2006). Radiation induced lipid peroxidation as reflected by the thiobarbituric acid reactive substance (TBARS) equivalents at various post-irradiation intervals have been studied to elucidate the amount of damage and protection provided by the melatonin. The biochemical endpoints to evaluate the protective role of melatonin against radiations were reduced glutathione, total protein, cholesterol and glycogen, which provide an appropriate scan of the tissue and radiation induced changes with and without melatonin.

The objective of the present invention is to test the reproductive performance of male mice with and without radiation vis-à-vis melatonin.

Survey of Literature

There is growing interest in using melatonin as a therapeutic agent for the treatment of a variety of medical conditions, including cancer, heart disease, glaucoma, stress, jet lag, and sleep disorders. However, use of melatonin as an antigonadal agent or contraceptive has been a controversial subject.

In order to test any possible adverse effects of melatonin on preimplantation embryos, Mc Elhinny *et al.* (1996) used the mouse as a model system. They used sufficiently high levels of melatonin to mimic the pharmacological concentration used in the oral contraceptive. It was found that there was no effect of melatonin on embryos from either mouse strain at any of the concentrations tested. The results suggest that if conception occurs while melatonin is being administered to treat a range of conditions, it would not adversely affect the embryo.

Further, it is unclear whether sex steroids influence melatonin secretion in the human. In an attempt to find an answer to this important question 36 women within an age range of 19 to 40 years were studied within a 3-month period under the following conditions: natural menstrual cycle, ovulation induction with gonadotrophins, early pregnancy, and intake of monophasic or triphasic oral contraceptives. Except in the case of pregnancy, repeated measurements in the same individual were done because of the well-known large inter-individual variations in melatonin secretion. Melatonin concentration was measured in plasma samples obtained at 4-hourly intervals in a 24h period and < 200 lux for all subjects studied. No consistent change in melatonin blood concentrations was demonstrated in response to the varying endogenous or exogenous concentrations of sex steroids. These observations suggest that circadian melatonin secretion is not significantly modulated by sex steroids (Delfs *et al.*, 1994).

The hypothalamic GnRH pulse generator activates the pituitary-gonadal reproductive axis, and contraceptive techniques have advanced to the point where GnRH analogues can block this effect. However, nature has an even finer form of contraception, whereby the GnRH pulse generator is activated or inactivated at different seasons of the year. Darkness affects the retino-pineal nervous pathway to cause the synthesis and release of melatonin from the pineal gland at night. The duration of the night time release of melatonin is longer in winter than in summer; and it is the prolongation in the duration of the nighttime release of melatonin, with the change of season from summer to winter, which acts as the endocrine signal for inactivating the hypothalamic GnRH pulse generator. Humans are not seasonal breeders, and evidence is presented to indicate that this is due to an impairment of the retino-pineal pathway rather than an impairment of melatonin hypothalamic function. Thus the way is open for utilizing melatonin as a human contraceptive, and a melatonin-based contraceptive is at present undergoing phase III clinical trials. The challenge is to develop more refined methods for administering (or releasing) melatonin, so that it has a night time amplitude and duration, which mimics that seen in long day breeders (Vanecek, 1995).

Melatonin and GnRH

An inhibitory effect of melatonin on GnRH-induced LH release from the neonatal rat anterior pituitary gland was first described by Martin and Klein (1976), who used cultured rat pituitary gland as a bioassay system for testing anti gonadotrophic factors in the pineal gland. The neonatal tissue was selected because small, immature tissues generally survive well in organ culture. The release of LH from gonadotrophs is low under resting conditions and is not affected by melatonin (Martin and Klein, 1976). The effect of melatonin on LH release has been confirmed using dispersed pituitary cells in culture (Martin *et al.*, 1982; Vanecek and Klein, 1992a,b).

Vanecek and Vollrath (1990) suggest that melatonin may block the GnRH induced Ca^{2+} increase by inhibiting GnRH-induced phospholipase C activity (Figure 25.1). Phospholipase C metabolizes phosphatidylinositol bisphosphate (PIP_2) into diacylglycerol and inositol trisphosphate (IP_3), which, in turn, induces the release of Ca^{2+} from the endoplasmic reticulum (Berridge, 1993). Melatonin inhibits the GnRH induced formation of diacylglycerol in the gonadotrophs (Vanecek and Vollrath, 1990) and preliminary observations indicate that it also inhibits GnRH-induced IP_3 accumulation.

Melatonin may inhibit GnRH-induced Ca^{2+} mobilization by inhibiting phospholipase C activity and inositol trisphosphate (IP3) formation. Inhibition of Ca^{2+} influx by melatonin may be due to hyperpolarization of plasma membrane, which closes voltage-sensitive Ca^{2+} channels. Hyperpolarization may be caused by the inhibitory effect of melatonin on Na^+ influx via an Na^+–Ca^{2+} exchanger or via cyclic nucleotide-gated channels. Alternatively, inhibition of adenylyl cyclase and decrease of cAMP by melatonin may decrease the permeability of the voltage-regulated Ca^{2+}-channels by preventing phosphorylation by cAMP-dependent kinase. AC, adenylyl cyclase; CNG, cyclic nucleotide-gated channel;

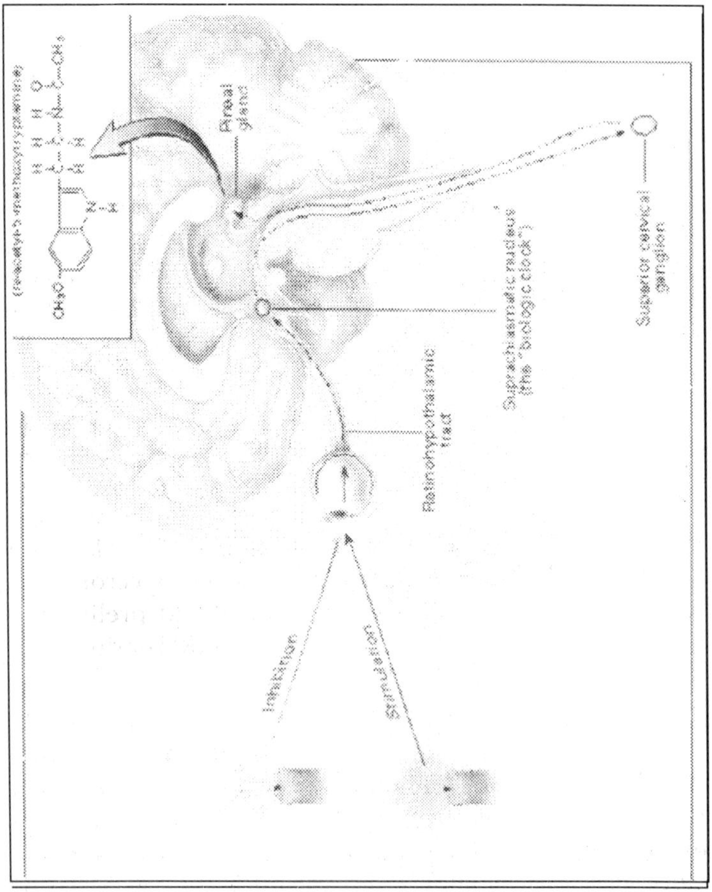

Figure 25.1: Position of Pineal Gland in Human Brain and its Stimulation

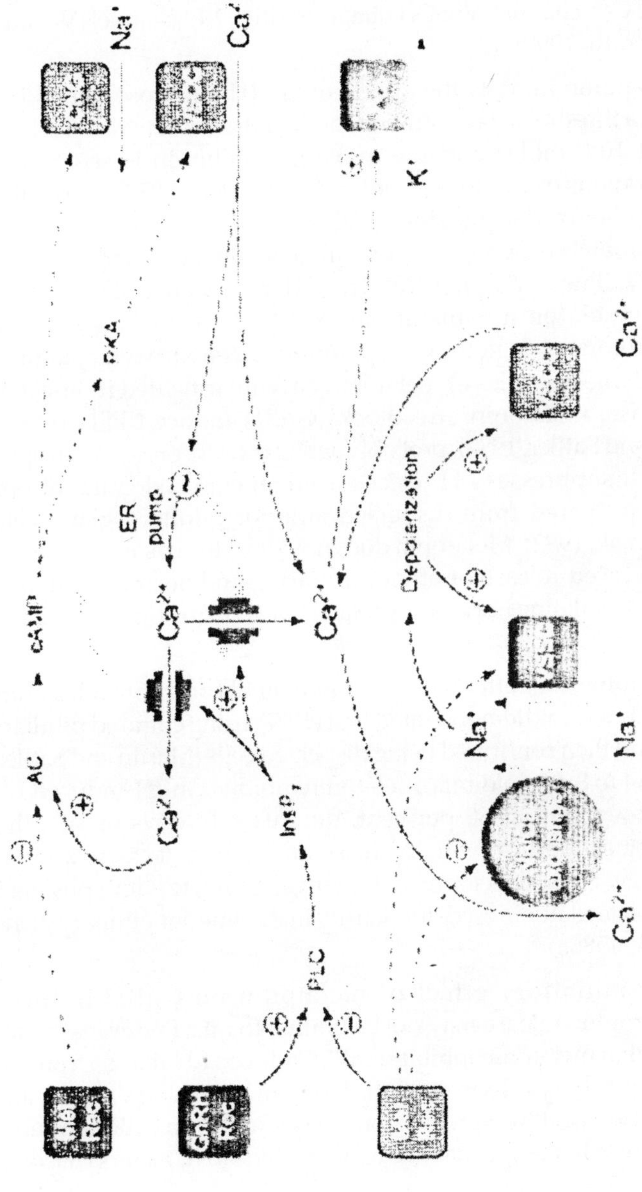

Figure 25.2: Proposed Mechanism of Inhibition by Melatonin of GnRH-induced LH Release from the Neonatal Rat Gonadotroph

ER, endoplasmic reticulum; GnRH Rec, GnRH receptor; KCa, Ca^{2+}-sensitive K^+ channel; Mel Rec, melatonin receptor; PKA, cAMPdependent kinase; PLC, phospholipase C; VSCC, voltage-sensitive Ca^{2+} channel; VSSC, voltagesensitive Na^+ channel (Vanecek and Vollrath, 1990).

Melatonin inhibits the effects of GnRH in a dose-dependent manner; a threshold concentration for inhibition of *in vitro* LH release is about 10^{-10} mol/l and the maximal inhibition is seen using concentrations of 10^{-8} to 10^{-7} mol/l (Martin *et al.*, 1977). This value correlates with the physiological concentrations of melatonin (approximately 0.5 nmol/l) found in rat serum at night (Wilkinson *et al.*, 1977; Illnerova *et al.*, 1978). The order of potency of the indoles is 2-iodomelatonin > melatonin > 6-hydroxymelatonin > N-acetylserotonin > 5-methoxytryptamine >> 5-hydroxytryptamine. Although most studies of melatonin have investigated its effect on LH release, melatonin also blocks GnRH-induced FSH release (Martin and Sattler, 1982), probably acting directly on gonadotrophs, because it suppresses LH release from an enriched gonadotroph fraction prepared from dispersed anterior pituitary gland cells (Martin *et al.*, 1982). Melatonin does not affect the basal or releasing factor-induced release of other pituitary gland hormones, that is, thyroid-stimulating hormone, prolactin or growth hormone (Martin and Sattler, 1982).

The inhibitory effects of melatonin on LH and FSH release are age-dependent, as demonstrated first (1977) using cultured pituitary glands and then confirmed using dispersed cells (Martin and Sattler, 1979). In 4 to 8 day-old rats, melatonin inhibits GnRH-induced LH release by about 50–60 per cent, but, after 10 days of age, this inhibitory action begins to decrease and is undetectable after 15 days of age. These developmental changes correlate with a postnatal decrease in melatonin receptor density in the anterior pituitary gland (Vanecek, 1988a).

The inhibitory effect of melatonin on GnRH-induced gonadotrophin release may not be limited to rats. Preliminary data indicate that melatonin inhibits GnRH induced LH release from the cultured pituitary cells in neonatal Siberian hamsters (Vanecek and Masson, 1988a). However, melatonin has no effect on GnRH-induced LH response in the pituitary of neonatal Syrian hamsters (Bacon *et al.*, 1981).

Melatonin Receptors

Melatonin receptor in the anterior pituitary gland has been investigated using ^{125}I-melatonin binding. The receptor has a high affinity ($Kd \sim 10^{-11}$ mol l^{-1}) and is located in the plasma membrane. Using RT-PCR, Mella receptor mRNA has been identified in the neonatal rat pituitary. Melatonin receptor is present in the pars distalis and pars tuberalis of the rat pituitary (Vanecek, 1988a; Williams and Morgan, 1988; Williams *et al.*, 1991). In the rat pars distal is the melatonin receptor density is age-dependent. On embryonic day 20 and postnatal day 1, the density is about 30 fmol mg^{-1} protein but by postnatal day 30; it decreases tenfold, that is, below 3 fmol mg^{-1} protein (Vanecek, 1988a). Similar developmental changes in melatonin receptor density have been described in the pars distalis of the Syrian hamster (Vanecek and Kosar, 1994). The concentration of melatonin receptor in the pars tuberalis does not change significantly as neonatal rats develop into adults. The factor(s) inducing the decrease in melatonin receptor concentration in the pars distalis between birth and 15 days of age are unknown. The decrease may be due to the decreased expression of the melatonin receptor, because the concentration of Mella mRNA in the rat pituitary decreases markedly between day 1 and day 30. The melatonin receptor density in the pars distalis of the pituitary increases twofold in postpubertal male rats 2–4 weeks after castration (Vanecek *et al.*, 1990). Melatonin receptor density may increase as a consequence of removal of gonadal steroids. In support of this view, an inhibitory effect of gonadal steroids on 125I-melatonin binding in caudal and cerebral arteries has been shown (Seltzer *et al.*, 1992). In rats, 125I-melatonin binding to the arteries is highest at dioestrus, when circulating oestrogen concentrations are low whereas, at pro-oestrus and oestrus, when oestrogen concentrations are high, melatonin binding capacity decreases. Therefore, it is possible that the increasing concentration of gonadal steroids in prepubertal rats causes the decrease of melatonin receptor expression in the anterior pituitary after 10 days of age.

The postcastration increase of GnRH secretion, which increases the number of gonadotrophs, is another factor that may account for the increase of the melatonin receptor density after castration (Hellbaum *et al.*, 1961; Sarkar and Fink, 1980; Clayton and Catt, 1981). Since available data indicate that melatonin receptor is located on gonadotrophs (Martin and Sattler, 1982), a GnRH-induced increase in the number of gonadotrophs may be responsible for the postcastration increase in melatonin receptor concentration in the pituitary. This mechanism may also explain the developmental changes of the pituitary melatonin receptor density, since the concentration of GnRH in amniotic fluid increases approximately tenfold between embryonic day (E) 12 and (E) 16 which corresponds with the time (E) 15 when the pituitary melatonin receptor is first detected (Williams *et al.*, 1991). After birth, the pars distalis is exposed to an abrupt decrease in GnRH, which may cause the post-natal decrease in melatonin binding site density (Vanecek, 1988a).

Function of the Inhibitory Effect of Melatonin on GnRH-Induced Gonadotrophin Release

The most important role of melatonin in mammals is the regulation of seasonal rhythms (Underwood and Goldman, 1987). Photoperiod drives seasonal breeding cycles in many mammalian species. Nocturnal secretion of melatonin from the pineal gland is prolonged in animals kept on short photoperiods and the duration of the melatonin pulse regulates the reproductive functions. The effects of melatonin on reproduction in adult animals are probably not mediated by the pars distalis of the pituitary, but through the mediobasal hypothalamus and pars tuberalis (Lincoln and Clarke, 1994; Maywood and Hastings, 1995). The physiological significance of the inhibitory melatonin effects on GnRH-induced gonadotrophin release from the pituitary of neonatal rats is not clear. It is possible that melatonin mediates the photoperiodic regulation of the timing of puberty. Adult laboratory rats are not photoperiodic, presumably because constant laboratory conditions remove the necessity for seasonal adaptations or as a result of artificial selection against seasonal breeding. However, prepubertal rats reared from birth on short photoperiods have smaller reproductive organs than those kept on long photoperiods (Vanecek and Illnerova, 1985). Although a clearly defined rhythm of pineal melatonin concentration starts from about 8 days of age (Tamarkin *et al.*, 1980; Pelisek *et al.*, 1994), maternal melatonin may carry the photoperiodic message to the fetuses and neonates. Melatonin crosses the placental barrier freely and is also secreted into the milk (Reppert and Klein, 1978; Illnerova *et al.*, 1993). Melatonin may exert the inhibitory effect on onset of puberty, at least partially, at the pituitary by inhibiting the GnRH induced gonadotrophin release. The removal of this inhibitory input due to the decrease of the melatonin receptors may be one of the events initiating puberty. It is not clear why rats have photoperiodically regulated puberty when the adults are not seasonal breeders, but it may be that the regulation of puberty is a remnant of the significant photoperiodic regulations existing in wild rats. In Siberian hamsters, in which photoperiod has marked effect on the timing of puberty, melatonin inhibits GnRH-induced LH release from the neonatal pituitary cells, but this effect has not been found in Syrian hamsters, in which the timing of puberty is not sensitive to photoperiod (Bacon, 1981; Sisk and Turek, 1983; Yellon and Goldman, 1984; Cherry, 1987).

Material and Methods

The present investigation has been done to evaluate the reproductive efficiency or reproductive performance of melatonin supplemented male mice in the presence and absence of radiation. For this, healthy male mice of 6–8 weeks age were divided into four groups. First group served as normal which did not receive any treatment. Second group was administered melatonin (0.1 mg/kg b.w./day) for 15 consecutive days. Third group was also administered melatonin as in group second but was exposed to 6Gy of gamma radiation after 30 minutes of last dose of melatonin. Fourth group received DDW for 15 consecutive days and then exposed to gamma radiation as in group third.

Group	No. of Female	No. of Male	Treatment
Normal	4	1	No treatment
Melatonin	4	1	Melatonin administration for 14 days
Irradiation	4	1	Exposed to γ-irradiation
Melatonin + Irradiation	4	1	Exposed to γ-irradiation after melatonin administration

For the breeding, estrous stage of female was confirmed by vaginal smears. Briefly, small volume of Mili Q water has been ejected into vagina through an eyedropper. Aqueous fluid containing epithelial cells has been sucked back and stages were confirmed by the microscopic examination of vaginal smear. After all 4 male mice were housed with one estrous female mouse until pregnancy was confirmed by the appearance of vaginal plug. The subsequent development of the embryos was brought up on the tap water. Variation in the litter size, weight of neonates, size of neonates and time of conception has been recorded in different group.

Results

It was observed that melatonin does not significantly alter the reproductive behavior of male mice whereas a dose of 4 Gy caused such an ill effect (Figure 25.3). Melatonin supplemented male mice though showed an urge towards female mice yet in sporadic cases it was not conspicuous due to drowsiness (perhaps) or seemingly an imperfect protocol for sexual urge. However, a positive response to female mice was evident while during maze learning. It was seen that male mice went to the goal box, reached with a time which is statistically not different from that of normal male mice (without any treatment). Furthermore the time taken by melatonin (only) supplemented male mice has also not been significantly different than those of normal mice.

Female used in the experiment were always normal Litter size did not show any significant alteration when melatonin was supplemented to male mice and females were fertilized by them. The variability of the time of conception had also greatly stabilized in the female mated with melatonin treated males (Table 25.1). Equal number of females conceived when fertilized with melatonin supplemented male mice.

It was recorded that the litter size got drastically reduced in those females, which were fertilized by irradiated males. Moreover the time of conception was also highly variable in matting with such irradiated males. Furthermore the growth of neonates born to such females after mating with irradiated males has also showed a considerable decline. In 15 days period the weight increased by only 51.32 per cent. However, melatonin supplementation had a positive effect (67.23 per cent growth) on the prevention of such growth retardation. The growth retardation is also conspicuous in term of length of neonates after 15 days.

Discussion

In present study, administration of melatonin to normal adult mice indicated that it does not significantly alter the reproductive

Table 25.1: Depicts Parturition of Normal Females, which were Mated with Experimentally Different Males. It also shows variation in the litter size, weight of neonates, size of neonates and time of conception after radiation exposure to only males, with and without melatonin administration (mated with normal females). Values±SD E_0 at the time of birth; E_{15}: 15 days post parturition.

Group	*Time taken for Conception*	*Litter Size*	*Size (cm) Neonates (E_0)*	*Size (cm) Neonates (E_{15})*	*Weight of Neonates (g) (E_0)*	*Weight of Neonates (g) (E_{15})*	*Per cent Growth*
Normal	1–2 days	7–10	$4.25 \pm 0.21^{a\#}$	$8.1 \pm 1.5^{a\#}$	7.34±1.9	$12.5 \pm 2.4^{a\#}$	70.30 per cent
Melatonin	1–2 days	6–10	4.20±0.20	8.6±1.1	7.12±1.8	12.3±2.2	72.75 per cent
Irradiated	3–26 days (highly variable)	0–1	$2.0 \pm 0.16^{b*}$	$4.9 \pm 1.0^{b*}$	4.56±0.9	$6.9 \pm 1.6^{b*}$	51.32 per cent
Melatonin+ Irradiation	2–3 days	3–5	$3.9 \pm 0.18^{c*}$	7.8±1.3	5.92±1.1	$9.9 \pm 1.7^{c*}$	67.23 per cent

a: Statistical difference with melatonin; b: Statistical difference with normal; c: Statistical difference with irradiated; *: $P < 0.001$; #: Non-significant; Per cent Growth: $A/E_0 \times 100$, where, $A = E_{15} - E_0$.

behaviour of male mice whereas a dose of 4 Gy caused such an ill effect (Figure 25.3). The time taken by melatonin supplemented male mice were also not significantly different than those of normal mice. While mating the melatonin treated males with normal females, the litter size did not show any significant alteration. (Table 25.1).

The physiological and pharmacological effects of melatonin seem to be mediated by activation of high-affinity receptors, but may involve different neurotransmitters in the brain (Gaffori and Van Ree, 1985a; Tenn and Niles, 1995). In particular, serotonin has been often considered as a possible mediator of melatonin effects (Gaffori and Van Ree, 1985b; Eison *et al.*, 1995). The main action of the hormone deals with adaptation of animal behavior, including sex behaviour, to the length of the dark period of the circadian cycle, hence to the seasons. Similar effects may be mimicked through the exogenous administration of melatonin. Animal studies show, for instance, that its chronic administration leads to inhibition of male sex behavior (Baum, 1968; Yamada *et al.*, 1992) and the implantation of melatonin-containing pellets near the SCN in female rats reduces their lordosis reflex (De Catanzaro and Stein, 1984). Besides, melatonin is involved in sexual cycle of the female rat. In particular, it was observed that melatonin administered by intraperitoneal (*i.p.*) injection produces an inhibiting effect on the nervous structures involved in the regulation of sexual cycle of female rats (Diaz Lopez and Fernandez, 1984). All these experiments, however, were made using rather high doses of melatonin. In present investigation low chronic administration of melatonin did not cause any significant effect on male sexual as well as reproductive performance.

Different Research findings indicated that melatonin inhibits GnRH-induced release of LH and FSH from the neonatal, but not the adult, rat anterior pituitary gland. This action of melatonin is mediated by the specific high-affinity membranebound receptors that are absent in adult rats. The intracellular mechanism of melatonin action involves a decrease in intracellular calcium [Ca^{2+}] in the gonadotrophs; melatonin inhibits GnRH-induced Ca^{2+} release from endoplasmic reticulum as well as Ca^{2+} influx through voltage-sensitive channels. Melatonin also inhibits GnRH-induced accumulation of cAMP, which may result in the decreased influx of Ca^{2+}, because cAMP, acting through protein kinase A, stimulates Ca^{2+} influx into the gonadotrophs. This age-dependent effect, of

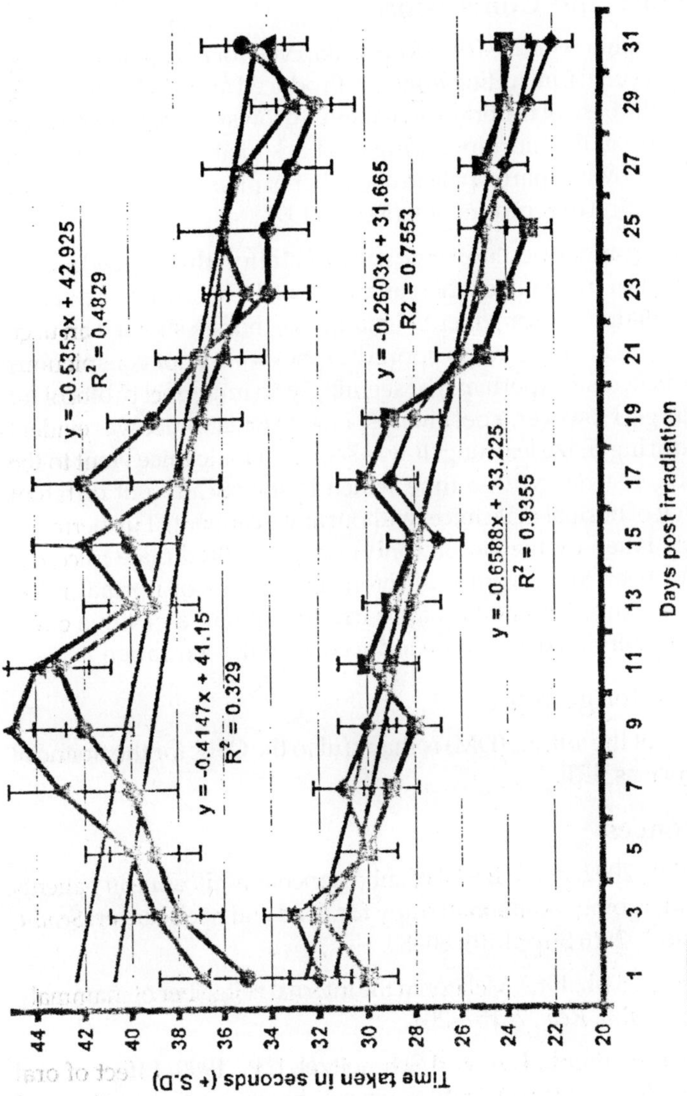

Figure 25.3: Reproductive Behavioural Performance in Male Swiss Albino Mice, with and without Supplementation of Melatonin

melatonin on gonadotrophin release from the pituitary may be involved in the timing of puberty (Vanecek, 1998, 1999).

Summary and Conclusion

Melatonin regulates the sleep-wake cycle, or circadian rhythms, in many animals, including humans. Produced at night by the pineal gland at the base of the brain, it makes us drowsy at night and, when levels drop in the morning, brings us back to alertness. Melatonin has been shown to participate in a number of physiological processes like regulation of reproduction (Reiter, 1998).

Melatonin does not significantly alter the reproductive behaviour of male mice whereas a dose of 4 Gy caused such an ill effect. Melatonin supplemented male mice though showed an urge towards female mice yet in sporadic cases it was not conspicuous due to drowsiness (perhaps) or seemingly an imperfect protocol for sexual urge. However, a positive response to female mice was evident while during maze learning. It was seen that male mice went to the goal box, reached with a time which is statistically not different from that of normal male mice (without any treatment). Furthermore, the time taken by melatonin (only) supplemented male mice has also not been significantly different than those of normal mice. Melatonin supplementation after irradiation had a positive effect on the prevention of radiation induced growth retardation.

Acknowledgement

One of the author (DM) is thankful to the CSIR for the financial assistance as SRF.

References

Anne, P.R., 2002. Phase II trial of subcutaneous amifostine in patients undergoing radiation therapy for head and neck cancer. *Semin. Oncol.*, 29(6 Suppl 19): 80–83.

Armstrong, S.M., 1989. Melatonin the internal zeitgeiber of mammals. *Pineal Res. Rev.*, 7: 157–202.

Ashley, P.F., Frank, L.A. and Schmeitzel, L.P., 1999. Effect of oral melatonin administration on sex hormone, prolactin, and thyroid hormone concentrations in adult dogs–Abstracts: recently published abstracts–Brief Article *J. Am. Vet. Med. Assoc.*, 215: 1111–1115.

Bacon, A., Sattler, C. and Martin J.E., 1981. Melatonin effect on the hamster pituitary response to LHRH. *Biology of Reproduction*, 24: 993–999.

Baum, M.J., 1968. Pineal gland: Influence on development of copulation in male rats. *Science*, 162: 586–587.

Berridge, M.J., 1993. Inositol trisphosphate and calcium signaling. *Nature*, 361: 315–325.

Bhatia, A.L. and Manda, K., 2004. Study on pre-treatment of melatonin against radiation-induced oxidative stress in mice. *Environmental Toxicology and Pharmacology*, 18: 13–20.

Cherry, J.A., 1987. The effect of photoperiod on development of sexual behaviour and fertility in golden hamsters. *Physiology and Behavior*, 39: 521–526.

Clayton, R.N. and Catt, K.J., 1981. Regulation of pituitary gonadotropin releasing hormone receptors by gonadal hormones. *Endocrinology*, 108: 887–889.

DeCatanzaro, D. and Stein, M., 1984. Suppression of the lordosis reflex in female rats by chronic central melatonin implants. *Horm. Behav.*, 18: 216–223.

Delfs, T.M., Baars, S., Fock, C., Schumacher, M., Olcese, J. and Zimmermann, R.C., 1994. Sex steroids do not alter melatonin secretion in the human. *Hum. Reprod.*, 9(1): 49–54.

Diaz, L.B. and Fernandez, B.M., 1984. Response to melatonin in brain areas implicated in the sexual cycle of the rat. *Brain Res.*, 296: 333–338.

Drago, F., Busa, L., Benelli, A. and Bertolini, A., 1999. Acute low doses of melatonin stimulate rat sex behavior: the role of serotonin neurotransmission. *Eur. J. Pharmacol.*, 385(1): 1–6.

Dubbels, R., Reiter, R.J., Klenke, E., Goebel, A., Schnakenberg, E., Ehlers, C., Schiwara, H.W. and Schloot, W., 1995. Melatonin in edible plants identified by radioimmunoassay and by high performance liquid chromatography-mass spectrometry. *J. Pineal Res.*, 18: 28–31.

Eison, A.S., Freeman, R.P., Guss, V.B., Mullins, U.L. and Wright, R.N., 1995. Melatonin agonists modulate 5-HT2A receptor-mediated neurotransmission: behavioural and biochemical studies in the rat. *J. Pharmacol. Exp. Ther.*, 273: 304–308.

Gaffori, O. and Van Ree, J.M., 1985b. Serotonin and antidepressant drugs antagonize melatonin-induced behavioural changes after injection into the nucleus accumbens of rats. *Neuropharmacology,* 24: 237–244.

Hellbaum, A.A., McArthur, I.G., Campbell, P.J. and Finerty, J.C., 1961. The physiological fractionation of pituitary gonadotrophic factors correlated with cytological changes. *Endocrinology,* 68: 144–153.

Illnerova, H., Backstrom, M., Saaf, J., Wetterberg, L. and Vangbo, B., 1978. Melatonin in rat pineal gland and serum: Rapid parallel decline after light exposure at night *Neuroscience Letters,* 9: 189–193.

Illnerova, H., Buresova, M. and Presl, J., 1993. Melatonin rhythm in human milk *Journal of Clinical Endocrinology and Metabolism,* 77: 838–841.

Konopacka, M., Widel, M. and Rzeszowska-Wolny, J., 1998. Modifying effect of vitamins C, E and beta-carotene against gamma-ray-induced DNA damage in mouse cells. *Mutat. Res.,* 417: 85–94.

Kuo, S.S., Saad, A.H., Koong, A.C., Hahn, G.M. and Giaccia, A.J., 1993. Potassium-channel activation in response to low doses of gamma-irradiation involves reactive oxygen intermediates in non excitatory cells. *Proc. Natl. Acad. Sci.,* USA, 90: 908–912.

Lincoln, G.A. and Clarke, I.J., 1994. Photoperiodically induced cycles in the secretion of prolactin in hypothalamo-pituitary disconnected rams: evidence for translation of the melatonin signal in the pituitary gland. *Journal of Neuroendocrinology,* 6: 251–260.

Manda, K. and Bhatia, A.L., 2003a. Melatonin-induced reduction in age-related accumulation of oxidative damage in mice. *Biogerontology,* 4(3): 133–139.

Manda, K. and Bhatia, A.L., 2003b. Prophylactic action of melatonin against cyclophosphamide-induced oxidative stress in mice. *Cell Biology and Toxicology,* 19: 367–372.

Martin, J.E. and Klein, D.C., 1976. Melatonin inhibition of the neonatal pituitary response to luteinizing hormone-releasing factor. *Science,* 191: 301–302.

Martin, J.E. and Sattler, C., 1979. Developmental loss of the acute inhibitory effect of melatonin on the *in vitro* pituitary luteinizing hormone and follicle stimulating hormone responses to luteinizing hormone-releasing hormone. *Endocrinology,* 105: 107–112.

Martin, J.E. and Sattler, C., 1982. Selectivity of melatonin pituitary inhibition for luteinizing hormone-releasing hormone. *Neuroendocrinology,* 34: 112–116.

Martin, J.E., Engel, J.N. and Klein, D.C., 1977. Inhibition of the *in vitro* pituitary response to luteinizing hormone-releasing hormone by melatonin, serotonin and 5-methoxytryptamine. *Endocrinology,* 100: 675–680.

Martin, J.B., McKeel, D.W. Jr. and Sattler, C., 1982. Melatonin directly inhibits rat gonadotroph cells. *Endocrinology,* 110: 1079–1084.

Maywood, E.S. and Hasings, M.B., 1995. Lesions of the iodomelatonin-binding sites of the mediobasal hypothalamus spare the lactotropic, but block the gonadotropic response of male Syrian hamsters to short photoperiod and to melatonin. *Endocrinology,* 136: 144–153.

McElhinny, A.S., Davis, F.C and Warner, C.M., 1996. The effect of melatonin on cleavage rate of C57BL/6 and CBA/Ca pre implantation embryos cultured *in vitro*. *J. Pineal Res.,* 21(1): 44–8.

Pang, S.F. and Allen, A.E., 1986. Extra-pineal melatonin in the retina: Its regulation and physiological function. *Pineal Res. Rev.,* 4: 55–95.

Pelisek, V., Kosar, E. and Vanecek, J., 1994. Effect of photoperiod on the pineal melatonin rhythm in neonatal rats. *Neuroscience Letters,* 180: 87.

Reiter, R.J., 1991. Pineal melatonin: Cell biology of its synthesis and of its physiological interactions. *Endocr. Rev.,* 12: 151–180.

Reiter, R.J., 1998. Oxidative damage in the central nervous system: Protection by melatonin. *Prog. Neurobiol.,* 56: 359–384.

Reppert, S.M. and Klein, D.C., 1978. Transport of maternal [3H] melatonin to suckling rats and the fate of [3H] melatonin in the neonatal rat. *Endocrinology,* 102: 582–588.

Rothkamm, K. and Lobrich, M., 2003. From the cover: Evidence for a lack of DNA double-strand break repair in human cells exposed to very low x-ray doses. *Proc. Natl. Acad Sci.*, USA, 100: 5057–5062.

Sarkar, D.K. and Fink, G., 1980. Luteinizing hormone releasing factor in pituitary stalk plasma from long-term ovariectomized rats: Effects of steroids. *Journal of Endocrinology*, 86: 511–524.

Seltzer, A., Viswanathan, M. and Saavedra, J.M., 1992. Melatonin-binding sites in brain and caudal arteries of the female rat during the estrous cycle and after estrogen administration. *Endocrinology*, 130: 1896–1902.

Sisk, C.L. and Turek, F.W., 1983. Gonadal growth and gonadal hormones do not participate in the development of responsiveness to photoperiod in the golden hamster. *Biology of Reproduction*, 29: 439–445.

Sminia, P., Vander Kracht, A.H., Frederiks, W.M. and Jansen, W., 1996. Hyperthermia, radiation carcinogenesis and the protective potential of Vitamin A and N-acetylcysteine. *J. Cancer Res. Clin. Oncol.*, 122: 343–350.

Tamarkin, L., Reppert, S.M., Orloff, D.J., Klein, D.E., Yellon, S.M. and Goldman, B.D., 1980. Ontogeny of the pineal melatonin rhythm in the Syrian (*Mesocricetus auratus*) and Siverian (*Phodopus sungorus*) hamsters and in the rat. *Endocrinology*, 107: 1061–1064

Tan, D.X., Manchester, L.C., Reiter, R.J., Qi, W., Hanes, M.A. and Farley, N.J., 1999b. High physiological levels of melatonin in the bile of mammals. *Life Sci.*, 65: 2523–2529.

Tan, D.X., Manchester, R.J., Reiter, R.J., Qi, W., Zhang, M., Weintraub, S.T., Cabrera, J., Sainz, R.M., and Mayo, J.E., 1999a. Identification of highly elevated levels of melatonin in bone marrow: Its origin and significance. *Biochim. Biophys. Acta*, 1472: 206–214.

Turek, F.W., Millan, J.P. and Menaker, M., 1976. Melatonin effects on the circadian locomotor rhythm of sparrows. *Science*, 194: 441–443.

Underwood, H. and Goldman, B.D., 1987. Vertebrate circadian and photoperiodic systems: Role of the pineal gland and melatonin. *Journal of Biological Rhythms*, 2: 279–315.

Ushakova, T., Melkonyan, H., Nickonova, L., Afanasyev, V., Gaziev, A. and Murdrik, N., 1999. Modification of gene expression by dietary antioxidants in radiation-induced apoptosis of mice splenocytes. *Free Rad. Biol. Med.*, 26: 887.

Vanecek, J., 1988a. The melatonin receptors in rat ontogenesis. *Neuroendocrinology*, 48: 201–203.

Vanecek, J., 1995, 1998. Melatonin inhibits increase of intracellular calcium and cyclic AMP in neonatal rat pituitary via independent pathways. *Molecular and Cellular Endocrinology*, 107: 149–153.

Vanecek, J. and Illnerova, H., 1985. Effect of short and long photoperiods on pineal N-acetyltransferase rhythm and on growth of testes and brown adipose tissue in developing rats. *Neuroendocrinology*, 41: 186–191.

Vanecek, J. and Klein, D.C., 1992a. Melatonin inhibits gonadotropin-releasing hormone-induced elevation of intracellular Ca^{2+} in neonatal rat pituitary cells. *Endocrinology*, 130: 701–707.

Vanecek, J. and Klein, D.C., 1992b. Sodium-dependent effects of melatonin on membrane potential of neonatal rat pituitary cells. *Endocrinology*, 131: 939–946.

Vanecek, J. and Klein, D.C., 1999. Melatonin inhibition of GnRH-induced LH release from neonatal rat gonadotroph: involvement of Ca^{2+} not camp. *American Journal of Physiology*, 269: 85–90.

Vanecek, J. and Kosar, E., 1994. Ontogenesis of melatonin receptors in anterior pituitary and pars tuberalis of golden hamsters. *Physiological Research*, 43: 379–382.

Vanecek, J. and Masson, J., 1988a. Melatonin binding sites. *Journal of Neurochemistry*, 51: 1436–1440.

Vanecek, J. and Vollrath, L., 1990. Melatonin modulates diacylglycerol and arachidonic acid metabolism in the anterior pituitary of immature rats. *Neuroscience Letters*, 110: 199–203.

Vanecek, J., Kosar, E. and Vorlicek, J., 1990. Daily changes in melatonin binding sites and the effect of castration *Molecular and Cellular Endocrinology*, 73: 165–170.

Wainwright, S.D., 1982. Role of pineal gland in the vertebrate master biological clock. *The Pineal Gland*, 23; 53–79.

Waldhauzer, F., Kovacs, J. and Reiter, E., 1998. Age related changes in melatonin levels in humans and its potential consequences for sleep disorders. *Exp. Gerontology*, 33: 759.

Wilkinson, M., Arendt, J., Bradtke, J. and De Ziegler, D., 1977. Determination of a dark-induced increase in pineal N-acetyl transferase activity and simultaneous radioimmunoassay of melatonin in pineal, serum and pituitary tissue of the male rat. *J. Endocrinol.*, 72: 243–244.

Williams, L.M. and Morgan, P.J., 1988. Demonstration of melatonin-binding sites on the pars tuberalis of the rat. *J. Endocrinol.*, 119: RI–R3.

Williams, L.M., Martinoli, M.G., Titchener, L.T. and Pelletier, G., 1991. The ontogeny of central melatonin binding sites in the rat. *Endocrinology,* 128: 2083–2090.

Yamada, K., Maruyama, K., Mogami, S., Miyagawa, N. and Tsuboi, M., 1992. Influence of melatonin on reproductive behaviour in male rats. *Chem. Pharm. Bull.*, 40: 2222–2223.

Yellon, S.M. and Goldman, B.D., 1984. Photoperiod control of reproductive development in the male Djungarian hamster (*Phodopus sungorus*). *Endocrinol.*, 114: 664–670.

Yu, H.S., Tsin, A.T.C. and Reiter, R.J., 1993. Melatonin: History, biosynthesis and assay methodology. In: *Melatonin, Biosynthesis, Physiological Effects, and Clinical Applications*, (Eds.) H.S. Yu and R.J. Reiter. CRC Press, Boca Raton, Florida, pp. 1–16.

Zhdanova, L.V., Cantor, M.L. and Leclair, A.D., 1998. Behavioral effect of melatonin treatment in non-human primates. *Sleep Res.* Online, 1: 114–118.

Chapter 26

Role of Pheromones and other Semiochemicals in Mammalian Reproduction and Behaviour

S.D. Pandey

Reproductive Endocrinology Laboratory, Department of Zoology, Christ Church College, Kanpur, U.P.

ABSTRACT

Chemical communication has been well recognized as operating in animals. For its survival and existence an organism must be able to perceive the changes in its environment and apparently respond to them. The perpetuation of a species is the ultimate aim of all organisms and it involves reproductive processes that may vary in their complexity and underlying intricate mechanism. Reproduction in animals occurs in harmony with existing dietary, physical and social conditions. To this end, natural selection has provided the mammals with variety of signaling systems, each of which couples environmental variations of same kind with appropriate endocrine responses. *Pheromones* are air-borne chemosignals which are wide spread throughout the animal kingdom as chemical communicants. In fact, the pheromonal communication is a world of gentle suggestions, urgent demands and seductive entreaties that occur in nature unnoticed to human observers. Since the specific olfactory signals like pheromones trigger quite selectively a cascade of biochemical processes in conspecifics, they have been variously called *viz. olfactory love letters or love letters in the air* or *smell of beauty* (see Rekwot, *et al.*, 2001). We have recently begun to understand the communication by pheromones in last 4-5 decades that have elapsed since the term pheromone (Greek pherin= to carry; harmon = to excite) was coined to describe the sex attractant of *Bombyx mori*, a silk moth (Karlson and Butenandt, 1959). Our knowledge has greatly increased in the field of chemical communication not only in the world of insects but also mammals, bird, reptiles and fishes. Communication among mammals has been well established and is the subject of many *reviews* (Parkes and Bruce, 1961; Whitten, 1966; Bruce, 1970; Bronson, 1971, 1974, 1979, 1982, 1985; Bronson and MacMillan, 1983; Dominic, 1978; Izard, 1983; Meredith, 1983; Vandenbergh. 1983 a, b; Johnston, 1983,1985; Pandey, 1989, 1992; Grammer and Jutte, 1997; Rekwot et al., 2001; Novotny, 2002; Archunan, 2003; Hurst and Beynon, 2004; Brennan and Keverne, 2004) and *books* (Shorey, 1976; Doty, 1976; Muller- Schwarze and Mozel, 1976; Brown and Mc Donald, 1985).

Definition

Pheromones are air-borne chemical substances emitted externally by an individual of a species and cause a specific response in another individual of the same species. Pheromones send chemical signals between individuals to elicit behavioral and endocrine responses in other members of the same species, unlike hormones which are secreted internally and cause a response within the body of same individual. Chemical communication with pheromones is thus one means of transmitting information at incredibly small quantities. Pheromones are released in urine of faeces or secrete from cutaneous glands and dissipated in the air.

Pheromones are believed to be chemical compounds containing 5-20 carbon atoms and having a molecular weight ranging from 80-300 (Wilson and Bossert, 1963). The important property of a compound used as pheromone in an appropriate volatility. Many different chemicals have been reported in pheromones from non-polar molecules like alkanes and alkenes to various polar compounds which may be acidic (*e.g.* acids, phenols) or basic (*e.g.* amines). Pheromones affect the behaviour or physiology of the recipient by ingestion, absorption or by olfaction (Bruce, 1970). Majority of pheromones so far recognized are olfactory in their mode of action. The most well known example is the sex attractant pheromone of insets. The queen bee substance of honey bee, secreted by mandibular gland of queen, which inhibits ovarian development in worker bees, is an example of ingestion pheromone. The best known example of a contact pheromone is the one produced by adult male locust which accelerates maturation of young and is also believed to be involved in the formation of migratory swarms. All mammalian pheromones so far known act through olfactory pathways.

Classification

Chemical substances which are employed in conveying massages between animals are termed *semiochemicals* (Regnier, 1971). Semiochemicals which are employed in intra-specific communications are called *pheromones* (Karlson and Butenandt, 1959). Semiochemicals used in inter-specific communications are known as *allelochemics*. Allelochemics are further divided into *allelomones* and *kairomones*. Allelomones induce effects which adaptably favour the emitter (*e.g.* spray of skunk, Musk shrew and repellant secretions of certain insects). Kairomones induce effects that adaptably favour the receiver (*e.g.* chemosignals of prey helping predators to locate it). Two more allelochemics have been described by Nordlund and Lewis (1976). These are *synomones* and *apneumones*. Synomones induce responses which are beneficial to both emitter and receiver (*e.g.* floral scents and nectars). Apneumones are emitted by nonliving materials that evoke a response adaptably favorable to recipient but detrimental to individuals of other species present on the nonliving material (*e.g.* decaying meet, food grains containing the prey), the attractant being emitted by the meat or food grains rather than by the prey.

Pheromones are further classified into *three types* on the basis of their functions: *Releaser pheromones, Primer pheromones* and *Imprinting pheromones* (Bruce, 1970; Dominic, 1978; Pandey, 1989, 1992).

Releaser Pheromones

Releaser pheromones are known to play an important role in intra-population communications. Releaser pheromones evoke an immediate and reversible response on the recipient and are mediated directly through the central nervous system or through rapidly acting neurohumoral channels. Some examples of the information transmitted through releaser pheromones are as under:

Species and Racial Identity

House mouse using olfactory signals can discriminate between conspecifics and other rodent species (Bowers and Alexander, 1967). Gerbil, vole and hamster have also been reported to distinguish the odours of conspecifics from those of other species (Eisenberg and Kleimen, 1972).

Identification of Sex and Reproductive State

Bowers and Alexander (1967) using a discriminating test, have shown that the mouse can identify the sex of the conspecifics by odour. Dagg and Windsor (1971) have described the same ability in gerbils. Similarly, a male dog can distinguish and prefers the odour of a receptive female to that of an unreceptive female (Beach and Gilmore, 1949). Michael, Keverne and Bonsall (1971) have found that olfactory signals are of major importance for male macaques in discriminating between the receptive and nonreceptive females.

Discrimination of Age

Most mammals can distinguish between the smell of prepuberal and mature individuals. Mykytowycz (1970) has found that the odour from the chin gland of the rabbit gets stronger as the animal ages. In male dogs and cats, the adult patterns of urination (leg-lifting and spraying backward) do not develop until the animals attain puberty.

Aggression and Dominance

Mugford and Nowell (1970) have reported that adult male mice produce a substance in its urine which increases the tendency to fight. Conversely, the female urine contains a factor which discourages aggression in males (Dixon and Mackintosh, 1971). It has been suggested by Ewer (1968) that a conspecific odour maintains a certain level of confidence in an animal as it traverses its territory. This increased confidence in the presence of personal odour accounts for winning of the fight by an individual in its home territory.

Scent Marking

The perception, production and deposition of scents are closely correlated in all mammalian species in which olfaction is one of the dominant sensory systems. In both males and females of any species, the odour of a foreign or familiar conspecific increases the frequency of scent-marking. Adult males especially mark in response to the odour of other adult males (Ralls, 1971). Territory marking is also common in many animals (*e.g.* dog, deer, lion etc) which frequently deposit their urine/faeces in their habitats so that they can easily make out the intruders.

Arousal

Most of the behaviour patterns increase in frequency in response to odour. It is probable that in communication some species-specific odours may act solely to arouse the recipient. Increased arousal can be expressed in a number of ways through scent-marking, aggression, grooming, etc.

Sexual Attraction

Vaginal secretions have been found to be the source of releaser pheromones for sexual attraction in sheep, dog, hamster and rhesus monkey (cf. Parkes and Bruce, 1961; Michael, 1975). Sexual attraction in male rhesus monkey is mediated by a releaser pheromone of vaginal origin. The active substance comprises volatile short-chain aliphatic acids called *copulins* which have also been demonstrated in vaginal secretions of human, even though their precise role has not been elucidated. In addition, secretions of many cutaneous glands in different mammalian species are also implicated in sexual attraction (see Ewer, 1968; Mykytowycz, 1970; Stoddart, 1974).

Immobilization Reflex

Saliva of certain species (*e.g.* boar) also contains releaser pheromones that induce the standing response in the sow called *immobilization reflex* during copulation. This is helpful in prolonged act of coitus (Clause and Alsing, 1976).

Maternal Pheromones

This pheromone is produced by females to maintain mother-young relationship in rats (Leon, 1974). It attracts mobile young and maintains cohesiveness of the litter. It is produced by the caecal bacteria of lactating females under the control of prolactin. This pheromone is emitted in large quantity in specific type of faeces called *caecotrope*. Similar reports are available for spiny mouse (Porter and Doane, 1976) and wild rabbit (Mykytowcyz and Goodrich, 1974).

Flehmen Reaction

Males of certain mammalian species exhibit a distinct grimance called flehmen following exposure to urine of conspecific females (see Estes, 1972). *Flehmen* seems to be a motor pattern associated with olfactory detection of certain substances (releaser pheromones), especially the urinary metabolites of sex hormones.

Alarming Pheromones

Releaser pheromones are also used to signal danger such as the alarm chemical emitted by a wounded minnow or aphid killed by predators alarms other members of its species for the danger.

Trailing Pheromones

Many insects like ants, termites and bark beetles etc. emit trailing pheromones which convey massage to other members of their species to follow, as if they are saying 'there is a party come and join me to enjoy it'. Tent caterpillars guide their tree mates to branches laden with tasty leaves by emitting trail pheromones.

Primer Pheromones

Primer pheromones demand prolonged stimulation and induce a delayed response in the recipients involving neuroendocrine pathways. These pheromones have been further categorized on the basis of their action on the oestrous cycle, pregnancy, and sexual maturation. Different primer pheromones and their functions are described as follows:

Oestrus-suppressing Pheromone

The oestrus-suppressing pheromone emanates from body of intact adult female mice. It is produced under the control of ovarian hormones (*e.g.* oestrogen) and is released in the urine. This pheromone suppresses the oestrous rhythm causing pseudopregnancy and/or anoestrus by involving neuro-endocrine pathways. Unisexual grouping of regularly cycling females in a cage results in disruption of the oestrous cycle in laboratory mice. There occurs an increased incidence of spontaneous pseudopregnancy in a small group of females (Van der Lee and Boot, 1955). This phenomenon is known as the *Lee-Boot effect*. They also noted that this effect is abolished following olfactory bulbectomy, indicating the involvement of olfactory chemosignals (Van der Lee and Boot, 1956). Whitten (1957) observed that housing the female mice in a larger group results in anoestrus (complete arrest of oestrous cyclicity) accompanied by decrease in ovarian and uterine weights. Anoestrus can also be mimicked by keeping females in cages recently soiled by grouped females (Champlin, 1971). Analogous studies have been carried out in deermice (Bronson and Marsden, 1964), voles (Richmond and Conaway, 1969) and wild mice (Pandey, 1976).

The female pheromone which mutually suppresses the oestrous rhythm in other females is excreted in the urine of adult female (Pandey and Pandey, 1986a) and its production is under the control of ovarian hormones (Pandey and Pandey, 1985a). It is interesting to note that castrated and oestrogenized males also develop the ability to suppress the cyclicity in females (Pandey and Pandey, 1987). This oestrus-suppressing pheromone acts through contact in wild mice (Pandey and Pandey, 1985b). The synergistic role of the physical contact in suppression of oestrus on unisexual grouping of females is also described in field mice (Dixit and Pandey, 2007a).Minimum grouping time required for mutual oestrus suppression in females is 3 hrs (Pandey and Pandey, 1986b).

Not much is known of the pathways of action of the oestrus-suppressing pheromone. The existence of a reciprocal relationship between prolactin and FSH-LH secretion is postulated (cf. Dominic, 1969b). If this concept is valid, the incidence of spontaneous pseudopregnancy in grouped mice is due to the increase in luteotrophic activity and decrease in gonadotrophic activity. In large groups the suppression of the gonadotrophic activity will be more profound resulting in anoestrus. The two types of responses seen in the small and large unisexual groups of female mice are believed to be related to differences in population density (Parkes and Bruce, 1961; Whitten, 1966). Pseudopregnancy in female group has been reported in both inbred and random-bred strains of mice, but the anoestrus response has not been found in inbred animals (Whitten an Bronson, 1970). Hence the genetic background of the females being a causative factor cannot be ruled out. Oestrus suppression appears to be a specific response to the female group situation and not a result of adrenal response to general stress (Whitten and Bronson, 1970; Bronson, 1971).

Oestrus-inducing Pheromone

The oestrus-inducing pheromone is emitted by adult intact male mice. Its production is under the control of androgens (*e.g.* testosterone) and it is released in the urine. This pheromone induces a new cycle in noncyclic females as well as synchronizes oestrus on day 3 or 4 depending upon the strain of the mouse. Whitten (1956) for the first time observed the induction and synchronization of oestrus in pseudopregnant or anoestrous female mice following exposure to adult males. The synchronization of oestrus followed by mating takes place on day 3 in laboratory mice. This is known as the *Whitten effect*. The chemosignal inducing oestrus is excreted in the male urine (Bronson and Marsden, 1964) and its production is under the control of androgens (Bronson and Whitten, 1968). The oestrus-inducing capacity of males vanishes following castration and restored by androginization of castrates of either sex (Pandey, 1976). The prepuberal administration of flutamide (a nonsteroidal antiandrogen) also inhibits the production of the oestrus-inducing pheromone in Indian field mice (Dixit and Pandey, 2007b). Androgenized females can also induce oestrus in noncyclic females (Dominic, 1969). Treatment with a steroidal and nonsteroidal antiandrogen inhibits the ability of males to induce oestrus in wild mice (Pandey and Dominic, 1978). The male pheromone could also induce oestrus in those females which were made noncyclic by treating them with psychotropic drugs like reserpine or chlorpromazine (Pandey, 1976). This shows that the male pheromone could override the effect of these drugs. However, artificial scents like *oil of wintergreen* or *kanta* masked the oestrus inducing ability of male wild mice (Dominic and Pandey, 1979b). Physical contact was found to be essential for the operation of this pheromone which is perceived through main olfactory system (Pandey, 1986). Occurrence of the Whitten effect has also been reported in other rodents like the deermouse, Peromyscus (Bronson and Marsden, 1964), wild mouse (Pandey, 1976), prairie vole, *Microtus ocharogaster* (Richmond and Conaway, 1969) and desert pocket mouse, *Prognathus pencillatus* (Wilkin and Ostwald, 1968).

Even though the neuroendocrine mechanisms involved in the Whitten effect have not been extensively investigated, it appears that the influence of the male odour stimulates gonadotrophic secretion in females. Since the effect of male on the oestrous cycle is not evident in either metaestrous or dioestrous females, it is presumed that the male pheromone initiates and/or accelerates follicular development by stimulating secretions of FSH (see Whitten, 1966) or of both FSH and LH (Zarrow *et al.*, 1970). The significant reduction in the pituitary acidophils in grouped females following exposure to males is consistent with the view that the female pheromone suppresses and male pheromone stimulates the secretion of FSH (cf. Dominic, 1976a).

Primer pheromones influencing oestrous cycle, analogous to that described in the laboratory mouse, operate in several mammalian species. The detailed investigations of Pandey (1976) have established the existence of a female-originating oestrus-suppressing pheromone and a male-originating oestrus-inducing pheromone in wild mouse. In the deermouse, *Peromyscus maniculatus* bairdii oestrus in females is suppressed by unisexual grouping and is induced in anoestrous females following exposure to males. The oestrus-inducing pheromone is present in the urine of males and its production is androgen dependent (see Bronson, 1968, 1974). Male induction of oestrus is also reported in sheep and goats. The sudden introduction of rams to ewes in the transition from non-breeding to breeding activity or to progestin-treated ewes results in a high degree of synchronization of oestrus (the ram effect) (cf. Dominic, 1969b; Bruce, 1970). The induction of oestrus and hence ovulation in response to a male or to bedding soiled by males has also been reported in the vole, *Microtus ochirogaster* (cf. Dominic, 1976a). Male influence on the oestrous cycle and ovulation also occurs in the pocket mouse, *Perognathus pencillatus*, the cuis, *Galea musteloid* (cf. Dominic, 1976a) and the bank vole, *Cletherionomys glareolus* (Clarke and Hellwing, 1977), even though in these cases physical contact is necessary. The proximity of a male has been found to be important in the induction of oestrus in certain species, *e.g.*, the green acouchi, *Mycoprocta prathi*, rabbit, Mongolian gerbil (Eisenberg and Kleimen, 1972); however, it is not known whether induction of oestrus in these species is accomplished solely through olfactory pheromone. In Wistar rats the oestrous cycle is shortened from 5 to 4 days as a result of exposure to male or female urine (Aron, 1973). This cycle shortening is related to an increase in the pituitary FSH content in late dioestrus. Castration of males or females does not suppress the production of the cycle-shortening urinary pheromones (cf. Aron, 1973). It thus appears that the rat has ability to respond to primer pheromones but long domestication and selective breeding has resulted in its reproduction not profoundly being influenced by them.

Pregnancy-blocking Pheromone

A third primer pheromone has been shown to terminate the pregnancy in newly inseminated female mouse following exposure to a 'strange' or 'alien' male, a male other than her stud one (Bruce, 1961). The female returns to oestrus and if mating is allowed, the pregnancy is sired by the second male. This phenomenon is known as the *Bruce effect*. The effectiveness of the strange male depends upon his genetic relationship with the stud, up to 80 per cent of pregnancies are blocked if strange male belongs to different strain, but the number is considerably lower if stud and strange males belong to same strain (Parkes and Bruce, 1961). The pregnancy-blocking factor is present in the urine of adult intact male and it vanishes following castration (Bruce, 1965) or by treatment with an antiandrogen (Bloch and Wyss, 1973). Anosmic females fail to react to the presence of males (Bruce and Parrot, 1960). It has been reported that both physical contact and vomeronasal organ (VNO) are necessary for the mediating of the pregnancy blocking pheromone (Rajendren and Dominic, 1984). As suggested by Bruce (1970), the pregnancy blocking factor induces the release FSH and LH and there is concurrent inhibition of the prolactin release resulting in block to ovoimplantation due to the failure of the development of functional corpora lutea. The Bruce effect has been demonstrated in *Peromyscus* (Bronson and Eleftherion, 1963), wild *Mus musculus* (Chipman and Fox, 1966) and a vole, *Microtus agrestis* (Cliulow and Clarke, 1968).

The Bruce effect provides one of the best examples of the physiological results of olfactory discrimination (Parkes and Bruce, 1961; Dominic, 1969b). Exposure of the female to the original stud male does not induce pregnancy block. Moreover, the efficiency of the blockade is increased if the stud and strange males are of different genetic strains (Parkes and Bruce, 1961; Dominic, 1966a). Thus the key concept would appear to be discrimination on the part of the newly mated female between odours of the two males, a discrimination results in her failure to react to the stud male after his original induction of her oestrous cycle but to respond to the new male by hormonal changes leading to the return of oestrus at the expense of pregnancy termination.

The neuroendocrine mechanisms involved in the Bruce effect have been studied in great detail (Dominic, 1966b). Administration of prolactin (Parkes and Bruce, 1961; Dominic, 1966b;, 1967b) or increased prolactin production induced by suckling in lactating females pregnant from postpartum conception (Parkes and Bruce, 1961) prevents the block; hence it is evident that the primary action of the male pheromone is to suppress prolactin secretion. Pregnancy block is also suppressed in intact newly mated females by the presence of a functioning ectopic pituitary graft in the renal capsule which secretes prolactin continuously (Dominic, 1966b, 1967a) or by administration of reserpine (Dominic, 1966b,c) which stimulates prolactin secretion by suppressing inhibitory centre in the hypothalamus controlling the release of this hormone. Pregnancy block is also prevented by administration of progesterone (Dominic, 1966b, d). The histological appearance of the corpora lutea in the ovary of the female after a blocked pregnancy also lends support to the view that the inhibition of implantation is due to the failure of luteal function (Dominic, 1966b, 1970). These studies support the view (Parkes and Bruce, 1961) that the immediate cause of the pregnancy block is the interference with prolactin secretion with consequent failure of corpus luetum development and return of oestrus.

Puberty-influencing Pheromones

A primer pheromone influencing sexual maturation in both male and female mice has been conclusively demonstrated (Drickamer, 1981; Vandenbergh, 1983; Pandey, 1986). Sexual maturation in young females is accelerated in presence of an adult male and suppressed in presence of the adult female (Cowley and Wise, 1972). The *puberty-accelerating* pheromone in young females is androgen dependent and is associated with male urinary protein (Vandenbergh, Whitsett and Lombardi, 1975). The odour of females wild mice has an acceleratory effect on maturation of young males (Pandey, 1986). On the other hand, a male-urinary cue accelerates the pubertal onset in young females (Pandey and Pandey, 1988a).

However, these olfactory cues from adult individuals have a suppressive effect on sexual maturation of young of the same sex.

The *puberty-delaying* pheromone is released in the urine of adult females after housing them together for 15 days and a daily exposure of subject females to the pheromone, at least for 2 hrs, is needed for effective delay of the puberty to occur (Pandey and Pandey, 1986c). The puberty acceleration is initiated only when the young females are exposed to adult male urine for a minimum of 6 days. However, the puberty delay in young females is apparent only after a long exposure of 9 days to female urine (Pandey and Pandey, 1988b). Effect of social odours on sexual maturation has also been described in prairie deermice (Terman, 1968) and the Mongolian gerbils (Payman and Swanson, 1980).

In contrast, a female-originating pheromone delays puberty in young females, presumably by partial inhibition of gonadotrophins release. The pheromone is present in the urine, since exposure to urine of adult virgin female retards sexual maturation in immature females (cf. Vandenbergh, 1973).

Imprinting Pheromones

Olfactory influences of the pre-weaning environment have been shown to affect reproductive behavior in certain laboratory rodents, particularly mice. Imprinting pheromones act at a critical period during development and bring about permanent modification in the behavior at adulthood. It was shown by Mainardi *et al.* (1965) that young female mice reared with both male and female parents eventually prefer to mate with males of a different strain when adult. However, in the absence of a male parent, the female mates with males of her own strain. This sexual preference and aversion to males of the two subspecies are not exhibited by females which are raised by their mother alone in the absence of the father and thus not exposed to the odour of the male parent or by females reared during the pre-weaning period with perfumed (with Parma violet) parents. According to the Yanai and Mc Clean (1972) these preferences do not exist in all mice strains; the 'father effect' is present in C57B1 mice while Balb mice mate randomly. Marr and Gardner (1965) deodorized rat pups and their mothers with chemical odours and found that as adults odour-reared animals preferred to stay near (in the same area of Y- maze) other rats bearing the odour they smelled in the litter. Male odour-reared rats were less sexually responsive to normal (unscented) females than were controls. Females segregated from males (but not males segregated from females) since birth respond differently to adult sex odours compared to animals reared in mixed litter (Carr *et al.*, 1970). Mc Carty and Suthwick (1977) reared litters of southern grasshopper mice (*Onycomus torridus*) and white-footed mice (*Peromuscus leucopus*) with natural parents or reciprocally cross-fostered to parent of the opposite species. After weaning, mice were allowed a choice between the soiled beddings of *Onycomus* and *Peromyscus* adults of the opposite sex. Compared to controls, cross-fostred mice of both species displayed decreased preferences for homo-specific odours. Cross-fostered *Peromyscus* males actually switched species preferences to *Onychomus* odours (Mc Carty and Southwick, 1977).

Human Pheromones

The existence of human pheromones has long been suspected with no conclusive evidence in its favour. Doty (1976) has indicated the involvement of human chemosignals in communication. It is suggested that human olfactory sensitivity is as strong as that of dog, but is suppressed in adult life due to psycho-sexual reasons. Comfort (1971) suggested that olfactory cues play an important role in infantile psychosexual development. Differences in the ability to smell certain substances exist between two sexes and their ability also varies during the menstrual cycle, pregnancy and following castration and/or exogenous administration of gonadal steroids (Doty, 1976). Human vaginal secretion contains volatile short-chain fatty acids, the amount of which varies during different stages of the menstrual cycle, reaching peak, as they do in nonhuman primates, near the time of ovulation (Michael, Bonsall and Warner, 1974). Women have olfactory sensitivity to boar taint in pork. The steroid-ketone responsible for the *boar odour* has been observed in auxiliary sweat of men (Clouse and Alsing, 1976), but its significance remains still obscure. Le Magnen (1952) found that only sexually mature women can perceive the musky odour of a synthetic perfume, exaltolide (15-pentadecanolide) and the sensitivity is highest at about the time of ovulation. This phenomenon provides a circumstantial evidence for the existence of an ancestral, musk-like sex pheromone that might have stimulated receptive women. A comprehensive account of pheromonal interaction in human has been given by Pandey (1977).

Recently, two pheromones have been identified in women giving evidence for potential pheromonal mechanism for synchronizing the menstrual cycles of women. For years women who live together and do not continuously consume oral contraceptives have noticed that their menstrual cycles become synchronized (McClintock, 1971). The reason for this phenomenon has been proven to be caused by the pheromones emitted from the female which lengthen and shorten the menstrual cycle of other women (Stern and McClintock, 1998). The study by Weller and Weller (1997) demonstrated menstrual synchrony in each of groups with high statistical confidence for two of the three months under optimal conditions. The two pheromones which cause mentrual synchrony are identified as *follicular pheromone* and *ovulatory pheromone* (Stern and McClintock, 1998). Follicular pheromone is released in the late follicular stage of the menstrual cycle and accelerates the pre-ovulatory surge of luteinizing hormone thus shortening the menstrual cycle. The ovulatory pheromone is released at ovulation from the axillary compounds of women, and causes a delay in luteinizing hormone surge, and thus lengthens the cycle.

Very recently, human pheromone called *androstenone* (a metabolite of androgens) which is produced by men and is now sold over the internet as *date-mate.* The spray of this pheromone guarantees to make men irresistible to women. In another study, women were found to choose the most attractive man on the sole basis of sweaty shirts that may contain *androstenol,* a male hormone related to androgen or some other chemical that may act like pheromone. When the compounds of the axillary odours from the arm-pits of male and females were examined, it was found that the major components were the same but in different concentrations (Zeng *et al.,* 1996). In males (E)-3-methyl-2-hexenoic acid is present in ratio of 10:1 to (Z)–3- methyl-2-hexenoic acid, which in females the ratio is 16:1. Female axillary odours also contains trace amounts of 2-piperidone, which is not seen in males, for an unknown reason. Female secretions also had no detectable amount of androstenone which is seen in male extract. A gene, named V1RL1 has been identified in humans. This gene codes for formation of a protein that is helpful in detection of pheromones. The receptor of this pheromone has been found to lie in the olfactory mucosal membrane lining of human nose.If the existence of sex attractant human pheromone were proven interest in other kinds of human pheromones, such as alerting, the interest in reproduction-repression would increase. Probably, the human reproduction-repressive pheromone would be used as the most efficient and harmless contraception method. Then the human could enjoy more romantic times without working about the consequences.

Mode of Action

Chemically pheromones range from ions to macromolecules and possibly all known chemoreceptor systems serve as pheromone detectors. Four to six different chemoreceptor systems have been reported to impart their role in pheromonal action. The number of these detecting systems depends upon the species and the type of chemosignals (see Meredith, 1983). Different chemoreceptors are:

1. The main olfactory system (MOS)
2. The accessory olfactory system (AOS) or Vomeronasal organ (VNO)
3. The septal organ
4. The trigeminal system
5. The nervus terminalis
6. The taste system.

The Main Olfactory System

The main olfactory system comprises olfactory receptors which are neurons with their apices in the mucus overlying the olfactory epithelium with axons passing back to the olfactory bulb. In mammals the receptors are located in a posterior dorsal pocket of the main nasal respiratory airway. The distributing concentration and flow rate of odourous substance across the olfactory mucosa are controlled by sniffing pattern (Marshall and Malton, 1977). Since the olfactory system can respond to almost all molecule that reach its receptors, it possibly serves as a specific detector for a particular chemical signal

The Accessory Olfactory System or Vomeronasal Organ

The accessory olfactory system consists of an elaborate crescent like structure connected to nasopalatine canal and is known as *vomeronasal organ* or *Jacobson's organ*. The inner wall of VNO is lined with sensory non-ciliated, pseudostratified epithelium (Monti-Bloch *et al.*, 1998a) that contains vomeronasal receptor neurons. The vomeronasal axons pass between main olfactory bulbs (MOB) and enter the accessory olfactory bulbs (AOB) at the posterior end of main bulbs. The entire vomeronasal lumen continuously bathes in the secretions of vomeronasal glands. As in the olfactory system this fluid may be important for determining the response properties of the system

Septal Organ

The septal organ consists of a path of tissue located on either side of nasal septum anterior and ventral to the main olfactory epithelium. It is also known as the *organ of Masera*. It comprises the ciliated receptor neurons like the olfactory receptors (Graziadei, 1977). This organ has been identified in mice, rat, hamster, guinea pig, rabbit and opossum (Meredith, 1983). The septal organ may detect chemical signals during non-sniffing respiration (Moulton, 1978).

Trigeminal System

The trigeminal is a somatic sensory third cranial nerve that has distribution to the epithelium inside the nasal cavity as well as to the skin of the head and cornea of the eye. The internasal and corneal fibers are sensitive to air-borne chemicals. The system also forms the chemosensory input for reflex changes in heart rate and respiration in response to strong irritating vapours (Stone *et al.*,1978) and probably also for the reflex changes in autonomic activity. Although this system is less sensitive than MOS for many odours but it responds to certain odours like phenyl ethanol at irritating concentration.

Nervus Terminalis

The nervus terminalis is present in all vertebrates but no function has yet been definitely assigned to it. This system comprises a network of cells and unmyelinated nerve fibers. The neurons of this system include both multipolar and bipolar nerves which enter the nasal epithelium without ending in specific structure. There are no electrophysiological or behavioral evidences in support of chemoreceptive functions of nervus terminalis but the possibility is not negligible.

Taste system

The taste receptor consists of specialized epithelial cells contained in the taste buds. These cells make synoptic contacts with gustatory nerve fibers whose cell bodies lie in cranial sensory ganglia and whose central processes enter the brain. In higher vertebrates odour secretion and excretion of other individuals are frequently licked off taken into mouth as well as sniffed (*e.g.* hamster, dog etc..). The participation of this taste system in pheromonal communication cannot be ruled out without experimental investigations. The human taste system is sensitive to protein and peptide which may carry much information (Beidler, 1977). Odorous material might stimulate the taste system, when they concentrate in oral or nasal fluids to allow identification by this system.

In mammals, two major systems impart their role in pheromonal communication. These include main olfactory system (MOS) and accessory olfactory system (AOS) which includes Vomeronasal organ (VNO). The MOS comprises the olfactory receptors in the form of ciliated neurons with their apices in the mucus overlaying the olfactory epithelium and with axons that pass back to the olfactory bulb. The MOS serves as a general molecular analyzer than a specific detector (Meredith, 1998). Signals from the MOS are transferred to anterior cortical nucleus, postero-lateral cortical nucleus and other peleo-cortical targets. The AOS consists of an elaborate crescent-like structure connected to the nasal cavity or naso-palatine canal. This system is known as *vomeronasal organ* or *Jacobson's organ* (Wysocki, 1979). Vomeronasal receptor neurons are expressed from two gene families, which express different *G-proteins*. A number of soluble small size pheromone binding-proteins participate in the so-called perireceptor events (Breer, 1994) to transport pheromones, assist receptors, deactivate the reception site, etc. Some carrier proteins called *lipocalins* are known to be expressed in nasal mucus of rodents while other similar proteins are excreted in urine of laboratory and wild mice (*e.g.* major urinary proteins, *MUPs*) (Finlayson and Baumann, 1958) and (*e.g.* α-2, 4 globulins) (Bayard *et al.*, 1996) and the vaginal secretion of the golden hamster (*aphrodisn*) (Korner *et al*, 1996). They all can bind small ligands, albeit in different tissues and different species, for different reasons. The AOS projects to the vomeronasal amygdla, the medial and postero-medial nuclei of the amygdla. This projects along with stria terminalis signals, which are sent to medial pre-optic area that is known as critical mating area.

The outputs from both MOS and AOS travel in separate pathways but converge on the medial nucleus of the amygdla, the bed nucleus of the stria terminalis (BNST) and the medial pre-optic area (MPOA) (Swann and Fiber, 1997). Once the pheromonal signal has reached the specific centre of the brain (*i.e.* hypothalamus), a change in hormone surge occurs resulting in release of gonadotropins releasing hormone (GnRH) from hypothalamus which acts on adenohypophysis to trigger the release of gonadotrophins (FSH/LH) followed by the gonadal secretions *e.g.* testosterone in case of male or oestrogen in case of female (Meredith, 1998). Thus in contrast to odour discrimination by the MOS, the pheromones are detected through the VNO resulting in behavioural and endocrine responses. Once the organism senses the odour or pheromone through AOS, these scents are then transmitted to brain for processing. The chemical messages of scents cannot directly enter the target cells, so they stimulate activation of other messengers like G-proteins and adenylic cyclase enzyme (Brown, 1994). The G protein functions in transduction of the signal from extra-cellular receptor complex to intra cellular second messenger system. The first messengers (odour/pheromone) thus activate second messengers like cyclic-AMP (cyclic adenosine mono phosphate). Odours bind to the receptors of main olfactory epithelium to induce the transduction cascade of a G-protein with adenylic cyclase-III which elevates cAMP level to open cyclic nucleotide gated ion channels, so as to transmit the signal (Berghard and Buck, 1996).

The mechanism of signal transduction to transmit volatile and non-volatile signals once they have been received on receptor neurons differs in MOS and AOS. The AOS (*e.g.* VNO) uses similar strategy for sensory signal transduction to the main olfactory process (Berghard and Buck, 1996). The pheromonal signal transduction is mediated by the principal secondary messenger, *inositol-(1, 4, 5)- tri phosphate* (VNO IP3), whose production is stimulated by the presence of pheromones (Wekesa and Anholt, 1997). The IP3 primarily replaces c-AMP in pheromone signal processing; however, cAMP may also play a role in VNO signal transduction. The IP3 responses involve G-proteins and by inference G-protein–coupled receptors. The two G-proteins highly expressed in VNO neurons are G_0 and

Gi2, which are found in different locations within the VNO neuroepithelium and are most likely the G-proteins involved in VNO sensory transduction (Berghard and Buck, 1996). Contrary to the main olfactory system, the vomeronasal system uses the adenylic cyclase type II (AC II) in sensory transduction cascade, as it is highly expressed in VNO neurons. It thus appears that in VNO signal transduction, there is a transduction cascade of G_0 and Gi2 with AC II which induces elevations in IP3 and somehow opens ions channels to transmit the signal (Berghard and Buck, 1996; Wekesa and Anholt, 1997). In hamsters the mounting pheromone (the aphrodisin) modulates IP3 production in villar membranes without a corresponding alteration in cAMP production (Korner *et al.*, 1996). Thus there is evidence that the while sensory transduction in the main olfactory and vomeronasal systems uses similar methods of sending signals they use different proteins and mechanisms.

Source of Mammalian Pheromones

The chief sources of mammalian pheromones are urine, faeces, saliva and secretions of cutaneous glands. The role of the urinary releaser and primer pheromones in behaviour and reproduction is well documented (Dominic, 1976a, b). It is noteworthy that hundreds even thousands of volatile and nonvolatile substances can be identified in the urine of most mammals. In guinea pigs, for example, numerous chemicals of a variety of molecular weights are involved in a male's discrimination between male and female urine and his preference for the latter. It is known that some of the steroids present in the urine have characteristic odours detectable even by the human nose (Doty, 1976). The regular sampling of the female's urine is a daily routine of males of many species (Estes, 1972).

There is ample information on the role of faecal odours in communication. The odour of faeces may be determined genetically or by environmental factors such as the type of food consumed and the action of bacterial flora. In the first case, the mixtures of secretion from glands located in the anal region contribute significantly to the individual characteristic of the faeces. In rabbits (Mykytowycz and Goodrich, 1974) and rats (Leon, 1974) the odour of the mother's faeces is of particular importance to the newborn because of its ability to regulate their movements. The role of maternal pheromone has already been discussed. In the domestic pig the main odorous component of saliva is 5-α-androst-16-en-3one with urine like smell. This steroid can be reduced to 5α-androst-16-en-3α-ol and 5α-androst-16-en-3β-01, both with a musk smell. The submaxillary glands act only as a reservoir for this substance which like the androgen is synthesized in the Leydig cells but does not possess androgen's activity (Claus and Alsing, 1976).

The odoriferous glands of mammals differ remarkably in origin and position and in the nature of their secretory products. Anatomically they may be found in any location (Mykytowycz, 1970). These glands are the main source of odour in mammals and even the characteristic smell of faeces may be modified by their secretion (see Mykytowycz and Goodrich, 1974). Scent glands have been identified in 15 of the 19 mammalian Orders, and on the basis of the location of body, as many as 40 different types of glands can be made out. A single species may show a varied assortment of glands. Lagomorphs, for example, have chin glands situated in the orbit (Mykytowycz, 1970). The scent glands may be composed exclusively of sebaceous elements (holocrine) as in the ventral gland of the Mongolian gerbil or sudoriferous elements (apocrine) as in the chin gland of European rabbit or both as in the gular gland of bat, *Taphozus longimanus* (Pandey, 1976). Various aspects of kin scent glands in some mammals have been worked out by Pandey (1976).

Even though eight Orders involving 55 species have scent glands and associated behaviour, correlated physiological and behavioral studies have been done only in some mammalian species such as Colombian ground squirrel, European rabbit, sugar glider, black tailed deer, pronghorn antelope, Maxwell duiker, Mongolian gerbil, golden hamster, guinea pig, marmoset monkey and lemur (see Thiessen and Rice, 1976). It is likely that future investigations will reveal many more species using scent glands for chemical communication.

Chemistry of Mammalian Pheromones

The identification of chemosignals is the most difficult aspect of research in chemical communication. Our knowledge is still meager about how the olfactory system works or how chemical signals are constituted. Much of the early work on chemical ingredients of mammalian scent glands was carried out in those species which have economic importance in perfume industry. These are; civets, musk deer, musk rat and beavers (Lederer, 1950; Kingston, 1965). Recently, with the advent of interest in chemical communication, the chemistry of variety of other scent gland secretions, vaginal secretions and urinary excretions has been worked out (c.f. Pandey, 1976; Muller-Schwarze and Mozel, 1977; Muller-Schwarze and Silverstein, 1980; Johnston, 1983). Available chemical studies suggest that most mammalian pheromones are complex substances. Unlike insect releaser model in which a single chemical constitutes the chemosignal and a specific receptor detects the molecule and odour quality, a different model has been suggested to operate in other animals. In this model the signal consists of a complex mixture of substances that is experienced by the organism as an odour quality or gestalt, which depends on the entire constituent for its essence. Among mammals, an additive but linear model has been suggested (Johnston, 1983). In this model, many components contribute to overall odour quality but it may not be the case for all. Pheromones thus mostly act in concert. It is the combination rather than the individual chemicals that carries the message. The chemical composition of mammalian pheromones, so far reported, is described in Table 26.1.

Table 26.1: Chemical Nature of Certain Mammalian Pheromones

Source	Species	Chemical Nature
Urine	Male Mouse	2-(Sec-butyl)4,5-dihydrothiazole (oestrus induction) and 3,4 Dehydroexobrevicomin (oestrus synchronization)
Preputial gland	Male Mouse	6-hydroxy-6-methyl-3-heptanone (puberty acceleration) E,E-α-farnesene and E-β-farnesene (opposite sex attraction)
Vaginal secretions	Male Red Fox	2-methylqunoline Dimethyl sulphide and 17 k Da
	Hamster	Polypeptide aphrodisn (mounting behavior)
	Rhesus monkey and other primates	Short chain aliphatic acids *e.g.* acetic, propionic, isobutyric, n-butyric and iovaleric acids
	Humans	C2-C5 aliphatic fatty acids (couplins)
	Domestic dog	Methyl-p-hydrobenzoate

Contd...

Table 26.1–Contd...

Source	*Species*	*Chemical Nature*
Tarsal gland secretions	Black tailed deer	Cis-4-hydroxydodec-6-enoic acid lactone
	Reindeer	Saturated alcohols and aldehydes
Subauricular gland secretions	Male pronghorn	Isovaleric acid as the most active component
Inter-digital gland secretion	Antelope	Methyl ketones
Anal gland secretion	Red Fox	Trimethylamine and a number of short chain aliphatic acids
	Mustelids	2, 2 dimethyl-thietane (mustelan) and di-isopentyl disulphide
	Mongoose	Lipids
Chin gland secretion	Rabbit	Protein and carbohydrate and 2-methyl-2-enal (nipple search in pups)
Inguinal gland secretion	Rabbit	Lipids
Ventral scent marking gland	Mongolian gerbil	Phenylacetic acid
Boar taint in pork	Boar	5α-androst-16-en-3-one and 3-hydroxy-5α-androt-16-ene (marketed as boar-mate)
Musk	Muskrat	Cyclopentadecanol and Cycloheptadecanol
Preputial secretions	Musk deer	Methylanrotenone
Side glands	Musk shrew	Lipids
Gular gland	Indian sheath-tailed bat	Lipids and Proteins
Skin scent gland	Marmoset	Saturated mono and diunsaturated fatty alcohols esterified with butaric acid and also squalene
Caudal organ	Apodemus (mouse)	Esters of long chain alcohols
Anal gland	Civet cat	Civetone
Anal sac	Mink	Diisopentyle disulphide
Skin gland	Skunk	trans-2-butane-1-thiol, 3-methyl-2-butanethiol, 3,2-methyl-1-trans-2-butanyl disulphide
Hormone metabolites in sweat	Humans	Androstenone (sold as Date-mate over the internet) and Androstenol (attraction to opposite sex)

Conclusion

Insects, mammals and possibly humans use chemical signals and pheromones to attract a member of opposite sex and signal the readiness to mate. The existences of pheromone have been documented in insects, mammals and humans. In insets, pheromones allow an individual to discover which other members of the same species are ready to mate. Mammals also use pheromones to communicate sexual behavior, to elicit precopulatory behavior, trigger the onset of puberty, synchronize oestrus and gestation periods, and facilitate mating by releasing LHRH. Thus insects and mammals have adapted the use of chemicals to communicate sexual intent and elicit behaviors in others. The two human pheromones are known to synchronize female menstrual cycle in a similar manner to other mammals, but signals in insect and mammals are processed through a accessory olfactory system (VNO), anatomically separate from the main olfactory system (nasal olfactory epithelium).These two pheromone receptor organs are similar to the main processing system, and they are highly specialized to different pheromone compounds, Since the human pheromone system closely resembles mammalian and insect systems, it is thus possible that humans also use pheromones to signal a mate through chemical attraction. The ability to attract a member of the opposite sex is of evolutionary significance to many mammalian species so that they can pass on their genes with the best mate. It is possible that genes coding for pheromones have been passed on through evolution, and may thus be present in humans too.

Acknowledgements

Parts of my own work described here were supported by HCS/ DST/1166/81, UGC/F3-87/86(SR-II) and CSIR/38/759/89, EMR-II. Working expenses were also defrayed, in part, from DST/SR/ SO/AS-09/2003. The notion of pheromonal regulation of reproduction in Indian field mice emerged during discussion with Professor C.J. Dominic (Former Head, Department of Zoology, Banaras Hindu University, Varanasi). I thank Mr. Himanshu Dixit, a SRF in DST project under my guidance, who was kind enough to help me in preparation of this manuscript. I also wish to extend my heartly thanks to Dr. P.E. Dean, Head of the Institution for providing required laboratory facilities.

References

Archunan, G., 2003. In: *Proceedings in the lecture of recent trend in ethology and Behaviour*, pp. 82–91.

Aron, C., 1973. *Arch. Hist. Embr. Norm. et. exp.*, 56: 209–216.

Beach, F.A. and Gilmore, R., 1949. *J. Mammal.*, 30: 391–392.

Bayard, C. Holmquist, L. and Vasterberg, O., 1996. *Biochim. Biophy. Acta,* 1290: 129–134.

Beidler, L. M., 1977. In: *Chemical Signals in Vertebrates*, (Eds.) D. Muller-Schwarze and M.M. Mozell. Plenum, New York, pp. 483–488.

Berghard, A. and Buck, L.B., 1996. *The Journal of Neuroscience,* 16(3): 909–918.

Bloch, S. and Wyss, H.I., 1973. *J. Endocr.*, 59: 365–366.

Bowers, J.M. and Alexander, B.K., 1967. *Science.* 158: 1208–1210.

Breer, H., 1994. *Semin. Cell Biol.*, 5: 25–32.

Brennan, P.A. and Keverne, E.B., 2004. *Current Biology,* 14: R81–R89.

Bronson, F.H. and Eleftheriou, B.E., 1963. *Gen. Comp. Endocr.*, 3: 515–518.

Bronson, F.H. and Marsdon, H.M., 1964. *Gen. Comp. Endocr.*, 4: 634–637.

Bronson, F.H. and Macmilan, B., 1983. In: *Pheromones and Reproduction in Mammals,* (Ed.) J.G. Vandenbergh, Academic Press, New York, pp. 175–197.

Bronson, F.H., 1971. *J. Reprod. Fert.*, 25; 279–282.

Bronson, F.H., 1974. In: *Pheromones,* (Ed.) M.C. Birch. Elsevier/Experta Medica/North Holland and Publishing Company, pp. 344–365.

Bronson, F.H., 1979. *Q. Rev. Biol.*, 54: 265–299.

Bronson, F.H., 1982. In: *Olfaction and Endocrine Regulation,* (Ed.) W. Breiphol. IRL, London, pp. 103–113.

Bronson, F.H., 1985. *Biol. Reprod.*, 32: 1–26.

Brown, R.E. and MacDonald, D.W. (Eds.) 1985. *Social Odours in Mammals,* Vols. 1 and 2, Calendron Press, Oxford.

Bruce, H.M. and Parrot, D.M.V., 1960. *Science*, 131: 1526.

Bruce, H.M., 1961. *J. Reprod. Fert.*, 2: 138–142.

Bruce, H.M., 1970. *Brit. Med. Bult.*, 26: 10–13.

Carr, W.J. Wylie, N.R. and Leob, L.S., 1970. *J. Comp. Physiol. Psychol.*, 72: 15–59.

Champlin, A.K., 1971. *J. Reprod. Fert.*, 27: 233–241.

Chipman, R.K. and Fox K.A., 1966. *J. Reprod. Fert.*, 12: 233–236.

Clarke, J.R. and Hellwing, S., 1977. *J. Reprod. Fert.*, 50: 155–158.

Clause, R. and Alsing, W., 1976. *J. Endocr.*, 68: 483–484.

Clulow, F.V. and Clarke, J.R., 1968. *Nature*, 291: 511.

Comfort, A., 1971. *Nature*, 230: 432–433.

Cowley, J.J. and Wise, D.R., 1972. *Anim. Behav.*, 20: 499–506.

Dagg, A.I and Windsor, D.E., 1971. *Can. J. Zool.*, 49: 283–285.

Dixit, H. and Pandey, S.D., 2007a. *J. Adv. Zool.*, 28(1): 39–43.

Dixit, H. and Pandey, S.D., 2007b. *Biochem. Cell. Arch.*, 7(2): 25–29

Dixon, A.K. and Machintosh, J.H., 1971. *Anim. Behav.*, 19: 138–140.

Dominic, C.J. and Pandey, S.D., 1979. *Annals d'Endocrinol.*, 40: 229–234.

Dominic, C.J., 1966a. *J. Reprod. Fert.*, 11; 407–414.

Dominic, C.J., 1966c. *Science*, 152: 1764–1765.

Dominic, C.J., 1966d. *Naturewissenschaften*, 53: 310–311.

Dominic, C.J., 1967a. *Nature*, 213: 1242.

Dominic, C.J., 1967b. *Ind. J. Expt. Biol.*, 5: 47–48.

Dominic, C.J., 1969a. *Gen. Comp. Endocr. Suppl.*, 2: 260–267.

Dominic, C.J., 1969b. *Indian Biologist*, 1: 1–18.

Dominic, C.J., 1970. *J. Anim. Morphol. Physiol.*, 17: 126–130.

Dominic, C.J., 1976a. Role of pheromones in mammalian fertility. In: *Neuroendocrine Regulation of Fertility*, (Ed.) T.C. Anand Kumar. S. Karger, Basel, pp. 226–245.

Dominic, C.J., 1978. *J. Anim. Morphol. Physiol. Silver Jubilee Vol.*, pp. 104–127.

Doty, R.L., 1976. In: *Mammalian Olfaction: Reproductive Processes and Behaviour*, (Ed.) R.L. Doty. Academic Press, New York, pp. 295–321.

Drickamer, L.C., 1981. In: *Environmental Factors in mammal Reproduction*, (Ed.) Gillmore and B. Cook. Macmilan, London, pp. 100–111.

Eisenbergh, J.F. and Kleiman, D.G., 1972. Ann. *Rev. Ecol. Syst.*, 3: 1–32..

Estes, R.D., 1972. *Mammalia*, 36: 315–341.

Ewer, R.F., 1968. *Ethology of Mammals*. Logos Press, London.

Finlayson, J.S. and Baumann, C.A., 1958. *Am. J. Physiol.*, 192: 69–72.

Graziadei, P.P.C., 1977. In: *Chemical Signals in Vertebrate*, (Eds.) D. Muller-Schwarze and M.M. Mozell. Plenum, New York, pp. 435–454.

Hurst, J.L. and Beynon, R.J., 2004. *BioEssay*, 26: 1288–1298.

Izard, M.K., 1983. In: *Pheromones and Reproduction in Mammals*, (Ed.) J.G. Vandenbergh. Academic Press, New York.

Johnston, R.E., 1983. In: *Pheromones and Reproduction in Mammals*, (Ed.) J.G. Vandenbergh. Academic Press, New York, pp. 3–37.

Johnston, R.E., 1985. In: *The Hamster: Reproduction and Behaviour*, (Ed.) H.I. Siegal. Plenum Press, New York, pp. 121–153.

Karlson, P. and Butenandt, A., 1959. *Ann. Rev. Entomol.*, 4: 39–58.

Kingston, B.H., 1965. *Experta Medica International Congress Series*, 83: 209–214.

Korner, C. Breer, H. Singer, A.G. and O'Conell, R.J., 1996. *Neuroreport*, 7(8): 2989–2992.

Le Magnen J., 1952. *Physiol. Behav.*, 6: 295–332.

Lederer, E., 1950. *Fortschr. Chem. Org. Naturstaffe*, 6: 87–153.

Leon, M., 1974. *Physiol. Behav.*, 13: 441–453.

McCarty, R. and Southwick, C.H., 1977. *Behav. Biol.*, 19: 255–260.

Mainardi, D., Marsan, M. and Pasquali, A., 1965. *Atti. Sco. Ital. Sci. Nat.*, 104: 325–338.

Marr, J.N. and Gardner, L.E., 1965. *J. Genet. Physiol.*, 107: 167–174.

Marshall, D.A. and Moulton, D.G., 1977. *Proc. Int. Symp. Olfaction Taste 6th*, 1976, p. 197.

McClintok, M.K., 1971. *Nature*, 291: 244–245.

McClintok, M.K., 1998. *Annals New York Acad. Sci.*, 855: 390–392.

Meredith, M., 1983. In: *Pheromones and Reproduction in Mammals*, (Ed.) J.G. Vandenbergh. Academic Press, New York, pp. 199–252

Meredith, M., 1998. *Annals New York Acad. Sci.*, 855: 349–361.

Michael, R.P., Bonsall, R.W. and Warner, P., 1974. *Science*, 186: 1217–1219.

Michael, R.P., Keverne, E.B. and Bonsall, R.W., 1971. *Science*, 172: 964–966.

Michael, R.P., 1975. *J. Steriod. Biochem.*, 6: 161–170.

Monti-Bloch, L., Diaz-Sanchez, V., Jennings-White, C. and Berliner, D.L., 1998a. *J. Steroid Chem. Mol. Biol.*, 65(1–6): 237–242.

Moulton, D.G., 1978. In: *Handbook of Behavioral Neurobiology: Sensory Integration*, (Ed.) R.B. Masterton. Plenum, New York, 1: 91–118.

Mugford, R.A. and Nowell, N.W., 1970. *Nature*, 226: 967–968.

Muller-Schwarze, D. and Mozell, M.M. (Eds.) 1976. *Chemical Signals in Vertebrates*. Plenum Press, New York.

Muller-Schwarze, D. and Silverstein, R.M. (Eds.) 1980. *Chemical Signals in Vertebrates*. Plenum Press, New York.

Mykytowycz, R. and Goodrich, B.S., 1974. *J. Invert. Dermatol.*, 62: 124–131.

Mykytowycz, R., 1970. In: *Advances in Chemoreception*, (Eds.) J.W. Johnson, D.G. Moulton and A. Turk. .Appleton-Century-Crafts, New York, pp. 327–360.

Nordlund, C.H. and Lewis, W.J., 1976. *J. Chem. Ecol.*, 2: 211–220.

Novotny, M.V., 2002. *Biochem. Soc. Trans.*, 31(1): 117–122.

Pandey, S.D. and Dominic, C.J., 1978. *Ind. J. Expt. Biol.*, 16: 887–889.

Pandey, S.D. and Pandey, C.J., 1985a. *Physiol. Behav.*, 35(b): 851–854.

Pandey, S.D. and Pandey, S.C., 1985b. *Ind. J. Expt. Biol.*, 23: 188–190.

Pandey, S.D. and Pandey, S.C., 1986a. *Acta Physiol.*, 67(3): 387–391.

Pandey, S.C. and Pandey, S.D., 1986b. *Ind. J. Expt. Biol.*, 24: 142–144.

Pandey, S.C. and Pandey, S.D., 1987. *Acta Zool. Bulgr.*, 33: 40–43.

Pandey, S.D. and Pandey, S.C., 1988a. *Acta Physiol.*, 72(2):165–170.

Pandey, S.C. and Pandey, S.D., 1988b. *Zool. Sci.*, 5: 153–157.

Pandey, S.C., 1986. *Ph.D. Thesis*, Kanpur University, Kanpur

Pandey, S.D., 1976. *Ph.D. Thesis*, Banaras Hindu University, Varanasi.

Pandey, S.D., 1989. *Dr. B,S. Chauhan, Comm. Vol.*, pp. 49–61.

Pandey, S.D., 1992. *Jour. P.A.S. Jaunpur*, (3): 176–186.

Parkes, A.C. and Bruce, H.M., 1961. *Science*, 144: 1049–1054.

Payman, B.C. and Swanson, H.H., 1980. *Anim. Behav.*, 28: 528–535.

Porter, R.H. and Dane, H.M., 1976. *Physiol. Behav.*, 16: 75–78.

Rajendren, G. and Dominic, C.J., 1984. *Arch. Biol.*, 95: 1–9.

Ralls, K., 1971. *Science*, 171: 443–449.

Rekwot, P.I., Ogwu, R.D., Oyedipe, E.O. and Seconi, V.O., 2001. *Anim. Reprod. Sci.*, 61(3–4): 157–170.

Regnier, F.E., 1971. *Biol. Reprod.*, 4: 309–326.

Richmond, M. and Conaway, C.H., 1969. *J. Reprod. Fertil.*, 6: 357–376.

Shorey, H,H, ed., 1976. *Animal Communication by Pheromones.* Academic Press, New York.

Stern, K. and McClintok, M.K., 1998. *Nature*, 392: 177–179.

Stoddart, D.M., 1973. *Nature*, 246: 501–503.

Stone, H., Williams, B. and Carregal, E.J.A., 1968. *Expt. Neurol.*, 21: 11–19.

Swann, J. and Fiber, J.M., 1997. *Brain Res. Bull.*, 44(4): 409–413.

Terman, C.R., 1968. *Ecology*, 49: 1169–1172.

Thiessen, D. and Rice, M., 1976. *Phychol. Bull.*, 83: 505–539.

Vandenbergh, J.G., 1973. *Physiol. Behav.*, 10: 257–261

Vandenbrgh, J.G., 1983a. In: *Hormones and Behaviour in Higher Vertebrates*, (Eds.) J. balthazart, E. Prove and R. Gilles. Springer-Verlog, Berlin, pp. 342–349.

Vandenbergh, J.G., 1983b. In: *Pheromones and Reproduction in Mammals,* (Ed.) J.G. Vandenbergh. Academic Press, New York, pp. 95–112.

Vanderbergh, J.G., Whitsett, J.M. and Lombardi, J.R., 1975. *J. Reprod. Fertil.,* 43: 515–523.

Van der Lee, S. and Boot, L.M., 1955. *Acta Physiol. Pharmacol. neerl.,* 4: 442–443.

Van der Lee, S. and Boot, L.M., 1956) *Acta Physiol. Pharmacol. neerl.,* 5: 213–214.

Wakesa, K.S. and Anholt, R.R.H., 1997. *Endocrinol.,* 138: 3497–3504.

Weller, A. and Weller, L., 1997. *J. Comp. Psycol.,* 111(2): 143–151.

Whitten, W.K and Bronsob, F.H., 1970. In: *Advances in Chemoreception,* (Ed.) J.W. Johnson, D.C. Moulton and A. Turk. Appleton-Century-Crafts, New York, pp. 309–325.

Whitten, W.K., 1956. *J. Endocr.,* 13: 399–404.

Whitten, W.K., 1957. *Nature,* 180: 1436.

Whitten, W.K., 1966. In: *Advances in Reproductive Physiology,* (Ed.) Mc Laren. Academic Press, New York, pp. 155–177.

Wilkin, K.K. and Ostwald, R., 1968. *J. Mammal.,* 49: 570–572.

Wilson. E.O. and Bossart, W.H., 1963. *Recent. Prog. Horm. Res.,* 19: 673–711.

Wysocki, C.J., 1979. *Neurosci. Biobehav. Rev.,* 3: 301–341.

Yanai, J. and Mcclean, G.E., 1972. *Nature,* 238: 281–282.

Zeng, X., Leyden, J.J., Spielman, A.L. and Preti, G., 1996. *J. Chem. Ecol.,* 22: 237–257

Zarrow, M.X., Estes, S.A., Danenbergh, V.H. and Clarke, J.H., 1970. *J. Reprod. Fert.,* 23: 357–360.

Chapter 27

Effect of Anosmia in Perception of the Male-Originating Primer Pheromones Involved in Oestrus Induction (The Whitten Effect) in *Mus booduga* Gray

*Himanshu Dixit and S.D. Pandey**

Reproductive Endocrinology Laboratory, Department of Zoology, Christ Church College, Kanpur – 208 001, U.P.

ABSTRACT

The primer pheromone of male origin induces and synchronizes oestrus in noncyclic unisexually grouped females. However, such oestrus induction was inhibited when females were intranasaly irrigated with zinc sulphate (ZS, 1.0mg in 0.05ml physiological saline). Present results suggest that irrigation of ZS causes complete blockade of olfactory perception rendering the females anosmic and hence they fail to perceive the primer pheromone of male origin. Those findings thus lend support to the concept that the oestrus-inducing male pheromone responsible for the Whitten effect is mediated through olfactory pathways.

Keywords: Anosmia; Mice, Oestrus, Pheromone, Zinc sulphate.

* Corresponding author: E-mail: sdpchchcollege@yahoo.co.in.

Pheromonal influences on mammalian reproductive functions and behaviour are well documented (Whitten, 1966; Bronson, 1971; Pandey, 1976, 1996; Meredith, 1983; Rekwot *et al.*, 2001; Novotny, 2002; Archunan, 2003; Brennan and Keverne, 2004; Hurst and Beynon, 2004; Soini *et al.*, 2005). Unisexual grouping of females disrupts the oestrous rhythm in many rodent species (Whitten, 1959; Bronson and Marsden, 1964; Zang, 2005). These irregularities in the oestrous cyclicity of grouped females vanish following exposure to adult males (Pandey, 1976; Whitten, 1956). A new cycle is initiated in majority of females, resulting in synchronization of oestrus. This male induced oestrus synchrony in noncyclic females is known as the Whitten effect.

The nature, source and hormonal regulation of the primer pheromones influencing oestrus has recently been worked out (Kimura, 1971; Clee *et al.*, 1975). Pheromonal perception in most mammalian species occurs through mode of olfaction (Whitten, 1966; Bronson, 1971; Dominic, 1976). Zinc sulphate (ZS) eliminates most of the main olfactory receptors in the form of ciliated neurons overlying the olfactory epithelium on ventral wall of the nasal passage and causes peripheral anosmia in many laboratory rodents (Lisk *et al.*, 1972; Albert and Galef, 1971; Thor *et al.*, 1976; Meredith, 1980; Pandey and Pandey, 1986). As there has been no report on the effect of zinc sulphate on olfactory responsiveness to the oestrus-inducing pheromone in Indian field mice, it was thought opportune to tackle this problem it in the present context.

Material and Methods

Animals employed in this investigation were trapped from the wild and maintained in the laboratory on socked Bengal gram (*Cicer araetinum*), boiled rice and milk. Water was available *ad libitum*. Their diet was also supplemented with paletted food. All animals used in this study were adults weighing 10-12g (10.1 ± 0.28). The Indian field mouse is a docile small rodent that can be handled easily and hence it serves a good experimental animal for pheromonal study. Thirty regularly cycling females employed in this study were first individually housed in galvanized steel cages, 45 × 15 × 10 cm for 21 days. They were then unisexually grouped together in a colony cage, 30 × 30 × 30 cm for another 21 days. Following this, these grouped females were exposed to four intact males, confined individually in wire mesh corrals, 8 × 6 × 6 cm for 7 days. This device prevented the physical contact between two sexes but facilitated the exposure of females to pheromones of males. The males were removed along with their corrals after 7 days of exposure and females remained unisexually grouped in the same colony cage for next 21 days. This followed the exposure of grouped females treated with ZS to four corralled intact males for 7 days. This time grouped females were intranasaly irrigated with 1 per cent sterile ZS solution (0.05 ml in physiological saline) under mild ether anaesthesia. ZS treatment was given 24 hrs prior to second exposure to males *i.e.*, on day 70. The experiment was terminated on day 77 (Table 27.1).

Table 27.1: Protocol of Experiment and Summary of Results (n=30)

Treatment	*Duration*	*Cycle history*
Isolation	Day 1 to 21	Regular
Unisexual grouping	Day 22 to 42	Anoestrus
Unisexual grouping + exposure to intact males	Day 43 to 49	Oestrus induction
Unisexual grouping	Day 50 to 70	Anoestrus
Unisexual grouping + ZS treatment (on day 70) + exposure to intact male	Day 71 to 77	Anoestrus

Body weights of females were recorded daily and vaginal smears from each female were also examined daily. A smear containing only cornified cells was taken to indicate the incidence of oestrus and one with only leucocytes as that of dioestrus. The presence of epithelial cells and leucocytes indicated the stage of pro-oestrus. Occurrence of such cellular picture at times during noncycling stage was considered 'attempted' oestrus (Bronson *et al.*, 1956).

Procedure for Treatment of Zinc Sulphate (ZS)

Sterile hydrated ZS solution (1 per cent) was prepared in physiological saline. Intranasal irrigation of the saline in females was done under mild ether anaesthesia. Each spray containing 0.05ml fluid was injected into nasal cavity through a 1ml hypodermic syringe with a blunt 23 gauge needle cut to a length of 2 cm and smoothed at the tip. The tip of the needle was inserted into each nasal passage for 0.3 to 0.5 cm to spray the solution. Following spraying, the pharynx was aspirated until the female regained consciousness and moved about normally.

Olfactory Impairment Test

Olfactory impairment in females treated with zinc sulphate was assessed by the procedure employed by Rowe and Smith (1972). Females were given two olfactory preference tests one 24 hrs after intranasal irrigation (ca. 30min prior to exposure to males) and the second on day 7 after removal of males. Olfactory preference was test for 5 min. Two identical tin cans, 9 × 6 cm, were placed at opposite corner of the colony cage, 30 × 30 × 30 cm. A small cup containing glucose biscuits (the baited cup) was placed inside one of the cans and the other can contained an identical cup but no biscuits (the neutral cup). Each of the females was left in the centre of the test cage and allowed to remain in the cage for 5 min. The females were observed from a distance of 1 m to avoid disturbance in their normal behaviour. The time spent by each female in sniffing the baited and neutral can was recorded. Females which spent more than 2–3 times investigating the baited cans than in investigating the neutral can were not considered to be anosmic and hence were discarded.

Results

All thirty females which exhibited regular oestrous cycle during the period of isolation attained anoestrus following unisexual grouping at both occasions (day 22–42 and day 50–70). Vaginal smears comprised exclusively cornified cells, at least twice, during the period of isolation where as leucocytes were common cells in smears of females during both periods of grouping. Mucus was also occasionally detected in smears of females during this period. Exposure of grouped females to intact males resulted in the prompt return of oestrus in majority (25/30) of females within 7 days, with a peak on day 4 (Table 27.2). Following removal of males grouped females again became noncyclic, as their smears contained the cells characteristic of anoestrus. In contrast to first exposure none of the grouped females treated with ZS return to oestrus during the period of second exposure to males. All females continued to exhibit only leucocytes in smears, indicating the suppression of oestrus (Tables 27.1 and 27.2). Smears from two females (one on day 5 and the other on day 6) contained epithelial cells along with few leucocytes, which indicated 'attempted' oestrus. Cornified cells were never detected in smears of grouped females during second exposure to males.

Table 27.2: Incidence of Oestrus in Grouped Females Before and After Treatment with ZS

Housing Condition/ Treatment	*Number of Females Returning to Oestrus within 7 Days*							*Total Number and Per cent of Females Returning to Oestrus within 7 Days*
	Day 1	*Day 2*	*Day 3*	*Day 4*	*Day 5*	*Day 6*	*Day 7*	
Unisexual grouping + exposure to males	0	2	5	8	6	3	1	25/30, 83.3 per cent
Unisexual grouping + ZS treatment + exposure to males*	0	0	0**	0	0**	0	0	25/30, 83.3 per cent

*: This trail excluded three females which were not rendered anosmic after ZS treatment.

**: One female on each of these days exhibited 'attempted' oestrus.

The length of the oestrous cycle in isolated females ranged from 6 to 9 days with a mean length 7.5 ± 0.09 days and the mean number of the oestrous cycle was 2.15 ± 0.08. Oestrus, which was completely suppressed during the period of unisexual grouping, was induced in most females following their exposure to intact males; the length of the cycle was also significantly shortened ($P < 0.001$), averaging 6.0 ± 0.28 days. Mean number of oestrus cycle was 3.05 ± 0.05. There was a significant reduction in body weight of about 1 g (0.98 ± 0.014) of all females on day 2, following intranasal irrigation with ZS. However, after this initial ponderal decrease, there was no further loss in body weight until the termination of the experiment on day 77. All females except 2 which were excluded from the trail were rendered anosmic at the beginning of the test (day 70). One female which was initially anosmic but was found to have regained normal olfactory acuity at the termination of the test (day 77) was also discarded in the analysis of the data.

Discussion

The data obtained in the present study are analogous to those reported in many rodent species, as regards the influence of olfactory stimuli on the oestrous cycle (Whitten, 1966; Meredith, 1983; Pandey, 1996; Rekwot *et al.*, 2001; Hurst and Beynon, 2004; Dominic, 1978). The male induced oestrus synchrony in the Indian field mouse is similar to that described in laboratory strains of mice (Whitten, 1956) and wild mice (Pandey, 1976). It is interesting to note that majority of females in the present study return to oestrus on day 4 as reported in wild mice, whereas this peak of oestrus return is observed on day 3 in laboratory mice (Dominic, 1966) and deer mice (Bronson and Marsden, 1964). This difference in the interval of the oestrus may be due to the incidence of an inherently longer oestrous cycle in the Indian field mice (Dixit and Pandey, 2005) and wild mice (Pandey, 1976; Chipman and Fox, 1966).

The failure of grouped females to return to oestrus following intranasal irrigation with zinc sulphate and subsequent exposure to intact males reveals that the treatment with ZS prevents the return of oestrus that normally occurs. This is suggestive of the fact that zinc sulphate may have caused peripheral anosmia, thereby eliminating the olfactory sensitivity of grouped females to the oestrus-inducing pheromone emanating from males. However, intranasal irrigation of the nasal cavity with physiological saline under mild ether anaesthesia did not alter the oestrus rhythm in female mice (personal observation). Present findings are in conformity with results reported Lisk *et al.* (1972) who also observed that intranasal spray of ZS results in peripheral anosmia in golden hamster, which subsequently fails to exhibit normal sexual behaviour mediated through olfactory pathway. It is suggested that zinc sulphate blocks olfactory sensitivity by coagulation necrosis of the nasal epithelium.

Certain anesthetics have also been reported to exert suppressive effects on the response of female mice to pheromones (Brown-Grant, 1966; Pandey and Dominic, 1980). The present findings thus support the view that the male-originating, oestrus-inducing primer pheromone is olfactorily mediated in Indian field mice.

Acknowledgements

The present investigation was financially sponsored by a major research grant (SR/SO/AS-09/2003) from Department of Science and Technology, Government of India, New Delhi.

References

Albert, J.R. and Galef, B.J. Jr., 1971. Acute anosmia in the rat: A behavioral test of a peripherally-induced olfactory deficit. *Physiol. Behav.*, 6: 619–621.

Archunan, G., 2003. Pheromones: Chemical signals for reproductive behavior. *Proc. 28th Conf. Ethol., India*, p. 38–42.

Brennan, P.A. and Keverne, E.B., 2004. Something in the air? New Insights into Mammalian Pheromones. *Current Biology*, 14: R81–R89.

Bronson, F.H. and Marsden, H.M., 1964. Male induced synchrony of oestrus in deermice. *J. Comp. Endocrinol.*, 4: 634–637.

Bronson, F.H., 1971. Rodent pheromones. *Biol. Reprod.*, 4: 344–357.

Bronson, F.H., Dagg, C.P. and Snell, G.D., 1956. Reproduction. In: *Biology of Laboratory Mouse*, 2nd Edn., (Ed.) E.L. Green. Mc Graw-Hill Book Co., New York, pp. 187–204.

Brown-Grant, K., 1966. The effect of anaesthesia on endocrine responses to olfactory stimuli in female mice. *J. Reprod. Fert.*, 12: 177– 181.

Chipman, R.K. and Fox, A.K., 1966. Oestrus synchronization and pregnancy blocking in the wild house mouse (*Mus musculus*). *J. Reprod. Fert.*, 12: 233–236.

Clee, M.D., Humphreys, E.M. and Russell, J.A., 1975. The suppression of the ovarian cyclical activity in groups of mice and its dependence on ovarian hormones. *J. Reprod. Fert.*, 45: 395–398.

Dixit, H. and Pandey, S.D., 2005. Evaluation of oestrus cyclicity in the Indian field mouse, *Mus booduga Gray* in proximity of adult male. In: *Abstract of Paper Presented in Inter. Conf. on Male Reproduction and Infertility*, September, 16–18, pp. 84.

Dominic, C.J., 1966. Observations on the reproductive pheromones in mice. I. Source. *J. Reprod. Fert.*, 11: 407–414.

Dominic, C.J., 1976. Role of pheromones in mammalian fertility. In: *Neuroendocrine Regulation of Fertility*, (Ed.) T.C. Anand Kumar, S. Karger, Basel, pp. 236–245.

Dominic, C.J., 1978. Mammalian reproductive pheromones. *J. Anim. Morphol. Physiol. Silver Jubilee Vol.*, pp. 104–127.

Hurst, J.L. and Beynon, R.J., 2004. Scent wars: The chemo biology of competitive signaling in mice. *Bioassay*, 26: 1553–1563.

Kimura, T., 1971. Modifications of the oestrous cycle in mice housed together with normal or ovariectomized females. *Sci. Pap. Coll. Gen. Ed. Univ., Tokyo*, 21: 161–166.

Lisk, R.D., Zeiss, J. and Ciaccio, L.A., 1972. Influence on olfaction on sexual behavior in male golden hamster (*Mesocricetus auratus*). *J. Exp. Zool.*, 181: 69–78.

Meredith, M., 1980. The vomeronasal organ and accessory olfactory system in the hamster. In: *Chemical Signals: Vertebrates and Aquatic Invertebrates*, (Eds.) D. Muller Schwarze and R.M. Silverstein. Plenum press, New York, pp. 303–326.

Meredith, M., 1983. Sensory physiology of pheromone communication. In. *Pheromones and Reproduction of Mammals*, (Ed.) J.G. Vandenbergh. Academic Press, New York, pp. 199–252.

Novotny, M.V., 2002. Pheromones, binding proteins and receptor responses in rodents. *Biochem. Soc. Trans.*, 31(1): 117–122.

Pandey, S., 1996. Social odours and reproduction in laboratory mouse: Primer pheromonal influences on the oestrous cycle. *Ph.D. Thesis*, Banaras Hindu University, Varanasi, India

Pandey, S.D. and Dominic, C.J., 1980. Effect of Nembutal anaesthesia on male induction of estrus (the Whitten effect) in the wild mouse *Mus musculus domesticus. India. J. Exp. Biol.*, 18: 445–447.

Pandey, S.D. and Pandey, S.C., 1986. Role of main olfactory system in perception of the oestrus-suppressing pheromones in wild mice. *Trends in Life Sciences* (*India*), 1(2): 145–148.

Pandey, S.D., 1976. Studies on mammalian pheromones. *Ph.D. Thesis*, Banaras Hindu University, Varanasi, India.

Rekwot, P.I., Ogwu, D. Oyedipe, E.O. and Sekoni, V.O., 2001. The role of pheromones and biostimulation in animal reproduction. *Anim. Reprod. Sci.*, 65: 157–170.

Rowe, F.A. and Smith, W.E., 1972. Effects of peripherally induced anosmia on mating behavior of male mice. *Psychon. Sci.*, 27: 33–34.

Soini, H.A., Weisler, D. Afelbach, R. Koning, P. Vassallivia, N.Y. and Novotany, M.V., 2005. Comparative investigations of the volatile urinary profiles in different hamster species. *Chem. Senses,* 31: 1125–1143.

Thor, D.H., Carty, R.W. and Flannelly, K.J., 1976. Prolonged peripheral anosmia in the rat by multiple intranasal applications of zinc sulphate solution. *Bull. Psyhonom.,* 7: 41–43.

Whitten, W.K., 1959. Occurrence of the anoestrus in mice caged in groups. *J. Endocrinol.,* 18: 102–107.

Whitten, W.K., 1966. Pheromones and mammalian reproduction. *Adv. Reprod., Physiol.,* 1: 155–177.

Whitten, W.K15. 1956. Modification of the oestrous cycle of the mouse by external stimuli associated with the male. *J. Endocrinol.,* 13: 399–404.

Zang, J.Z., Soini, H.I., Bruce, K.K., Weisler, D., Woodley, S. K., Boum, M.J. and Novotny, M.V., 2005. Putative chemosignals of ferrets (*Mustella furo*) associated with individual and gender recognition. *Chem. Senses,* 30: 727–737.

Previous Volumes

— Volume 1 —

2002, xiii+308p., figs., tabls., ind., 23 cm Rs. 650

ISBN 81-7035-272-X

Section I: Animal Ecology

Section II: Animal Reproduction

— Volume 2 —

2004, xi+387p., figs., tabls., ind., 23 cm Rs. 850

ISBN 81-7035-322-X

Section I: Animal Ecology

— Volume 3 —

2006, xvii+443p., figs., tabls., ind., 23 cm Rs. 1150

ISBN 81-7035-424-2

Section I: Animal Ecology

Section II: Animal Reproduction

— Volume 4 —

2007, xiv+550p., figs., tabls., ind., 23 cm **Rs. 1400**

ISBN 81-7035-459-5

Section I: Animal Ecology

Section II: Animal Reproduction

— Volume 5 —

2008, xiv+476p., col. plts., figs., tabls., ind., 23 cm Rs. 1600

ISBN 81-7035-563-X; 978-81-7035-563-2

Section I: Animal Ecology

Section II: Animal Reproduction

Index

E

G

H

I

P

Q

R

S

X

Y

Z